A	Number of cases in the upper left hand cell of a 2×2 frequency matrix $(f_{1,2})$. See section 7.5
B	Number of cases in the upper right hand cell of a 2×2 frequency matrix $(f_{1,2})$. See section 7.5
b_i	Total number of cases in the ith block. See section 5.3
C	Number of cases in the lower left hand cell of a 2×2 frequency matrix $(f_{2,1})$. See section 7.5
d_i	Difference between a pair of ranks, used by Spearman in his method for computing ρ. See section 11.2
D	Number of cases in the lower right hand cell of a 2×2 frequency matrix $(f_{2,2})$. See section 7.5
E, E_{ij}	Expected number of cases in the cell in the ith row and the jth column of a frequency table when H_0 true. See section 9.8
$E(L)$	Expected value of L when H_0 true. The average of L over all permutations of the ranks. See section 3.2
f_{ijq}	Number of cases in the qth category of the dependent variable in the jth sample in the ith block. See sections 5.6 and 10.7
H	Statistic measuring the dispersion of the sample rank means; approximately distributed as chi-square. See section 3.3
H'	H when corrected for ties. See section 3.3
H_0	Null hypothesis. See section 2.9
H_1	Specific alternative hypothesis. See section 2.10
i	Index normally reserved for the block number (e.g. b_i)
j	Index normally reserved for the sample number (e.g. R_j)
k	Number of samples. Note that k is lower case
K	Statistic reflecting differences between the rank means. See section 3.3. Note that K is upper case
L	Statistic measuring agreement between data and a specific hypothesis. See section 3.2.
M	Median; a value exceeded by half of the scores in the data set. See section 2.9
m	Number of blocks. See section 5.3
m'	Number of blocks which are not completely tied. See section 8.2
n_{ij}	Number of cases in the cell in the ith block of the jth sample
n_j	Number of cases in the jth sample $\left(\text{i.e. } n_j = \sum_{i}^{m} n_{ij} \right)$
n	Number of cases in each cell when cell frequencies are all equal
N	Total number of cases $\left(\text{i.e. } N = \sum_{j}^{k} n_j \right)$
O, O_{ij}	Observed (as opposed to expected) number of cases in the cell in the ith row and jth column of a frequency table. See section 9.8
p	Probability; usually the probability of obtaining the results under discussion, or similarly extreme results, when H_0 is true
q	Index normally reserved for the number of a category of the dependent variable. See section 8.9

Q	Number of categories of the dependent variable. See section 8.9
r	The rank of a particular score
r_q (r_{iq})	The rank shared by all scores in the qth category of the dependent variable (in the ith block). See section 6.4
R_j	The sum of all ranks in the jth sample. See section 3.2
$\bar{R}_j$	The mean rank in the jth sample. See section 3.2
R_+	The sum of the ranks of positively signed differences; used in the Wilcoxon matched pairs test. See section 8.3
R_-	The sum of the ranks of negatively signed differences; used in the Wilcoxon matched pairs test. See section 8.3
$S_i.$	The number of 1s in the ith block when the data consist of only 0 and 1 (dichotomy). See section 10.4
$S._j$	The number of 1s in the jth sample when the data consist of only 0 and 1 (dichotomy). See section 10.4
t, t_q	Number of cases in the qth category of the dependent variable (or the qth row or column total). See section 5.6
Σt_q	Cumulative total of cases in the qth *and lower* categories of the dependent variable. See section 6.4
$T, (T_i)$	Correction factor for ties (in the ith block). See section 3.2
$\text{var}(L)$	Variance or dispersion of all possible values of L when H_0 true (i.e. overall permutations of the ranks). See section 3.2
W	Kendall's coefficient of concordance. See section 11.4
Z	Statistic for evaluating L. Z is approximately normally distributed. See section 3.2
Z'	Z corrected for ties. See section 3.2
z	Critical value of the normal distribution
χ^2	(pronounced kigh square) Critical value of the chi-square distribution. See section 3.3
λ_j	(pronounced lambda) Coefficients which represent numerically the specific alternative hypothesis. See section 3.2
ρ	(pronounced roe) Spearman's coefficient of rank correlation. See section 11.2
$\bar{\rho}$	Average value of ρ over all possible pairs of variables in a concordance analysis. See section 11.4
$\displaystyle\sum_i^m$	Instruction to add all the values to the right of the symbol by systematically changing the index i from a starting point of 1 to finish at m
σ_L	Standard deviation of L; the square root of $\text{var}(L)$. See section 3.2

$X > Y$	X is greater than Y
$X \geqslant Y$	X is greater than or equal to Y
$X < Y$	X is less than Y
$X \leqslant Y$	X is less than or equal to Y

Statistics Using Ranks

A Unified Approach

Ray Meddis

Basil Blackwell

© Ray Meddis 1984

First published 1984

Basil Blackwell Publisher Ltd
108 Cowley Road, Oxford OX4 1JF, UK

Basil Blackwell Inc.
432 Park Avenue South, Suite 1505,
New York, NY 10016, USA

British Library Cataloguing in Publication Data

Meddis, Ray
    Statistics using ranks.
    1. Statistical hypothesis testing
    2. Ranking and selection (Statistics)
    I. Title
    519.5'6   QA277
    ISBN 0-631-13788-2

Library of Congress Cataloging in Publication Data

Meddis, Ray
    Statistics using ranks.

    Bibliography: p. 444
    Includes index
    1. Social sciences – Statistical methods. 2. Ranking
and selection (Statistics) I. Title
HA29.M43885 1984   519.5   84-11087
ISBN 0-631-13788-2

Typeset by Unicus Graphics Ltd, Horsham, West Sussex
Printed in Great Britain by Bell and Bain Ltd, Glasgow

Contents

Preface

The seed idea for this book came in 1972 when my colleague John Valentine pointed out that some of the nonparametric statistical tests which we were teaching as quite separate were, in fact, very similar under the skin. I became obsessed by the problem of proving the similarities, and when this had been done for the handful of tests in question I set out in search of more tests to bring under the umbrella of the system. It was an exciting and rewarding obsession because new tests came quickly and the system grew. Much later I became aware of the similar work of Benard and Van Elteren (1953), but by then the division into specific and nonspecific statistical tests had already formed itself and correlation/concordance tests had also joined the fold. When it later became clear that frequency tables could also be processed by the same methods, I knew that my private obsession could have public benefits and a paper on the topic was published (Meddis, 1980).

This was to have been the end of the story, but a computer program which I had written to demonstrate the system was beginning to be requested for routine laboratory use by researchers around the world. This brought with it a stream of further requests for clarification of many points and it was clear that a textbook was needed. My classroom teaching was also suffering because the statistical ideas in my head were increasingly out of step with the available textbooks, and my presentation became full of parentheses and qualifications which made little sense to beginners. A book which took them straight to where I was obviously going was clearly long overdue.

While these events were taking place, I moved from Bedford College, London University, to the University of Technology, Loughborough. This was a most fortunate move since my new colleagues were very supportive of the new approach to statistics using ranks. We had many discussions and the new methods were tried out with various groups of undergraduates and postgraduates. Although support was widespread in the department, I would particularly like to thank Murray Sinclair, John Atha and Mike Cooke for their help. The mathematics department also gave me invaluable help in the form of Tony Pettit, an experienced innovator himself in the field of nonparametric statistics. I have so often gone to him for help with papers which I could not fully understand that

I blush now to recall these occasions. He also was a constant source of useful perspectives in an area which was relatively new to me. I never really worked out what he thought of my obsession with the ergonomics of statistics, but his grasp of the general problem was such that I never managed to present him with any theoretical surprises.

I determined from the outset that the book would use examples from real projects in disciplines which dealt with human beings, though at that time I did not appreciate what a considerable administrative chore that decision entailed. Fortunately, I had little difficulty with the researchers themselves who were generous with their data and with their time. They were not at all defensive about having their numbers crunched again in public with the possibility that I would not agree with their analysis. I acknowledge a considerable debt of gratitude to them.

Lastly I must acknowledge the people without whom none of this would have been possible – the typists, who struggled for years with the endless stream of handwritten hieroglyphs which were never written with the humble typewriter or currently available word processors in mind. Nevertheless, Beryl Bruce, Bron Hunt and Jeanne Preston undertook the long uphill struggle with a steady professionalism which inspired my admiration and deepened my already considerable gratitude. I am also grateful to Dave Easom who did all of the draft artwork and helped generally in so many other ways.

1 Introduction

This book presents a unified collection of statistical tests for rankable data. These tests are especially applicable in those sciences which deal with people and other living things, because many of our measurements (e.g. school examinations, depression rating scales, health questionnaires, consumer preferences) typically assess individuals *relative* to other individuals. Sad to say, our measures often lack the elegance, regularity and absolute qualities of such measures as temperature, length, mass and energy which are the privileged domain of the 'hard scientist'. Sometimes our investigations go beyond relative measures but typically they do not, and tests based on rank order information can then prove essential.

Ranking tests form a subcategory of the class of *distribution-free* or *non-parametric* statistical procedures. Many such tests have been published, mainly in the last 40 years. They have come in many shapes and sizes, often with idiosyncratic techniques of application and very different mathematical justification. The purpose of this book has been to pick through the jungle of tests, select those which have a common theoretical basis and then present them anew with a single rationale, a common notation and common computational plans. The theoeretical basis of this exercise is given in chapter 14 for the advanced reader. The beginner need simply note that the pages which follow offer a unified and clarified version of a long story which is told piecemeal elsewhere in many books and papers.

Except for the largely ignored pioneering efforts of Benard and Van Elteren (1953) this is the first serious attempt to unify these tests in this way, and I want the fruits of my efforts to reach a wide range of potential users. The book should be suitable for use by undergraduates and researchers whose statistical expertise is merely competent rather than sophisticated. (After all, they are the people who justified the effort toward unification and simplification; statisticians can look after themselves.) However, it is also important to justify and advertise the system among more seasoned statistical campaigners who have served their time with the wild variety of nonparametric tests. But this presents very special problems of presentation. If we attempt to start our story at the beginning and work slowly and completely through to the end, many will fall by the wayside with little benefit before the end of the road is reached. As a consequence, a linear approach has been abandoned. Instead the book has been divided into three parts. The first looks in a straightforward manner at some important conceptual issues. The second presents the mechanics of the administration of the tests. The third deals with the mathematical justification of the procedures. It is assumed that the completely naïve user will have an instructor to direct his attention to the relevant part of the test. As in many other respects, the structure of the book has been influenced by Siegel's (1956) seminal text which did so much to introduce nonstatisticians to the possibilities of non-parametric statistics.

Apologies are therefore due, in advance, to true statisticians trained in the mathematical tradition. The transfer of statistics from the realms of absolute, rational truth to the day-to-day exigencies of the laboratory can be a most painful transition. All rigour seems lost, concepts appear subtly devalued, nice distinctions are ignored and the mathematics is reduced to mere arithmetic. It can be a sad thing to see one's beautiful daughter on the stage. A balance has been sought between the requirements of clear communication and the need for statistical rigour. This was no easy task and I have no illusions of unqualified success.

Part I (chapters 2-5) summarizes in common sense terms many of the conceptual issues surrounding hypothesis testing using ranking techniques. Much of this will only make sense after the reader has developed some familiarity with the mechanical aspects of carrying out the tests. The basic thrust of these chapters is that we can make a small number of concepts work for us in a large number of research situations. This represents a saving on current state-of-the-art approaches to the teaching of nonparametric statistics where tests are supported by a bewildering wide variety of arguments. The emphasis on plain English rather than mathematical terminology is only partly an acknowledgement of the limited mathematical training of many test users. It is also an assertion of my firmly held belief that many key statistical ideas can be handled in the 'common sense domain' as aids to *understanding* the tests even if they cannot be used to prove the validity of the tests.

Part II (chapters 6-13) presents the tests purely as arithmetical procedures. Every effort has been made to make the techniques clear and simple to use. Each design and subspecies of design is illustrated with a worked example. The unification of rank sum techniques means that, for the first time, we can use a common notation and that we need only two basic computational plans for a very wide range of experimental designs. A separation of one- and two-sided tests into the two different computational plans should reduce the perennial confusion in this area, allow the two-sample techniques to be readily generalized to many-sample designs, and prepare the test user for the niceties of *post hoc* and multiple hypothesis testing methods. Frequency table techniques can also be embraced by the same two computational plans that are used for dealing with individual rankable scores.

The author has a long-standing interest in the general aspects of teaching and has many years' experience of the tribulations of teaching statistics in particular. Much of the interest in writing this book has, therefore, come from long discussions and experiments concerning various alternative ways of streamlining and clarifying existing procedures. Nonparametric statistical techniques have multiplied dramatically in recent years in a largely uncoordinated way, producing many dilemmas for any presenter of a unified approach. No single computational plan can be devised which preserves every test originator's recommended procedure. Many decisions among alternatives have therefore had to be made. Part II can be read by seasoned statistics teachers as an anthology of these decisions. The computational plans offered here are, ergonomically, a considerable improvement on the nonparametric status quo even if they are still found to fall far short of the ideal.

Part III contains two chapters. Chaper 14 aims to show how the computational formulae in part II can be derived from two master formulae. It also shows how the analysis of variance by ranks procedure comes to embrace a very wide range of existing nonparametric procedures; it puts the book into its historical context. So far, many of these tests have been (very properly) known by the names of their originators (e.g. Friedman, Kruskall, Wilcoxon and Page) and we intend no disrespect by abandoning this practice in the body of the text. However, the use of a different name for each test emphasizes the differences rather than the similarities between them. This practice may alone have been responsible for the slowness of the emergence of a nonparametric system to rival the flexible parametric analysis of variance procedures. Chapter 14 (see also chapter 5) is meant to go part of the way towards a full acknowledgement of our debt to these many authors. It should also help orient experienced researchers who were educated in the tradition of many individual distribution-free tests.

Chapter 15 is a BASIC computer program which will handle most of the examples given in the book.

1.1 How to use this book

Beginners

Beginners will normally be using this book under the guidance of an instructor who will introduce the basic statistical concepts as necessary in a manner best suited to the class. Most of the concepts which are directly pertinent to rank sum statistical analysis should be easily found in the chapters of part I. However, the development of competence will be a mixture of practice of the arithmetical procedures of part II and discussion of their underlying rationale. Neither approach precedes the other successfully and happiest progress comes by shuttling between the two.

Researchers

Researchers who use tests only occasionally but already have a grounding in statistical methods may benefit by reading quickly through part I to get a broad overview and then trying to apply the computational procedures of part II to their own data. Once the basic system has been grasped, the usual shuttle between parts I and II can then continue indefinitely as and when the reader requires refreshing on particular issues. Two points of departure should be noted by the experienced test user. Firstly, this book contains many popular tests in an unfamiliar disguise. These can be unearthed by looking for the traditional name in the index. With very minor exceptions the techniques in this book should give *exactly the same answers* as the original tests. Secondly, this book has abandoned the system of one- and two-tailed tests for two-sample designs in favour of a more explicit distinction between procedures which test a specific alternative hypothesis and procedures which test a nonspecific alternative hypothesis. When we have only two samples, the former correspond to one-tail (one-sided) tests whereas the latter correspond to two-tail (two-sided) tests. The change is intended to clarify this vexed business and to allow the distinction to generalize to many-sample problems where trends and contrasts will normally belong to the class of one-tailed (one-sided) tests.

Statisticians

Statisticians, who may look at this book as something of a curiosity, should begin with part III where the substance of the unification idea is expressed tolerably succinctly. They may also be prepared to master the computational plans in part II so that they are able to advise researchers who come seeking their advice. Part I, however, may take some swallowing. A separate statistical language used among nonstatistician researchers has evolved which looks funny and debased to the expert just as pidgin English looks childlike to the speaker of original English. It is, however, a flexible and useful means of communication and (where frank errors have been avoided) not to be decried for its simple innocence. Part I is written in this language of the statistical underworld. Perhaps it might serve as a source book when contemplating the difficulties of communication across the great divide between lofty expert and humble practitioner – a divide which, in my experience, is successfully bridged only too rarely. Finally, statisticians may be interested in the exact tables in the book which have all been recomputed and some of which appear here for the first time.

Part I

2 Logic of decision making in statistics

The logical basis of decision making in statistics is as simple as it is ingenious. It is also a source of endless difficulty to students and researchers coming to statistics for the first time. For many, the complex and ritualized arithmetic procedures of the tests themselves are relatively simple compared with discussion of what the tests are achieving at a conceptual level. The purpose of this chapter is to review in plain English some of the more important ideas which underpin the methods to be presented later in the book.

The exposition is nonmathematical, not because the reader is assumed to be uneducated but because the ideas are largely common sense. To prevent a discussion in the abstract, we shall begin with an illustrative example which we can refer to throughout the chapter. After that we have a quick overview of the logic of statistical hypothesis testing, followed in turn by a detailed individual look at some specific concepts. Many of the notions to be introduced have a fairly general application throughout the study of statistics, but it must be remembered that this is an introduction to statistics using ranks. Whenever a detailed example is required, we shall take it from the domain of rank sum analysis. Chapter 4 discusses the relationships between rank sum tests and other varieties of tests.

2.1 Illustrative example

This example is fictitious but is loosely modelled on an actual experiment (see section 7.7). The sample sizes are greatly reduced here to make it easy to follow. At the time of writing, the issue is of considerable medical and social importance although still very controversial.

For many years, psychiatric patients suffering from severe and prolonged depression have been liable to treatment by electroconvulsive therapy (ECT), which involves applying an electric shock to the patient's head while the patient is under anaesthesia. Patients may be given a course of such shocks over a number of weeks. It is commonly observed that the depression becomes gradually less severe following the treatment.

Opponents to the therapy claim that the patients would have recovered even if they had not received the shocks. Depression is well known to be subject to spontaneous recovery even though this may take months or even years. The issue, then, is whether ECT promotes a greater degree of recovery than would occur in the absence of the treatment.

In our experiment we take a group of ten patients, all of whom are diagnosed as suffering from severe depression. We assign five patients at random to the treatment (ECT) group and the remaining five to the control (no treatment) group. This random assignment is achieved by drawing names from a hat. All patients continue to receive their regular drugs and psychotherapy sessions. The patients and all staff who deal with them regularly are led to assume that they are *all* being treated by ECT. All patients are anaesthetized and removed to the treatment room as usual, but only the patients in the treatment group are given shocks.

For the purposes of our experiment the progress of the depression is monitored by the nursing staff, who rate the patients' behaviour on a number of standardized scales. The combined ratings yield a depression score on a scale where a high score indicates more depression. Table 2.1 gives the depression ratings for our ten patients one month after the end of their treatment (or pseudo treatment).

2.2   Overview of the decision making procedure

The control group of patients has a higher average depression score. Does this mean that the treatment was effective in alleviating depression? Should psychiatrists now be confident that ECT does relieve severe depression? Unfortunately, the difference in the means of the two samples is rather small and our researcher

Table 2.1   Depression ratings for two groups of patients one month after treatment

ECT treatment group	Control no treatment group
16	18
14	24
15	26
17	19
21	20
mean 16.6	21.4

will need to decide whether the expense, the distress and the side effects of the procedure are justified by an expected five-point drop in the depression rating.

Even if we accept that a five-point difference is enough, we are left with the more difficult problem of deciding whether the lower scores in the treatment group were caused by the treatment or by other uncontrolled factors. For example, the patients were assigned at random to the two samples. It could be that some of the more severely depressed paients were assigned to the control group purely by chance. Other uncontrolled factors may have been at work too. Perhaps one or two of the patients in the control group were suffering from a virus infection on the day the depression ratings were made. One of the patients in the treatment group may have been cheered by the recent visit of a loved one. Even the rating scales are subject to inaccuracies if the nurses are in a hurry or fail to notice a symptom. It is possible that uncontrolled fluctuations in rating error happened to work in favour of the treatment group on this occasion.

Clearly, the experimenter is forced into a decision concerning two possible explanations of the lower depression levels of his treatment group. Firstly, it may be that the treatment is responsible for the difference in the two mean scores. Alternatively, the difference may simply be the result of uncontrolled influences. After all, even if ECT were totally ineffective, we would not expect the two groups to have *exactly* the same mean depression rating – some fluctuations of the individual scores are inevitable.

These two explanations are the two competing hypotheses at the heart of the logic of hypothesis testing. The aim will be to accept one hypothesis and reject the other using the data as our only guide. We will, unfortunately, not be able to choose with absolute certainty. Whichever way we choose, we will run some risk of having made the wrong decision. All we can do is make the choice which seems most likely to be correct. One of the jobs of hypothesis testing statistics is to assess the degree of error associated with our choice.

Surprisingly, our methods concentrate on the second explanation or hypothesis – that the result obtained was due to uncontrolled influences alone. The main thrust of our calculations is to assess how likely it is that these influences should have combined to generate our result. If it seems unlikely, we reject the hypothesis, and by default we are left with the explanation or hypothesis which attributes our result to the effectiveness of the treatment. Common sense would dictate that we address ourselves directly to the main hypothesis concerning the value of the treatment. However, the logic of hypothesis testing looks in a different direction. It does not seek to prove that the treatment was or was not effective but simply assesses the likelihood that other uncontrolled factors were responsible for the difference. If, in the end, we accept that the treatment was effective, we do so merely because we have rejected the hypothesis that uncontrolled factors were at work and not because we have proved, in any positive way, that the alternative hypothesis is correct.

This may seem to be an unsatisfactory approach, too indirect and too inconclusive. Whatever the merits or demerits of this procedure, it is important that the reader grasp the logic for what it is. Much unnecessary confusion can be avoided if this crucial aspect of the logic is accepted and understood. Clearly it would be preferable to address our efforts directly to assessing the truth or falsehood of our theory but so far it has not been possible to devise satisfactory methods to do so, and we are left with this oblique approach. In practice, most scientists are satisfied with this method despite its strangeness on first encounter.

The history of statistics may illuminate the situation. Early students of this arcane art were interested in gambling and the laws of chance. As a result, we have accumulated a powerful collection of procedures for modelling the operation of uncontrolled (random) influences. This means that we can do a good job of assigning a probability statement to the explanation that the results of our experiment could have arisen simply as the result of random influences. To the statistician the problem is the same as that of detecting cheats in gambling. It is possible for a player to be lucky but, when his luck continues, the question arises as to whether he is cheating. The casino detective may look for positive evidence in the form of underhand acts. The statistician, however, operates only on the account of winnings and losses to produce a statement of the likelihood that the gambler's success could be attributed solely to chance. When this likelihood gets very low we may reject the 'luck' hypothesis and, by default, accept the alternative hypothesis that he is in fact a cheat.

The basic idea throughout the decision making process, whether in the casino or in the laboratory, is that the random influence explanation becomes less and less likely as the main effect under study gets larger. The idea that the gambler is just lucky grows less and less acceptable to the management as his winnings increase. The idea that lower depression scores in the treatment groups could be explained in terms of random effects such as rating inaccuracy, birthdays, visits and sampling fluctuations obviously grows less and less likely as the difference between the mean scores of the two groups gets larger.

The business of statistics is to convert a statement of the size of the effect under study into a statement of the likelihood that such a result could have occurred by chance factors alone. Once we have such a statement, we can then make a decision to accept or reject the idea that our result could be attributed to uncontrolled factors. If we reject this idea, then we have some grounds for favouring the alternative idea that the effect was brought about by the factors under study – in our case the shock therapy.

The rest of the book is concerned with procedures to allow us to make these probability statements on which our decisions will be based. Before becoming immersed in these we will need to acquire a suitable vocabulary and a set of useful concepts and distinctions for dealing with the wide variety of issues raised by their application to the data arising from experiments and surveys.

2.3   Variables

Every situation we study is characterized by variables, only some of which may be of interest to the experimenter. The following is a small sample of the variables in our illustrative example:

(a) depth of each patient's depression
(b) whether or not they were treated by ECT
(c) whether or not they knew they were subjects of an experiment
(d) sex
(e) religion
(f) season of the year
(g) nurses' morale and general demeanour.

The experimenter can do different things to the variables; he can manipulate them, measure them, seek to control them by preventing them from varying, or ignore them altogether. We classify the variables according to how he treats them.

Independent variables

Independent variables are aspects of the situation which are deliberately manipulated in order to observe the effect on other variables. If we apply different types of fertilizer to strips of a farmer's field to see which promotes the most vigorous growth of potatoes, then 'fertilizer type' is an independent variable. In our psychiatry example, the independent variable is whether or not the patients received shock therapy. Many experiments have more than one independent variable. In addition to trying various fertilizers we might repeat the process in various kinds of soil (clay, sandy, loamy etc.). We would then have two independent variables, 'fertilizer type' and 'soil type'. Most of the designs studied in this book have only one independent variable.

Dependent variables

Dependent variables are aspects of the situation which the experimenter measures and which are potentially influenced by the independent variable. In the fertilizer example, the weight of the crop of potatoes would be a dependent variable. The depth of the depression of our patients is the dependent variable in our psychiatry example.

Blocking variables

Blocking variables are qualities which cannot be controlled but must be taken into account even though they are not specifically relevant to the hypothesis under examination. If we wish to evaluate a new method for teaching children to

Table 2.2 Hypothetical reading improvement scores for 36 children from three schools in an evaluation of a new method for teaching reading

	New method	Old method
School A	27	14
	13	17
	19	13
	21	10
	14	8
	16	8
School B	17	6
	19	10
	13	12
	8	6
	14	6
	10	7
School C	9	2
	6	4
	8	8
	4	4
	8	8
	11	3

read, we should ideally try the method out in a number of different schools. We are not specifically interested in the schools themselves but the schools are known to vary in quality of staff and students. For various reasons, we need to keep the data from the different schools separate. Table 2.2 illustrates the situation. The schools form separate blocks. In this example the independent variable is 'teaching method', the dependent variable is 'reading improvement' and the blocking variable is 'school'. If we were to repeat our psychiatry example one year later with different nursing staff and a different psychiatric team, then the two trials would constitute separate blocks.

Controlled variables

Controlled variables concern those aspects that are kept constant across all of the subjects under study. Our depressed patients were all given to understand that they were receiving the same treatment. In this way, their expectations were not allowed to influence the dependent variable (depression level). Good experimental practice seeks to control as many variables as possible. The 'hard' sciences, such as physics and chemistry, are able to exercise considerable control over their experimental subject matter. In the life sciences and social sciences this is rarely so easy. In many cases it may even be undesirable on the grounds that control, when too severely exercised, can distort the situation so much that it is no longer the one which the researcher wanted to study. If our depressed patients were obliged to eat the same food, receive visits from the same standardized visitor, read the same books, take the same drugs etc., the situation would no longer be one that would interest psychiatrists. The several control procedures introduce new and irrelevant pressures on the patients and may create unwanted effects.

Uncontrolled variables (biased and unbiased)

Uncontrolled variables are all those aspects of the situation that the researcher neither manipulates nor measures. Usually these cannot be avoided entirely, and it is their presence which makes the art of hypothesis testing so essential. They constitute the pool of alternative explanations to the main hypothesis under consideration.

They occur in three broad categories; uncontrolled events, measurement error and sampling error. Uncontrolled events are most easily understood. In the psychiatry example they include such occurrences as birthdays, visits from relatives, virus infections, disputes among patients etc. Measurement error involves the variation in scores attributable solely to the weaknesses of the measuring techniques. Sampling error is the consequence of assigning subjects at random to the samples. It may be the best technique we have for producing unbiased samples but it does not guarantee equality of the samples.

A crucial distinction for the experimenter is whether uncontrolled variables are biased or unbiased with respect to the independent variable. Statistical methods for hypothesis testing can deal with unbiased uncontrolled variables but biased variables present special difficulties. For example, if the patients are assigned at random to the samples then the samples may not be perfectly matched, but we have no reason for suspecting a *bias* which might put the less depressed patients in the treatment group. Similarly, there is no reason why patients in the treatment group should be more likely to have a birthday than those in the other sample group. Both sampling error and the uncontrolled occurrence of birthdays are therefore *unbiased uncontrolled variables*.

Bias among the uncontrolled variables might occur if the nurses knew who was being treated with ECT and who not, and as a consequence allowed this knowledge to influence their assessment of the patients. Biased variables can render experiments completely worthless and often do so. It is an important skill in research to be able to design studies which are free of bias because, once the data have been collected, there is often little that statistics can do to remedy the damage caused by the presence of an unbiased uncontrolled variable.

2.4 Statistical hypotheses

Null hypothesis (H_0)

The central idea in statistical hypothesis testing is the possibility that the independent (or deliberately manipulated) variable does *not* influence the dependent variable, and that any apparent correlation between them can be attributed to the operation of uncontrolled variables. This idea is given the name 'null hypothesis', and the purpose of our procedures is to specify how likely it is that we could have obtained the results that we did if the null hypothesis were in fact true.

In our psychiatry example, the null hypothesis is that treatment by ECT does *not* relieve depression. We therefore seek the probability of obtaining by chance a difference of at least 4.8 points between the treatment and control groups if the treatment is totally ineffective. In other words we want to know how often such a result could arise simply as the result of the action of unbiased, uncontrolled variables.

If the probability is extremely low, we reject the null hypothesis. If the probability is not low, then we may decide not to reject the null hypothesis. By convention, we *assume* that the null hypothesis is true until rejected on the strength of the evidence. Failure to reject the null hypothesis is not the same as proving it to be true – just as failure to prove a man guilty is not the same as proving him to be innocent. Nevertheless, we assume the man to be innocent until proved guilty.

We shall look more formally at the nature of the null hypothesis following presentation of other major concepts.

Alternative hypothesis (H_1)

The hypothesis in which the researcher is mainly interested appears to play a passive role in our little drama, and appears to be accepted or rejected merely as a consequence of what happens to the null hypothesis. It serves only as an *alternative hypothesis*, which is never proven to be true or false. It is accepted if the null hypothesis is rejected and is rejected if the null hypothesis is itself not rejected. However, we shall see later that alternative hypotheses take on various forms and their nature influences the way in which the null hypothesis is put to the test. Their role is therefore not as passive as might at first appear.

In our psychiatry example the alternative hypothesis is that ECT really is effective in relieving severe depression in patients. The hypothesis could take a number of detailed forms. For example, we might be suggesting that all depressed patients are equally relieved by the treatment or we might be suggesting that not necessarily all but at least some of the patients are relieved by the treatment. A more general hypothesis might simply be that ECT has some effect, without specifying whether it relieves depression or makes it worse. These possibilities will be elaborated later.

2.5   Statistics for making decisions

Our first step in deciding whether to reject the null hypothesis is to decide on a measure of the degree of the relationship between the independent and dependent variables. In the psychiatry example we need a measure of how big the treatment effect is on the depression ratings. One obvious measure of the effect is the difference between the mean depression scores for the two groups. Clearly the larger the difference, the bigger the supposed effect.

We could use the difference between the median depression score for each group instead. In fact there are many alternative measures of the effect. Rank sum procedures tend to concentrate on the ranks of the individual scores, and a useful measure here might be the difference between the mean ranks for the two groups. This is illustrated in table 2.3, where the difference is found to be 3.8. In practice the actual statistic used varies from one situation to another, but the principle remains the same.

These measures of the effect under study are called 'statistics'. A statistic is defined as a numerical quantity which reflects some quality in which we are

Table 2.3   Ranks of data in table 2.1 used to establish the difference between the mean ranks of the two samples. The ranks have been assigned to the scores irrespective of the sample in which they occur

ECT treatment group			*Control* no treatment group	
Score	*Rank*		*Score*	*Rank*
14	1		18	5
15	2		19	6
16	3		20	7
17	4		24	9
21	8		26	10
rank sum	R_1 18			R_2 37
mean rank	$\bar{R}_1$ 3.6			$\bar{R}_2$ 7.4

difference between the mean ranks $7.4 - 3.6 = 3.8$

interested. The benefit of generating a statistic is that it allows us to express our main problem in a manner which is amenable to mathematical analysis, for example: 'If the null hypothesis is in fact true, what is the probability of obtaining, by chance, a value of our statistic greater than or equal to that actually obained?' In the case of our psychiatry example, we can express the matter precisely: 'When drawing two samples of size 5 at random from the ten ranks, one to ten, what is the probability of obtaining a difference between the mean ranks of the two samples of 3.8 or greater?' It may seem to be a mouthful but it turns out to be a simple exercise in probability mathematics.

2.6   Decision errors

When the mathematics are complete, we are left with a probability statement and the need to decide whether or not to reject the null hypothesis H_0. If the probability is very low (i.e. that our result is very unlikely to have occurred by chance if the null hypothesis had been true), then we may wish to reject the null hypothesis. Conversely, if the probability is high we will want to accept the null hypothesis. Either way we run the risk of making an error. The two important kinds of error are:

Type I   error (H_0 rejection error), which occurs when we reject H_0 when in reality it is true.

Type II   error (H_0 acceptance error) which occurs when we accept H_0 when in reality it is false.

The terms 'type I error' and 'type II error' are sanctioned by traditional usage but they are difficult for beginners to use, and I prefer instead the more memorable alternative names of 'H_0 rejection error' and 'H_0 acceptance error' The H_0 rejection error is the probability that we were discussed above. Thus, if our result could have occurred by chance with a probability of 0.02 (two in a hundred) when the null hypothesis were true, when we run a 0.02 probability of making an H_0 rejection error (type I) should we decide to reject the null hypothesis. If, in the face of the evidence, we decide to accept the null hypothesis, we run the risk of making an H_0 acceptance error (type II), but it is usually difficult to specify just how large this error is.

2.7   Statistical significance

If the probability of an H_0 rejection error is very low and we therefore decide to reject the null hypothesis, the convention is to say that the results obtained are 'statistically significant'. This is of course a value judgement, and researchers will differ in terms of what they find statistically significant. It is, therefore, conventional also to refer to the probability value (of an H_0 rejection error) when making this statement so that people can judge for themselves. We might say, for example, that our results are statistically significant at a probability level of 0.02. We could paraphrase that statement thus: 'Our calculations have shown that, given H_0 true, the likelihood of obtaining such favourable results (or better) by chance is only 0.02'.

The term 'statistical significance' should not, of course, be confused with the practical significance of the result. It is not uncommon, especially when dealing with large samples, to have trivially small effects in the practical sense which are nevertheless still statistically significant in the sense that they were unlikely to have occurred by chance if the null hypothesis were true.

Significance levels – rejection regions

It is often thought to be pretentious to specify the exact probability value associated with a result since it suggests an irrelevant degree of precision. Instead, researchers often merely specify a region in which the probability lies. The regions used in this book are:

(a) greater than 0.05   not significant at the 5 per cent level
(b) less than 0.05   significant at the 5 per cent level
(c) less than 0.025   significant at the 2.5 per cent level
(d) less than 0.01   significant at the 1 per cent level
(e) less than 0.001   significant at the 0.1 per cent level.

To say that our results are significant at the 0.025 level will then usually imply that the exact probability lies between 0.025 and 0.01. More commonly, we convert the probabilities to percentages·to create a less formal communication. We would then say that our result was significant at the 5 per cent (2.5, 1, 0.1 per cent) level of significance.

These are called rejection regions because they represent significance levels at which researchers would typically want to reject the null hypothesis. However, these are arbitrarily chosen levels and no hard and fast rules are implied. The actual significance level should be quoted in all communications so that the reader can decide for himself what the data really imply.

2.8   Statistical power

It is obviously desirable to use a test which will be as sensitive as possible to a situation where the null hypothesis is false. It is only by rejecting the null hypothesis that the researchers can establish support for his alternative hypothesis. A sensitive test, therefore, is one which creates only low H_0 acceptance error rates (type II errors). In other words we want to avoid accepting the null hypothesis when it is in fact false. We can increase the sensitivity of a test in many ways: by using larger samples, by reducing uncontrolled variation, by making more accurate measurements and by devising mathematical procedures which make most use of the information available in the data we collect. We can also increase sensitivity by making our alternative hypothesis as specific as possible.

All these points will be discussed in detail later, but the reader should be aware of the controversies among statisticians as to which mathematical procedures are more sensitive. These discussions go under the general heading of 'power' considerations of a statistical test. Power is taken here as synonymous with sensitivity. A more powerful test is one which, for a given data set, yields a lower rate of H_0 acceptance error (type II error). In practice, however, choosing a more powerful statistical test is often a less effective or less practicable method of reducing H_0 acceptance error rates than, say, increasing sample sizes or taking more care in making accurate measures or reducing bias in uncontrolled variables. Despite the arguments for and against different statistical tests in terms of their relative efficiency, the best way to increase sensitivity is to collect more data and to collect it more carefully.

2.9 Formal aspects of the null hypothesis

So far we have restricted our definition of the null hypothesis to the specification that there is no relationship between the dependent and the independent variable *in our study*. It is worth noting, however, that the null hypothesis carries an implication for future studies too. It implies that, *in general*, there is no relationship. To accept the null hypothesis now is to believe that future experimentation will also fail to demonstrate a relationship between the crucial variables.

A more particular deduction is that a pair of individuals selected in future becauser they differ in terms of the independent variable will not necessarily show any directional bias with respect to the dependent variable. In terms of our psychiatry example, the null hypothesis says that we can make no predictions concerning the relative depression ratings of two patients of whom one has recently been treated by ECT and the other has not. More formally we might attempt to express it thus:

$$p(d_t < d_c) = p(d_t > d_c) \qquad (2.1)$$

where d_t is the depression rating of a randomly selected, recently treated patient, d_c is the depression rating of a randomly selected, untreated (control) patient, $p(d_t < d_c)$ reads 'the probability that d_t is *less then* d_c', and $p(d_t > d_c)$ reads 'the probability that d_t is *greater than* d_c'. In other words, treatment makes no difference to the expected depression rating of a patient.

This may seem to be a somewhat formal approach in an introductory presentation, but expression (2.1) captures the essence of the reasoning of rank sum analysis. It pinpoints the central concern with the *relative* values of data points rather than their absolute values. In the psychiatry example we want to know whether to expect a treated patient to be less depressed than an untreated patient. Rank sum analysis does not consider the actual degree of reduction of depression but concentrates on the certainty of finding some reduction, however small. The two ideas are of course linked in that we will be more sure of finding a reduction if the size of the effect is greater. Nevertheless, rank sum analysis works, in the first instance, on the consistency of the effect rather than its magnitude.

Medians, mean ranks, and probabilities

It will be clear already that the probability expression (2.1) is unwieldy and not the most readily assimilated of statements. Often we talk about population medians instead and rephrase the null hypothesis thus:

$$H_0: M_1 = M_2 \tag{2.2}$$

where M_1 and M_2 are the medians of the populations from which the samples were drawn. Since the median of a population specifies the midpoint score it is clear that (2.2) amounts to much the same thing as (2.1). This would not be so true if we substituted means for medians in (2.2) and, as a consequence, it is usually preferable to specify the medians of samples when carrying out rank sum analysis. Often it makes little difference but it is not always possible to compute a mean (for example, if the scale used is merely a rank order scale such as 'improved/same/worse'). As a result, the median is the preferred index of central tendency (average value) in this book.

Often we are dealing with more than two samples and we need to recast the null hypothesis appropriately. For example, for three samples the null hypothesis would be

$$H_0: M_1 = M_2 = M_3$$

and for k samples

$$H_0: M_1 = M \ldots = M_j = \ldots = M_k \tag{2.3}$$

where M_j is the median of the jth population.

Unfortunately medians are not easily subject to direct statistical testing, and the relevant computational procedures have concentrated on the mean rank of the scores in the samples. The scores are ranked independently of the sample to which they belong. Table 2.3 illustrates this procedure for the psychiatry example. Since the relative ordering of the sample mean ranks will usually reflect the ordering of the medians we can recast (2.3) thus:

$$H_0: E(\bar{R}_1) = E(\bar{R}_2) = \ldots = E(\bar{R}_j) = \ldots = E(\bar{R}_k) \tag{2.4}$$

where $\bar{R}_j$ is the mean rank of the jth sample and $E(\bar{R}_j)$ is the expected or most likely value of $\bar{R}_j$.

We talk about the expected mean rank of the sample because it makes little sense to talk about the mean rank of a population which might be infinitely large. If the medians of the populations are equal, then we can express this in terms of an expectation that random samples drawn from these populations will have equal mean ranks.

Strictly speaking, we cannot seriously expect them to be exactly equal. We really intend to say that we do not expect any one sample mean rank to be either greater or less than any other sample mean rank. This is a small but important point, since it highlights the crucial difficulty facing the researcher that a difference between the medians (or mean ranks) of samples is not in itself proof of a similar difference between the medians of the populations from which the data points have been taken.

2.10   Formal aspects of the alternative hypothesis

Nonspecific alternative hypothesis

If the null hypothesis states that the medians of the populations studied are equal, then the most obvious alternative is that they are not. To be more precise, if the null hypothesis is untrue, then it follows that at least two of the populations sampled must have different medians. This is a nonspecific alternative to the null hypothesis.

Our deduction does not allow us to say which two populations are different (unless, of course, there are only two), and we must not make the elementary error of assuming that all the populations are different. This is a weak conclusion to draw, although it is often perfectly satisfactory for analysing two-sample experimental designs. In the psychiatry example, the nonspecific alternative hypothesis could be expressed as

$$H_{ns}: M_t \neq M_c$$

where M_t is the median of the treated population, M_c is the median of the untreated (control) population and H_{ns} denotes nonspecific hypothesis. The symbol $\neq$ denotes 'is not equal to'.

Specific alternative hypothesis

More usually, a researcher wants to draw a more specific conclusion than this. This is possible if he tailors his analysis to a specific alternative hypothesis which specifies some expected pattern among the sample rank means. In our preliminary example, he will want to conclude that treated patients produce a lower median depression rating than do untreated (control) patients. His hypothesis therefore specifies the ordering of the medians of two populations:

$$H_1: M_t < M_c$$

He will be able to accept this specific hypothesis on rejection of the null hypothesis only if his computational procedures were tailored to incorporate this hypothesis. Clearly we need two sets of techniques, one for analysing data using a nonspecific alternative hypothesis and another when a specific hypothesis is proposed. These two kinds of technique can be applied to the same data but, to avoid the confusion and suspicion which often arise in this context, they are kept well separated in this book.

Two-sided tests

The distinction between nonspecific and specific alternative hypotheses is less crucial in the case of two samples since the nonspecific alternative hypothesis can be decomposed into two specific alternative hypotheses:

$$H_1: M_1 \neq M_2 \text{ is the the same as } M_1 > M_2 \text{ or } M_1 < M_2$$

In words: to say that the two medians are not the same is to say that either the first is greater than the second or the first is less than the second. This has led to the traditional practice of using the same computational technique for assessing specific and nonspecific hypotheses in the two-sample cases. The only difference is a last-minute adjustment to the significance level – usually doubling it for the nonspecific case. The terminology used is 'one- and two-tailed' or 'one- and two-sided' testing according to whether the alternative hypothesis is specific or nonspecific.

The terminology is sanctified by tradition but it does give rise to considerable confusion, especially among beginners. Students are unhappy about the confusion of the two ideas and suspicious of the last-minute change of the significance level according to whether or not the researcher claims to have predicted the outcome. The confusion becomes worse, however, when dealing with many-sample designs where the distinction between specific and nonspecific hypotheses remains but the analogy with one- and two-tailed testing is less than helpful. The bold decision has therefore been made to keep the two varieties of test quite separate to avoid confusion and to stress the conceptual continuity across the two-sample/many-sample divide.

Trends and contrasts

There are two important types of specific alternative hypotheses. In one case, the researcher specifies a rank ordering of the population medians:

$$H_1: M_1 < M_2 < \ldots < M_k$$

These hypotheses are known as *trends*. In the other case, the researcher specifies a relationship between two subsets of the population medians:

$$H_1: (M_1, M_2) < (M_3, M_4)$$

Such a hypothesis is known as a *contrast*. The expression $(M_1, M_2) < (M_3, M_4)$ can be understood loosely to mean that the medians of populations 1 and 2 are expected to be less than the medians of populations 3 and 4. More specifically it often represents the idea that the *combined* first and second populations are expected to have a lower median than the combined third and fourth populations. Trends and contrasts are analysed using very similar procedures. In the two-sample design, trends and contrasts are obviously the same thing.

A priori *and* post hoc *hypotheses*

When a researcher generates a specific hypothesis on the basis of some theory or previous experience and he produces it before he collects the data, we say that the hypothesis is *a priori*. A single *a priori* hypothesis is dealt with in a straightforward manner and gives rise to few problems. In our illustrative example, we have the simple *a priori* specific hypothesis that ECT will reduce depression more quickly than the control condition. If there is a large enough difference in favour of the ECT group between the median depression levels for the two samples, we shall reject the null hypothesis and accept our specific hypothesis as probably true.

Complications arise when the researcher wishes to propose a hypothesis only after he has seen the data. Such hypotheses are called *post hoc* (after the event). Normally we are suspicious of wisdom after the event but it does have its use. Imagine, for example, that the ECT patients were unexpectedly more depressed than the control patients. In this case we might want to conclude that ECT makes you worse – not better – than no treatment at all. This hypothesis is called *post hoc* and is analysed using slightly different methods. The main difference is that a test of a *post hoc* hypothesis is treated as statistically less significant than a test of an *a priori* hypothesis. Extra credit is given for saying which way the wind will blow *before* it starts blowing!

A *post hoc* hypothesis has much in common with a nonspecific hypothesis. In both cases, the direction of the result is not specified in advance of data collection.

Multiple hypotheses

Sometimes a researcher has a number of prior specific alternative hypotheses and a decision has to be made as to whether the multiplicity of hypotheses has created a *post hoc* testing situation. After all, if the list of so-called prior hypotheses contains all possible trends and contrasts, then this is in no useful way different from the situation where we wait until the data have been collected and summarized before generating our hypotheses.

The test which is normally applied here specifies that the hypotheses in the list must be 'statistically independent' if they are to continue to be treated as *a priori*. Two hypotheses are statistically independent if the truth of one hypothesis in no way affects the likelihood of the other being true. Clearly opposite hypotheses are not independent. The following two hypotheses are independent, however:

$$H_1: (M_1, M_2) < (M_3, M_4)$$
$$H_2: (M_1, M_3) < (M_2, M_4)$$

If we interpret (M_1, M_2) as 'the median of the first and second populations', and so on, then we can see that *both* H_1 and H_2 can be true when

$$M_1 < M_2 < M_3 < M_4$$

Alternatively, if

$$M_2 < M_1 < M_4 < M_3$$

then H_1 is true but H_2 will be false. In this way H_1 being true does not imply the truth of H_2; they are independent.

Multiple hypothesis testing must, therefore, not be confused with *post hoc* hypothesis testing. If the hypotheses are statistically independent, they can be assessed individually as prior hypotheses. If they are not statistically independent, we must think again. One form of multiple hypothesis testing involves testing all samples against a single control sample. These hypotheses are not independent since an extremely low (or high) control sample median will clearly influence all of the hypotheses equally. The degree of independence of the hypotheses can be measured, however, and taken into account using special procedures which have been devised. Otherwise, the researcher may have to fall back on *post hoc* procedures as a conservative approach even though this may be expensive in terms of a rise in the H_0 acceptance error rate.

Sensitivity and the alternative hypothesis

The nature of the alternative hypothesis determines the kind of statistical computation. This point will be elaborated in chapter 3, but we may anticipate the discussion with the generalization that the sensitivity of the test procedures is greatly enhanced by a specific prior alternative hypothesis. It is as if we are more likely to find something if we are clear, in advance, what we are looking for. In our case we are more likely to reject the null hypothesis when it is false if we know what we are looking for by way of an alternative. *Post hoc* hypotheses suffer a similar loss of power or sensitivity as nonspecific alternative hypotheses for the same reason. Although a *post hoc* alternative hypothesis looks specific enough on the surface, the fact that it was thought of only after the data were inspected means that the situation is basically the same as not having a hypothesis. As a general guide, nonspecific or *post hoc* alternative hypotheses are less sensitive (have lower 'power') than specific, prior alternative hypotheses. This effect becomes stronger as the number of samples increases since the number of possible alternatives to the null hypothesis increases too. A single, accurate, specific hypothesis greatly narrows the search and permits a more certain discovery.

3 Evaluating alternatives to the null hypothesis

3.1 Introduction

The logic of hypothesis testing requires that we begin our statistical analysis with a null hypothesis and some suitable alternative. Next we compute a statistic which measures the agreement between the data we have collected and our alternative hypothesis. In this book, the statistic will typically get larger as the agreement gets better. Finally, we try to work out how often such a good agreement would occur if the null hypothesis were in fact true. If the likelihood of getting such a good agreement by chance alone is very small, we reject the null hypothesis and, by default, accept the alternative hypothesis. This chapter looks at the general principles surrounding the computation of an appropriate statistic and the procedures used to convert this statistic to a suitable probability statement.

There are two kinds of alternative hypothesis, the specific and the nonspecific, and it will be convenient to deal with them separately to avoid confusion. The basic principles are the same for both; they differ in points of detail only. Ironically, it is because they are similar, and therefore so readily confused, that they need to be kept apart in two sections. To begin, we will consider procedures for dealing with specific alternative hypotheses. Later, we will present a similar but abbreviated treatment for the nonspecific alternative to the null hypothesis.

3.2 Evaluating a specific alternative to the null hypothesis

Lambda coefficients to represent the hypothesis

A well formed specific alternative hypothesis indicates a predicted ordering of the sample rank means. To compute a statistic of agreement, we need a number which will reflect the success of the prediction; this number will be high if the actual order of the sample rank means follows the predicted pattern, and low if the order is different. As a first step, we phrase our hypothesis in terms of a string of numbers which reflect the predicted pattern of sample rank means ($\bar{R}_j$). For example, if we expect our means to go in descending order, thus:

$$\bar{R}_1 > \bar{R}_2 > \bar{R}_3$$

we can express our hypothesis in terms of the following three numbers:

3   2   1

which have the important properties that the first is greater than the second and the second is greater than the third. Notice that the first coefficient refers to the first sample rank mean, the second coefficient refers to the second sample rank mean, and so on.

These three numbers are called lambda coefficients. 'Lambda' is the Greek letter λ, corresponding to the Roman L, and 'coefficient' implies that we will be soon using them as multipliers. There are exactly as many lambda coefficients as there are sample rank means. We have considerable freedom in assigning actual values to the coefficients so long as the basic pattern remains unchanged. The same hypothesis can be represented by a number of different sets of lambda coefficients, all of which are equally successful in representing the same hypothesis (H_1):

	H_1:	$\bar{R}_1 > \bar{R}_2 > \bar{R}_3$	
	λ_1	λ_2	λ_3
set 1	3	2	1
set 2	2	1	0
set 3	1	0	−1
set 4	6	4	2

Notice that, for each set, all three values form a descending, equally spaced series, and *that this is all that matters* since our hypothesis merely asserts a descending series. Therefore, any of the four sets of coefficients will serve our purpose of representing the hypothesis. In fact, all will give exactly the same answer at the very end of the statistical exercise!

Calculating L, our statistic of agreement

Now that we have expressed our hypothesis in terms of a string of numbers, we are well placed to evaluate their agreement with the sample rank means since they too constitute a string of numbers. We are now looking for a correlation between two strings of numbers, which is easily found. In fact, it turns out that we do not need to compute the sample rank means but can work directly with the

Table 3.1   Ranks, rank sums and lambda coefficients for a three-sample design experiment with equal sample sizes: computation of L

	Sample 1	Sample 2	Sample 3
	6	2	1
	8	5	3
	9	7	4
R_j (rank sum)	23	14	8
λ_j (lambda coefficient)	3	2	1
$\lambda_j R_j$	69	28	8
$L = \sum_{j=1}^{3} \lambda_j R_j = 69 + 28 + 8 = 105$			
$\bar{R}_j$ (rank mean)	7.67	4.67	2.67

sample rank sums (even when the sample sizes are unequal). This cuts one whole step in the arithmetic. Table 3.1 gives a set of rank sums and their corresponding coefficients. Notice that the rank sums form a descending series, as predicted. The correlation with the lambda coefficients looks good; all we need do now is quantify it.

A useful statistical procedure for measuring the agreement between two strings of numbers is the method of product moment. This involves multiplying the elements of corresponding pairs together and adding the products to form a single value, which we shall call L. This value will be relatively large when the correlation is great and relatively small when the correlation is poor or even negative. You can test this procedure for yourself by changing the order of the coefficients, repeating the calculations and noting how the value of L rises and falls together with the agreement between the order of the two sets of figures.

Fortunately, our statistic L is not merely sensitive to the relative ordering of the sample rank means and the hypothesis lambda coefficients. It also responds to the degree of spread among the means; so the value of L is at its highest when the means are most widely separated. For example, if the sample rank sums were 24, 15, 6 this would represent the same ordering as the figures in table 3.1 but this time the means are more widely spread. Consequently, we expect and get a higher value of L (see table 3.2). Our value of L has risen from 105 to 108.

In summary, our statistic L is a measure of the success of our alternative hypothesis in two respects. Firstly, it reflects success in predicting the *order* of the sample rank means. Secondly, it represents the tendency of the means to be

Table 3.2 Computation of *L*. Compare with table 3.1: in this example the rank means are more widely spaced

	Sample 1	Sample 2	Sample 3
	7	4	1
	8	5	2
	9	6	3
R_j	24	15	6
λ_j	3	2	1
$\lambda_j R_j$	72	30	6
$L = \Sigma \lambda_j R_j = 72 + 30 + 6 = 108$			
$\bar{R}_j$ (rank mean)	8	5	2

widely separated. Both properties, when present, represent evidence in favour of the alternative hypothesis and make the null hypothesis less likely to be true. Clearly, *L* is a most useful statistic and we can formulate the general rule that *as L becomes larger, so the null hypothesis becomes an increasingly less likely explanation of our results*.

The letter *L* was chosen to represent our statistic as a tribute to Page (1963), who used the same letter in a similar procedure for evaluating trends across a number of samples of ranked data. The techniques in this book represent a considerable enhancement of his simple test but the basic principle remains the same.

More about lambda coefficients

The cumbersome expression 'lambda coefficients' is taken from analysis of variance terminology and is intended to highlight a structural similarity in the two methods of analysis of trends and contrasts. Rank sum procedures, however, are *not* subject to the restrictions usually imposed when using analysis of variance. In particular, we do not require that the sum of the coefficients be zero. Nor is it required that the number of scores in each sample be equal.

Our freedom to choose coefficients can be very useful when trying to minimize arithmetical effort. Keeping the coefficients small keeps the arithmetic manageable; 3, 2, 1 is therefore better than 6, 4, 2. Having a zero in the list can also produce great savings in terms of effort; 2, 1, 0 is therefore better than 3, 2, 1. If you are at home with negative numbers, you can meet both objectives by straddling the coefficients across zero, thus: 1, 0, −1. In practice, these are

B

particularly convenient since

$$L = \Sigma \lambda_j R_j = 1 \times R_1 + 0 \times R_2 + -1 \times R_3$$
$$= R_1 - R_3$$

This represents an arithmetical saving on 3, 2, 1 which involves the following calculations:

$$L = \Sigma \lambda_j R_j = 3R_1 + 2R_2 + R_3$$

Both will ultimately give the same result when making a decision about the null hypothesis.

When we have only *two samples*, we can take full advantage of this freedom. We use the coefficient 1 for the sample with the predicted higher mean and 0 for the sample with the predicted lower mean. Thus if we predict that sample 1 will have the higher rank mean we proceed thus:

$$L = \Sigma \lambda_j R_j = 1 \times R_1 + 0 \times R_2$$
$$= R_1$$

which is a useful simplification.

It was stated earlier that we could use the sample rank sums in our computations rather than the rank means, even though the hypothesis in question was specifically concerned with the rank means. Since this principle applies even when the sample sizes are unequal, some explanation is clearly required. The basic idea of the technique can be best understood in terms of the individual ranks in each group. We must consider our hypothesis as saying that individual scores in sample 1 will be higher than scores in sample 2 which, in turn, will be higher than scores in sample 3. It therefore follows that we can compute L by applying the lambda coefficients to the individual ranks (see table 3.3). This gives the same answer as using the rank sums but by a longer route. Therefore, when we use rank sums, we are evaluating a statement about typical scores within the samples and this amounts to the same thing as a statement about the sample rank means.

Our statistic L is also affected by the absolute size of the coefficients and the number of scores in each sample. We cannot begin to decide whether L is relatively large or not until we know these details and can take them into account. Therefore, whenever L is reported, the lambda coefficients used must be reported as well as the sample sizes. In our example we might say $L = 105$ ($\lambda_j = 3, 2, 1$; $n_j = 3, 3, 3$). When we have only two samples, it will always be assumed that the coefficients 1, 0 were used and that the sample with the highest predicted sample rank mean was assigned the nonzero coefficient. No reporting of the coefficients is therefore required in this case.

Table 3.3 Computation of L using individual ranks: r_{ij} is the ith rank in the jth sample. See table 3.1 for comparison

$\lambda_1 = 3$	$\lambda_2 = 2$	$\lambda_3 = 1$
Sample 1	*Sample 2*	*Sample 3*
λ r	λ r	λ r
$3 \times 6 = 18$	$+2 \times 2 = 4$	$+1 \times 1 = 1$
$+3 \times 8 = 24$	$+2 \times 5 = 10$	$+1 \times 3 = 3$
$+3 \times 9 = 27$	$+2 \times 7 = 14$	$+1 \times 4 = 4$

$$L = \sum_{i=1}^{3} \sum_{j=1}^{3} \lambda_j r_{ij}$$

$$= 18 + 24 + 27 + 4 + 10 + 14 + 1 + 3 + 4 = 105$$

Distribution of L using randomization experiments

There are three main methods of finding out the values which our statistic L typically assumes when the null hypothesis is true; the randomization method, the exact method and the large-sample approximation method. Each of these methods has its advantages and disadvantages, but all three deal with exactly the same problem and are three different ways of arriving at a similar answer. We use these techniques when we have a particular value of L and we want to know how often such a value of L (or larger) would occur if the null hypothesis were true. In the previous section we found a good agreement between our hypothesis and the observed sample rank means, and this was reflected in a value of L equal to 105. Although acknowledging that such a good agreement is unlikely to happen if the null hypothesis were true, we really need to have a good estimate of just how great this likelihood is. We need to quantify the likelihood so that we can communicate to other researchers the level of significance at which the null hypothesis was rejected.

When the null hypothesis is true, there are no systematic influences at work on the sample rank means. Any observed differences result from the unbiased and essentially random action of uncontrolled variables. We can therefore mimic the situation where the null hypothesis is true by rearranging the ranks entirely at random. In our working example we have a design with three samples and three scores in each sample. Table 3.4 illustrates the design but leaves blanks where the scores should be. These blanks will need to be filled by the ranks 1 to 9 – *but filled entirely at random*.

Table 3.4 Basic design for our working example. The blanks indicate where the ranks will be inserted

Sample 1	Sample 2	Sample 3
—	—	—
—	—	—
—	—	—

Table 3.5 Computation of L given ranks assigned at random to the samples

	Sample 1	Sample 2	Sample 3
	9	5	6
	2	1	7
	8	3	4
R_j	19	9	17
λ_j	3	2	1
$R_j\lambda_j$	57	18	17

$$L = \sum_{j=1}^{3} R_j\lambda_j = 57 + 18 + 17 = 92$$

In table 3.5 the blanks have been filled in by drawing the numbers 1 to 9 from a hat and replacing the blanks column by column from top to bottom. We can now compute L by finding the rank sums and applying the coefficients in the manner described above in the previous section. Our new value of L is 92 and is considerably less than the value (105) which we obtained earlier. This result supports the idea that our original result is unlikely to have occurred if the null hypothesis were true.

Of course, a single randomization experiment is not enough. We need to repeat the exercise many times to produce a clear picture. Table 3.6 illustrates a second randomization experiment. In this case, the resulting value of L is 75. Both values of L (75 and 92) resulting from randomization experiments are low compared with the value of L (105) currently being considered.

Computer-assisted example. To proceed further, either we need many willing hands to work in parallel to repeat the exercise many times, or we can use a computer to do the work for us. Table 3.7 summarizes the output of a computer

Table 3.6   Second randomization experiment (see text)

	Sample 1	Sample 2	Sample 3
	3	6	9
	2	5	7
	4	1	8
R_j	9	12	24
λ_j	3	2	1
$R_j\lambda_j$	27	24	24
$L = \Sigma R_j\lambda_j = 27 + 24 + 24 = 75$			

Table 3.7   Results of 100 randomization experiments using three samples each with three values and using the lambda coefficients $\lambda_j = 3, 2, 1$; (a) gives the 100 computed values of L, in order of occurrence, and (b) summarizes the results in the form of a cumulative frequency table. See text for further details

(a)

88	94	87	87	89
86	88	95	87	94
95	94	82	83	93
96	98	92	84	91
88	87	85	86	102
85	94	87	83	88
93	103	82	97	80
89	97	79	78	77
80	92	80	76	89
99	80	94	88	88
88	97	104	89	86
94	87	106	86	80
91	77	94	96	94
95	88	98	81	89
84	103	91	84	98
98	97	104	93	99
88	94	80	88	87
85	88	92	103	84
88	85	97	85	98
105	79	84	91	78

Table 3.7　continued

(b)

1 L	2 Frequency	3 Cumulative frequency	4 Cumulative probability
106	1	1	0.01
105	1	2	0.02
104	2	4	0.04
103	3	7	0.07
102	1	8	0.08
99	2	10	0.1
98	5	15	0.15
97	5	20	0.2
96	2	22	0.22
95	3	25	0.25
94	9	34	0.34
93	3	37	0.37
92	3	40	0.4
91	4	44	0.44
89	5	49	0.49
88	12	61	0.61
87	7	68	0.68
86	4	72	0.72
85	5	77	0.77
84	5	82	0.82
83	2	84	0.84
82	2	86	0.86
81	1	87	0.87
80	6	93	0.93
79	2	95	0.95
78	2	97	0.97
77	2	99	0.99
76	1	100	1

program which simply repeats, as often as we require, the exercise we have already carried out in tables 3.5 and 3.6. Although the computer has no hat to draw numbers from, it can achieve a similar effect by using a random number generator.

One hundred values of L generated by the program are given in the top half of the table beginning with 88, 94, ... Each value of L is the result of a separate randomization experiment. The largest value of L obtained using this method is 106, just one more than the results of our real experiment. In fact a value of 105 occurs only once. We can, therefore, say that our value of L (105) was *equalled or exceeded* twice in 100 randomization experiments. Thus, even when the null hypothesis is true we expect to get a value of L equal to or greater than 105 on roughly 2.0 per cent of occasions. This is the answer we have been seeking.

The bottom half of table 3.7 provides a summary of the 100 randomization experiments. Column 1 gives the different values of L, starting with the highest since these are of most interest to the researcher. Column 2 specifies how often that particular value of L occurred. Column 3 specifies how often that value of L was *equalled or exceeded*. We call this the cumulative frequency. This column is of interest to us because it indicates how often (when the null hypothesis is true) we would expect to get results *as high or higher than* those obtained in our research. Column 4 expresses the cumulative frequency in terms of a cumulative probability. This column indicates our significance level. For $L = 105$ we can read off our probability as 0.02 or a significance level of 2.0 per cent.

The need for long computer runs. For all its simplicity and intuitive appeal, the randomization method has an important weakness; every time we repeat the exercise we get slightly different estimates of our significance level. These fluctuations can be quite large if we carry out only a small number of randomization experiments. To counter this objection we can run the computer program for longer. Table 3.8 summarizes the results of 1000 randomization experiments. The individual results have, of course, been omitted. For $L = 105$ the significance level is now 1.6 per cent, whch we must assume is a more accurate assessment. Clearly, the longer we run the program the more accurate our estimates become. Only the user can weigh up the costs and benefits in terms of computer time and marginal increases in accuracy. As a rough guide, we can expect generally acceptable accuracy at the 1 per cent level of significance after 10 000 randomizations. However, 1000 is quite satisfactory as an informal method of getting a rough impression.

As we shall see later, there are often quicker and easier ways of estimating the significance levels for a particular experiment. This does not mean that the randomization method should be ignored. From time to time every researcher gets himself into a position where no exact or approximate procedure can help him. When this happens, a return to first principles may prove the only way to save the day. It has an equally important function, however, in the classroom, where it provides a direct and concrete introduction to the meaning of significance levels and offers conceptual pegs for anchoring the more abstract notions of exact and approximate distributions of statistics. It can also supply a useful introduction to the use of computers in statistics.

Table 3.8   Summary of 1000 randomization experiments using three samples. Lambda coefficients are 3, 2, 1. Sample sizes are 3, 3, 3. Compare table 3.7

1	2	3	4
		Cumulative	*Cumulative*
	Frequency	*frequency*	*probability*
107	1	1	0.001
106	3	4	0.004
105	12	16	0.016
104	11	27	0.027
103	15	42	0.042
102	16	58	0.058
101	17	75	0.075
100	15	90	0.09
99	13	103	0.103
98	35	138	0.138
97	37	175	0.175
96	36	211	0.211
95	43	254	0.254
94	50	304	0.304
93	65	369	0.369
92	59	428	0.428
91	51	479	0.479
90	40	519	0.519
89	59	578	0.578
88	56	634	0.634
87	44	678	0.678
86	60	738	0.738
85	55	793	0.793
84	38	831	0.831
83	31	862	0.862
82	34	896	0.896
81	27	923	0.923
80	24	947	0.947
79	14	961	0.961
78	13	974	0.974
77	10	984	0.984
76	8	992	0.992
75	1	993	0.993
74	5	998	0.998
73	1	999	0.999
72	1	1000	1

Distribution of L using the exact method

We can evaluate the distribution of L more directly by enumerating all possible arrangements of the ranks across the samples. This commonly used mathematical device is based on the idea that each arrangement of the ranks is equally likely to occur when the null hypothesis is true. We therefore construct a distribution which is based upon a single example of every possible arrangement. According to well established mathematical principles we may expect the resulting distribution to give us an unbiased picture of the distribution of L which we might obtain following an infinity of randomization experiments.

We need not examine every possible permutation of the ranks but only what is technically known at every possible partition of the ranks. The following two permutations of nine ranks in three samples:

| 1 | 2 | 3 | | 4 | 5 | 6 | | 7 | 8 | 9 |
| 3 | 2 | 1 | | 6 | 4 | 5 | | 9 | 7 | 8 |

are equivalent partitions from our point of view, since the samples contain the same scores. The use of partitions rather than permutations introduces a considerable computational saving because there are many fewer partitions than permutations. In this particular example, it makes the calculations 218 times quicker!

This is an important consideration since enumerating the possibilities is a lengthy business. There are a total of 1680 partitions in our simple example, and the number increases dramatically with increasing sample size. Exact distributions are usually evaluated by professional statisticians who publish tables for the use of others. Even so, exact tables are usually only available for experimental designs with small sample sizes because of the computational labour involved. In some simple cases, such as a 2×2 frequency table, the tables are easily generated by computer program and then the restrictions on length are simply the cost of printing and publishing.

The reader is unlikely ever to be required to generate his own exact distribution of any statistic, but he may find the following example instructive. Once again, we shall use the earlier example featuring three samples of three ranks in each sample. We must evaluate each of 1680 possible partitions and compute the value of L for each partition separately. Our first partition might be the perfect arrangement shown at the top of table 3.9 with the computation of L, which is 108. Two more partitions are shown lower in the same table with two more values of L, 107 and 106. The complete process cannot be shown but table 3.10 summarizes the outcome.

We can see with a glance at table 3.10 that our result of $L = 105$ is equalled or exceeded on 18 of the 1680 partitions. Expressed as a proportion, this is 0.011 or 1.1 per cent of the total. This is a little less than our estimate ($p = 0.016$) using the method of randomization. The exact method should be treated

Table 3.9   First three of 1680 partitions of nine ranks into three equal sized samples with associated computation of the L statistic, using lambda coefficients 3, 2, 1

Partition 1	9	8	7			6	5	4			3	2	1	
R_j			24					15						6
λ_j			3					2						1
$\lambda_j R_j$			72					30						6

$\Sigma \lambda_j R_j = 72 + 30 + 6 = 108$

Partition 2	9	8	6			7	5	4			3	2	1	
R_j			23					16						6
λ_j			3					2						1
$\lambda_j R_j$			69					32						6

$\Sigma \lambda_j R_j = 69 + 32 + 6 = 107$

Partition 3	9	8	5			6	7	4			3	2	1	
R_j			22					17						6
λ_j			3					2						1
$\lambda_j R_j$			66					34						6

$\Sigma \lambda_j R_j = 66 + 34 + 6 = 106$

and so on

as the more precise, however, since it gives a more faithful picture of what will happen in the very long run.

Clearly, the distribution of L in table 3.10 applies only to the very specific situation where we have three samples, with three scores in each group, and where we use the lambda coefficients 3, 2, 1 for the first, second and third samples respectively. For any alteration of these details we will need a whole new distribution. Accordingly, every table must be accompanied by these appropriate identifying features and the user of the tables must be very careful to check that these features are exactly right for his needs.

There are some advantages to the method of presentation used in table 3.10, but it is very extravagant in terms of space and new users experience difficulty in using such tables quickly. One way of presenting the essential information economically and clearly is given in table 3.11. This method isolates the four values of L which need to be *equalled or exceeded* to be significant at the indicated level. They are the *critical values* of L. Thus to be significant at the 5 per cent

Table 3.10   Exact distribution of L for three samples of size three using lambda coefficients 3, 2, 1

L	Frequency	Cumulative frequency	Cumulative probability
108	1	1	0.000 59
107	2	3	0.001 8
106	5	8	0.004 7
105	10	18	0.011
104	13	31	0.018
103	20	51	0.030
102	24	75	0.045
101	32	107	0.064
100	34	141	0.084
99	42	183	0.110
98	51	234	0.140
97	64	298	0.178
96	68	366	0.218
95	78	444	0.264
94	78	522	0.311
93	90	612	0.364
92	88	700	0.417
91	102	802	0.477
90	76	878	0.523
89	102	980	0.583
88	88	1068	0.636
87	90	1158	0.689
86	78	1236	0.736
85	78	1314	0.782
84	68	1328	0.823
83	64	1446	0.861
82	51	1497	0.891
81	42	1539	0.916
80	34	1573	0.936
79	32	1605	0.955
78	24	1629	0.970
77	20	1649	0.982
76	13	1662	0.990
75	10	1672	0.995
74	5	1677	0.998
73	2	1679	0.999
72	1	1680	1.000

Table 3.11   Critical values of *L* taken from table 3.10. *L* must be *equal to or greater than* a critical value to be significant at the corresponding significance level. $\lambda_1 = 3, \lambda_2 = 2, \lambda_1 = 1$

Sample size	Significance level			
	5%	2.5%	1%	0.1%
3, 3, 3	$L \geqslant 102$	104	106	108

level *L* must be *at least* 102; to be significant at the 1 per cent level *L* must be *at least* 106; and so on.

In our worked example, *L* is 105 and is clearly significant at the 2.5 per cent level (critical value 104) but not at the 1 per cent level (critical value 106). We therefore say that our results were significant at the 2.5 per cent level. In fact, we know from table 3.10 that the probability associated with our statistic under the null hypothesis is 0.011, which is only very marginally short of the 1 per cent level. Nevertheless it is not quite good enough and the brief summary table relegates it to the level below. What this method of presentation gains in terms of economy it sometimes loses in terms of subtlety.

The advantages of the exact method over the randomization method are improved precision and greater economy of effort (since the tables are available and the user simply looks up the answer). The main disadvantage lies in the fact that the user is totally dependent on the statistician to prepare the tables. These are typically available only for the most popular experimental designs and then usually only for restricted sample sizes. In addition, the tables do not allow for the possibility of tied scores and, when ties do occur, the user is reduced to using the tables as if no ties were present. For small numbers of ties, the error introduced is typically small but it is uncontrolled and therefore unsatisfactory.

Distribution of L using the approximate method

When we have no exact tables, we must resort to the approximate method. This is often called a 'large-sample approximation' because the approximation is at its best when large samples are used. This is convenient since the approximate method is most likely to be needed with large samples when exact tables are not available. However, the approximation is applicable whatever the sample size; it is merely less accurate with smaller samples. In general, however, the exact method is always to be preferred when it is feasible.

Figure 3.1 shows the distribution of *L* given in table 3.10 in histogram form. The height of each column represents the frequency of each particular value of *L*. The shape of the distribution shows a family resemblance to the normal

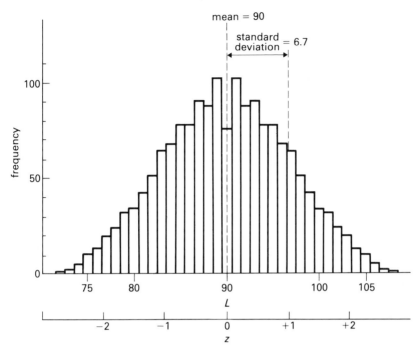

Figure 3.1 Exact distribution of L for $\lambda_j = 3, 2, 1$ and $n_j = 3, 3, 3$, showing standard scale. Based upon table 3.10

distribution, which is given in figure 3.2. If we change the sample sizes or the lambda coefficients, the distribution of L will change in terms of its average value and degree of spread but the overall shape will remain approximately normal. We can use this property of the shape of the distribution to discover the critical values of L irrespective of sample sizes or coefficients used.

The normal distribution itself has been thoroughly studied by statisticians, and tables already exist to allow us to discover the probability that a score chosen at random from a normal distribution will exceed any critical value. Table A in the appendix is an example of such a table. To simplify discussions, statisticians prefer to talk about a standard version of the normal distribution which has a mean of zero and a standard deviation of one. Figure 3.2 illustrates this concept. The horizontal axis of the figure gives standard scores or z scores as they are more popularly known. It can be seen from the figure that z values greater than 1.64 involve only 5 per cent of the total population. The z values which correspond to our four significance levels are given in table 3.12. All we need do is convert our value of L into a standard score, then check against table 3.12 to get an approximate guide to the significance level.

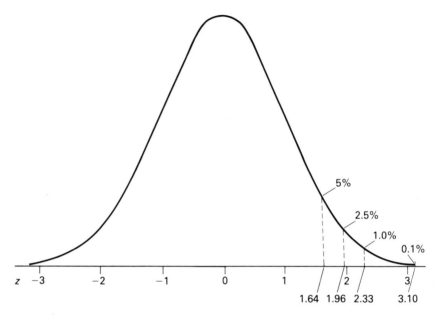

Figure 3.2   Normal distribution showing critical values of z at four significance levels

Table 3.12   Critical values of z at four levels of significance. Z must be *equal to or greater than* a critical value to be significant at the corresponding significance level

Significance levels	5%	2.5%	1.0%	0.1%
Standard scores	$z \geqslant +1.64$	$+1.96$	$+2.33$	$+3.10$

Converting L to a standard score. Figure 3.1 indicates that the mean of the distribution is 90 and the standard deviation is 6.7. The mean is an indicator of the middle point of a symmetrical distribution. We call it the 'expected value' of L and use $E(L)$ as the shorthand. Therefore, in our case,

$$E(L) = 90$$

The standard deviation is a measure of the spread of the distribution and can be thought of as a 'typical distance from the mean' indicator. We use the Greek letter sigma (σ) as the shorthand. Therefore, in our case,

$$\sigma_L = 6.7$$

We convert any value of L to a standard score by calculating how many standard deviation units it is from the mean:

$$Z = \frac{L - E(L)}{\sigma_L}$$

(Note that we use an upper case Z to indicate the conversion of L to a standard score and a lower case z to refer to the standard scores of the normal distribution.)

In our worked example we have a score of $L = 105$. We must convert this to a standard score thus:

$$Z = \frac{L - E(L)}{\sigma_L} = \frac{105 - 90}{6.7} = 2.24$$

If we refer to the critical values in table 3.12 we can see that our value of Z is significant at the 2.5 per cent significance (critical value $Z = 1.96$). Notice that the Z score must be equal to or larger than the value in the table. It is not, however, significant at the 1 per cent level (critical value $Z = 2.3$) although it is almost so. This is a similar result to that given by both the exact method and the randomization method.

Calculating mean and standard deviation of L. The mean and standard deviation of the distribution of L vary according to (a) the sample sizes, (b) the coefficients used and (c) even the basic design of the experiment. Throughout the book a number of formulae are given for calculating $E(L)$ and σ_L according to particular circumstances. These are derived and summarized in chapter 14. However, for the sake of completeness, the calculations for our particular example will be shown here. When we have a simple design (only one block) with more than two samples we use the following two formulae:

$$E(L) = \frac{1}{2}(N + 1) \sum_{j=1}^{k} n_j \lambda_j$$

$$\text{variance } (L) = \frac{1}{12}(N + 1)\left[N \sum_{j=1}^{k} n_j \lambda_j^2 - \left(\sum_{j=1}^{k} n_j \lambda_j \right)^2 \right]$$

$$\sigma_L = \sqrt{\text{variance (L)}}$$

The meanings of the symbols and their actual values in our example are:

N total sample size $= 9$

n_j individual sample sizes $= 3, 3, 3$

λ_j lambda coefficients $= 3, 2, 1$

k number of samples $= 3$

$\displaystyle\sum_{j=1}^{k}$ summation sign: sum all k instances of ...

We now have

$$E(L) = \tfrac{1}{2}(9 + 1)(3 \times 3 + 3 \times 2 + 3 \times 1) = 90$$

$$\text{variance } (L) = \tfrac{1}{12}(9 + 1)[9(3 \times 9 + 3 \times 4 + 3 \times 1)$$

$$-(3 \times 3 + 3 \times 2 + 3 \times 1)^2]$$

$$= 45$$

$$\sigma_L = \sqrt{\text{var}(L)} = \sqrt{45} = 6.71$$

These are the values used earlier in the calculation of Z.

Usefulness of the approximate method. The accuracy of the approximate method is often discussed in terms of how large the sample size needs to be before the approximation is 'valid'. This is somewhat misleading since the approximation is always useful: it simply becomes gradually less accurate as the sample sizes get smaller. A few comments are in order, however, to give the user a feel for the situation. These comments are based on a systematic exploration of the approximation to all of the exact tables of the distribution of L given in this book.

The most important point to bear in mind is that the user of this book is mainly concerned with decisions concerning the viability of the null hypothesis as an explanation of his results. He is, therefore, mainly interested in the accuracy of the probability estimates below the 5 per cent level of significance. Within this region the inaccuracy is very rarely greater than ± 1 per cent, even for very small samples. Therefore, for most *informal* purposes, the approximation is a very good basis for making a decision to accept or reject the null hypothesis. If the researcher believes that he really does need more accuracy than this, he should seek expert advice. The largest errors occur close to the 5 per cent region where the approximation almost always underestimates the true probability. Errors are at a minimum in the 0.03–0.01 (1–3 per cent) region, but become small and negative as the significance level gets less. Table 3.13 shows a typical pattern of errors for our worked example.

Correction for ties

So far our discussion has assumed that the ranks used in our computations are the integers from 1 to N, where N is the total sample size. Unfortunately, this

Table 3.13   Comparison of exact and approximate probability values based on table 3.10 and tables of the normal distribution

L	Exact probability	Approximate probability	Error (exact − approx.)
108	0.0006	0.0037	−0.0031
107	0.0018	0.0057	−0.0039
106	0.0047	0.0087	−0.0040
105	0.0107	0.0127	−0.0020
104	0.0185	0.0185	0.0000
103	0.0304	0.0263	+0.0041
102	0.0446	0.0367	+0.0078
101	0.0637	0.0505	+0.0132

assumption is not always justified. Very often some of the raw scores to be ranked are tied and we need to assign the same shared rank to a number of scores. The problem is particularly acute when we assign ranks to scores in frequency tables where all the scores in a given category share the same rank. When ties are present the distribution of L is different, and this fact needs to be taken into account in our procedures.

In general, ties make our basic test conservative; the more extensive the ties the more conservative the test becomes. This means that the true critical values of L are always lower when ties are present than the critical values of L for the situation where no ties are present. As a consequence, *we know that if our results are significant at a given level before correcting for ties then they will remain significant at that level after correction and may even be significant at a higher level.* We need only correct for ties if we are dissatisfied with the obtained significance level and wish to know the true (i.e. higher) level.

The effect of correcting for ties is typically very small except in situations involving the most extensive ties. By scanning through the examples in chapters 6 to 10, the reader will notice how small the effect is. However, ties can be very extensive in frequency table examples where many data points lie in the same category. In such circumstances, the correction for ties may make the difference between a significant and a nonsignificant result. In general, however, the beginner can afford to ignore the complications of correcting for ties in his first pass through the book.

The *randomization method* of evaluating L is easily adapted to cope with tied data. Whereas we normally put the numbers 1 to N in our hat, we now simply put the *actual* rankings (i.e. including the shared ranks) into the hat and proceed as usual. An analogous procedure can be used in a computer program, where the array of ranks to be sampled must now contain a list of the actual rankings based on the raw scores in the experiment.

The *exact method* of evaluating L is not easily adjusted to suit tied data. Most users of this method rely upon published tables and it would be very difficult to create and publish every variety of exact table adjusted for every conceivable configuration of tied rankings. Even when the user is able to evaluate his own exact tables, it is difficult to imagine many situations where the effort would be worth while. When using the exact method, it is usual to ignore the presence of ties except to acknowledge that this procedure is *conservative* and that the true significance level is certainly a little more impressive than that given by the exact tables.

The *approximate method* copes with the evaluation of L by using a computed correction factor (T) which compensates for the reduced variance of L in the presence of ties. Ties do not affect $E(L)$ but they limit the spread of individual values of L in a manner which can be computed thus:

$$\sigma'_L = \sigma_L \sqrt{T}$$

where σ'_L is the standard deviation of L *corrected for ties* and T is a correction factor. The correction factor is discussed more fully in chapter 6. T is never greater than one but is equal to one when there are no ties present. As a result, when there are no ties present, the correction factor can be ignored since it does not affect the result. For a further, more technical discussion of ties, the reader should begin with Bradley (1968, pp. 49-54).

3.3   Evaluating a nonspecific alternative to the null hypothesis

Measuring disagreement with the null hypothesis

The nonspecific alternative to the null hypothesis is simply that the medians of the populations (from which our samples were drawn) are *not* all equal. This alternative hypothesis does not specify which medians will be different, nor does it say anything about the ordering of these medians. It can be vindicated by any significant departure from equality. It is clearly going to be difficult to find any statistic to measure agreement between our obtained results and the nonspecific alternative hypothesis.

Fortunately, we can turn the problem on its head and look for a measure of disagreement between our data and the null hypothesis. When the null hypothesis is true, we expect our sample rank means to be approximately equal. We do not expect to find any large differences between them. In other words, we do not expect the sample rank means to vary to any great extent. Therefore any measure of the variance of the sample rank means is expected to be low. An indication of substantial variance between the sample rank means could therefore be used as evidence for rejecting the null hypothesis.

Calculating K, our statistic of disagreement

We could measure the variance of the sample rank mean in a formal way, but the following indicator of variability works just as well and is much simpler to compute:

$$K = \sum_{j=1}^{k} \frac{R_j^2}{n_j}$$

where R_j is the sum of the ranks in the jth group, n_j is the number of ranks in the jth group, $\sum_{j=1}^{k}$ means sum all k possible values of the following expression, and k is the number of samples. Our statistic K is small when the sample rank means are similar but is large when they are dissimilar. A large value of K might be offered as evidence against the null hypothesis.

The following two examples illustrate the calculation of K and show how it reflects the variability of the sample mean. Table 3.14 shows the computation of K for an example where the data have been arranged to show no variability of the sample rank means; the sample rank mean is 5.0 for all three samples. Table 3.15, however, shows the computation of K for an example where the data have been arranged to show the greatest possible spread among sample rank means.

Table 3.14   Computation of K for an example with minimum variability of sample rank means

	Sample 1	Sample 2	Sample 3
	1	2	3
	5	6	4
	9	7	8
R_j (rank sum)	15	15	15
R_j^2	225	225	225
n_j	3	3	3
$\dfrac{R_j^2}{n_j}$	75	75	75
$K = \sum_{j=1}^{3} \dfrac{R_j^2}{n_j} = 75 + 75 + 75 = 225$			
$\bar{R}_j$ (rank mean)	5.0	5.0	5.0

Table 3.15 Computation of K for an example with maximum variability of sample rank means

	Sample 1	Sample 2	Sample 3
	1	4	7
	2	5	8
	3	6	9
R_j	6	15	24
R_j^2	36	225	576
n_j	3	3	3
$\dfrac{R_j^2}{n_j}$	12	75	192

$$K = \sum_{j=1}^{3} \frac{R_j^2}{n_j} = 12 + 75 + 192 = 279$$

$\bar{R}_j$ (rank mean)	2.0	5.0	8.0

Our statistic goes from 225 at minimum variability to 279 at maximum variability. The reader is invited to try the computations for the same design with some other arrangement of the ranks. Since this will represent an intermediate degree of spread, he should expect a value of K between 225 and 279.

In table 3.16 we have computed K for this chapter's standard example. Our result is $K = 263$. This is, as expected, between the minimum of 225 and the maximum of 279. We now need to know how often a value of 263 or greater is expected when the null hypothesis is true.

Distribution of K using randomization experiments

We can evaluate the statistical significance of K using the same three methods as we used when evaluating L in section 3.2: the randomization method, the exact method and the approximate method. The basic principles are the same for all three as for the corresponding methods used in evaluating L, and will not be presented again in the same detail.

Table 3.17 illustrates the result of 100 randomization experiments for the design with three samples and three scores in each sample. The 100 values of K are printed to three decimal places at the top of the table. The summary table at the bottom omits the decimal places for clarity. We can see, at once, from the cumulative frequency column that our result of $K = 263$ was equalled or

Table 3.16 Computation of K for this chapter's standard example. Compare with tables 3.14 and 3.15

	Sample 1	Sample 2	Sample 3
	6	2	1
	8	5	3
	9	7	4
R_j	23	14	8
R_j^2	529	196	64
n_j	3	3	3
$\dfrac{R_j^2}{n_j}$	176.33	65.33	21.33
$K = \sum\limits_{j=1}^{3} \dfrac{R_j^2}{n_j} = 176.33 + 65.33 + 21.33 = 263$			
$\bar{R}_j$ (rank mean)	7.67	4.67	2.67

exceeded 13 times out of the 100 trials. Since this corresponds to a significance level of 13 per cent, it can hardly be offered as evidence against the null hypothesis. We normally require a value less than at most 5 per cent.

The randomization method is prone to error with only a small number of experiments. For serious work we need at least 10 000 experiments. However, for a useful informal estimate 1000 experiments are often quite adequate. Table 3.18 shows the summary table only of the results of a randomization experiment using 1000 experiments. On this occasion, our result of $K = 263$ is equalled or exceeded 78 times in 1000 or 7.8 per cent. This is much smaller than our previous estimate and illustrates the danger of using only a small number of randomization experiments. Nevertheless 7.8 per cent is still too large a value of the likelihood of K being equalled or exceeded to warrant rejection of the null hypothesis. Our result remains nonsignificant.

Distribution of K using the exact method

For the exact method we calculate K for every possible set of partitions of the ranks. The details of this technique are the same as for the evaluation of the distribution of L and are discussed in section 3.2. For our example we need to calculate K for 1680 different sets of three partitions of nine ranks. This was achieved using a computer program and the results are summarized in table 3.19.

Table 3.17 Results of 100 randomization experiments using three samples with three ranks in each sample

K

229.667	239	233.667	227.667	225.667
263	231	249.667	243.667	243.667
265.667	231	245.667	231	225.667
239	227.667	263	231	253.667
227.667	257.667	233	239	239
259.667	231	231	233	229.667
227	265.667	225	235.667	233.667
241.667	243	233.667	233.667	245.667
253.667	239	229.667	269.667	265.667
265.667	245.667	245.667	249.667	229.667
239	231	235.667	245.667	243
265.667	251	239	241.667	233.667
233.667	227.667	265.667	229.667	267.667
235.667	253.667	265.667	253.667	239
227.667	231	239	229.667	239
267	249.667	227.667	243	243
249.667	251	233	239	243.667
227	233.667	239	233.667	227
243.667	237.667	249.667	243.667	237.667
253.667	233.667	273.667	235.667	227

K	Frequency	Cumulative frequency	Cumulative probability
273	1	1	0.01
269	1	2	0.02
267	2	4	0.04
265	7	11	0.11
263	2	13	0.13
259	1	14	0.14
257	1	15	0.15
253	5	20	0.2
251	2	22	0.22
249	5	27	0.27
245	5	32	0.32
243	9	41	0.41
241	2	43	0.43
239	12	55	0.55
237	2	57	0.57
235	4	61	0.61
233	12	73	0.73
231	8	81	0.81
229	6	87	0.87
227	10	97	0.97
225	3	100	1

Table 3.18   Summary of 1000 randomization experiments using three samples with three ranks in each sample

K	Frequency	Cumulative frequency	Cumulative probability
279	3	3	0.003
273	5	8	0.008
269	11	19	0.019
267	20	39	0.039
265	23	62	0.062
263	16	78	0.078
259	18	96	0.096
257	37	133	0.133
253	33	166	0.166
251	24	190	0.19
249	63	253	0.253
245	54	307	0.307
243	70	377	0.377
241	21	398	0.398
239	51	449	0.449
237	66	515	0.515
235	30	545	0.545
233	121	656	0.666
231	51	717	0.717
229	97	814	0.814
227	112	926	0.926
225	74	1000	1

We can see from the table that $K = 263$ is equalled or exceeded on 144 out of 1680 occasions. This corresponds to a probability of 0.086. At 8.6 per cent our result is reasonably similar to the 7.8 per cent result of the randomization experiment, although it must be emphasized that the exact method is the preferred method. At 8.6 per cent our result remains nonsignificant.

In table 3.20 we have selected critical values of K which are significant at our four significance levels. Notice that we chose values of K associated with occurrences equal to *or less* than the corresponding percentage. This means that a value of K found to be significant at a given level is *at least* significant at that level; the actual percentage may be lower than that specified. No value of K occurs on less than 0.1 per cent of occasions, so no entry is made in the table. To be significant at the 5 per cent level we need a value of K equal to or greater than 267. Our

Table 3.19   Exact distribution of K for three samples of size three

K	Frequency	Cumulative frequency	Cumulative probability
279	6	6	0.003 6000
273.667	12	18	0.010 7143
269.667	24	42	0.025 0000
267.667	6	48	0.028 5714
267	36	84	0.050 0000
263.667	36	120	0.071 4286
263	24	144	0.085 7143
259.667	24	168	0.100 0000
257.667	54	222	0.132 143
257	12	234	0.139 286
253.667	48	282	0.167 857
251	32	314	0.186 905
250.667	16	330	0.196 429
249.667	60	390	0.232 143
249	36	426	0.253 571
245.667	72	498	0.296 429
243.667	72	570	0.339 286
243	36	606	0.360 714
241.667	36	642	0.382 143
239	96	738	0.439 286
237.667	120	858	0.510 714
235.667	54	912	0.542 857
233.667	144	1056	0.628 571
233	60	1116	0.664 286
230	96	1212	0.721 429
229.667	180	1392	0.828 571
227.667	84	1476	0.878 571
227	84	1560	0.928 571
225.667	108	1668	0.992 857
225	12	1680	1

result of $K = 263$ clearly does not meet that requirement, and we conclude that our result is not significant at the 5 per cent level.

Distribution of K using the approximate method

The approximate method for the evaluation of K depends upon the similarity of the exact distribution to the chi-square (χ^2) distribution. Figure 3.3. shows our

Table 3.20   Critical values of K extracted from table 3.19. K must be equal to or larger than the critical value to be significant at the corresponding significance level

Sample size	Significance level			
	5%	2.5%	1%	0.1%
3, 3, 3	$K \geqslant 267$	269.7	279	–

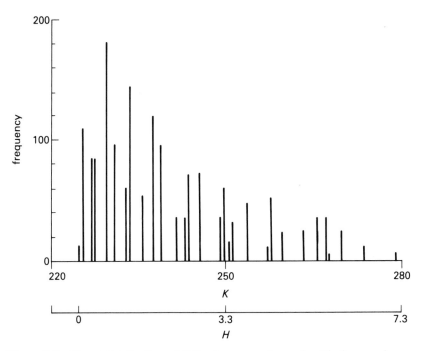

Figure 3.3   Exact distribution of K for three samples each with three ranks. Values are taken from table 3.19. The H values are rescaled K values to make use of the approximation to the chi-square distribution

exact distribution for three sample sizes and three ranks in each sample. Figure 3.4 shows examples of the chi-square distribution. Notice that the shape of the distribution changes according to the number of samples we have used. When using the chi-square distribution we need to use only that distribution which caters for the number of samples in our design. The exact distribution shown in

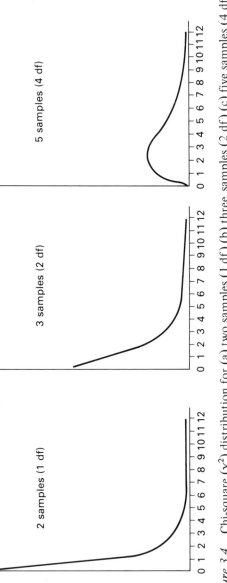

Figure 3.4 Chi-square (χ^2) distribution for (a) two samples (1 df) (b) three samples (2 df) (c) five samples (4 df)

figure 3.3 is supposed to approximate the chi-square distribution for three samples (figure 3.4b). The similarity is not very impressive for small samples, however.

To use the chi-square distribution we need to rescale our value of K so that it corresponds to the scale of the parent distribution. Different formulae are available for different designs. To illustrate, we can use the formula appropriate to a design with three samples:

$$H = \frac{12K}{N(N+1)} - 3(N+1)$$

where H is the rescaled value of K and N is the total sample size.
For $K = 263$ and $N = 9$,

$$H = \frac{12 \times 263}{9 \times 10} - 3 \times 10 = 5.07$$

We now need to know how often a value of 5.07 is equalled or exceeded in the chi-square distribution corresponding to the three-sample design. Table 3.21 gives some critical values, and we can see that a χ^2 value of 6.0 is needed to be significant at the 5 per cent level when we have three samples. Our value of $H = 5.07$ is therefore not significant at the 5 per cent level. This is the same result that we obtained using both the exact and randomization methods.

Correction for ties

The presence of ties in the data has the effect of reducing the spread of the distribution of K. This means that a correction factor needs to be applied when using the approximation to the chi-square distribution. In this book we use a correction factor which can be applied as the final step in the calculations:

$$H' = Hm/T$$

Table 3.21   Critical values of the chi-square distribution for two, three and four samples. χ^2 must be equal to or greater than the critical value to be significant at the corresponding significance level

Number of samples	Degrees of freedom	5%	2.5%	1%	0.1%
2	1	$\chi^2 \geqslant 3.9$	5.0	6.6	10.8
3	2	6.0	7.4	9.2	13.8
4	3	7.8	9.3	11.3	16.3

where H is the uncorrected value, H' is H corrected for ties in the data, T is the correction for ties and m is the number of blocks of data. Formulae for evaluating T are given in chapter 6.

The effect of the correction is *always* to increase the value of H, but often by only a very small amount unless ties are very extensive. As a result the test user may wish to examine his uncorrected value of H to decide whether the extra computations are necessary. *If the results are already significant at a satisfactory level or are far from significance then the correction for ties may be ineffective in bringing about any substantial change and may be omitted.*

Accuracy of the chi-square approximation

The approximation of the chi-square distribution to the exact distribution of H is, unfortunately, not very good and certainly not as good as the normal approximation to the distribution of L. The test user is therefore well advised to refer to exact tables whenever these are available. However, the general tendency is to be conservative for significance levels below 2.5 per cent, i.e. the estimate of the probability is typically greater than the true   probability for values less than 0.025. It is not possible to make categorical statements since the pattern of errors varies from design to design and we do not have many exact tables on which to base our comparisons.

3.4   Comparison of power of L and K procedures

Throughout this chapter we have repeatedly used the same data to illustrate the procedures under discussion. Nevertheless, we obtained quite different significance levels for the tests using specific and nonspecific alternatives to the null hypothesis. Using a specific alternative hypothesis and computing L, we achieved significance at the 2.5 per cent level (almost 1 per cent, in fact). However, with only a nonspecific alternative hypothesis and computing K we failed to achieve significance at even the 5 per cent level. This illustrates the following general rule:

> *If the null hypothesis is false, and if the only specific alternative hypothesis is true, then a statistical test based upon the specific alternative will always be more sensitive (i.e. more likely to reject the null hypothesis) than a test based on a nonspecific alternative hypothesis.*

In statistical jargon we say that a test based upon a true specific alternative hypothesis is more *powerful* than a nonspecific test. The matter has already been discussed briefly in chapter 2, where it was stressed that we can only legitimately take advantage of this increased power if the alternative hypothesis is generated *a priori* (on the basis of a valid argument before the collection of the data). Of course, specific tests are not always more powerful; they are only superior if the alternative hypothesis is true. If the alternative hypothesis is in fact false, then the test can be many times less sensitive than a nonspecific test.

3.5   More about chi-square (χ^2)

Some supplementary comments are required concerning chi-square to avoid confusions and misunderstandings. These concern the relevance of chi-square frequency tests, the meaning of the phrase 'degrees of freedom' (see table 3.21), and the close conceptual link between specific and nonspecific tests when only two samples are involved (one- and two-tailed tests).

Traditional chi-square tests for frequency tables (homogeneity tests)

It is unfortunate that certain well known tests for analysing frequency tables have come to be known as 'chi-square tests'. This gives the impression that they are the only tests which can deal with frequency tables using the chi-square distribution as the basis for an approximate method for estimating the significance level. This book introduces different procedures for analysing frequency tables and these new techniques also use the chi-square distribution. The two methods differ in that homogeneity chi-square techniques do not require that the categories of the dependent variable be ordered with respect to one another. Rank sum procedures for dealing with frequency tables can, however, only be used with rankable categories of the dependent variable. In this way, they take advantage of the order information available in the data and produce a more sensitive (powerful) set of procedures. When there are only two categories of the dependent variable the two types of test generate the same outcome, but when there are more the results are quite different (see chapter 4).

Clearly, the reader should not imagine that all tests which resort to the chi-square distribution are the same test, either superficially or structurally. To minimize confusion – since it cannot now be avoided altogether – we shall reserve the phrase 'traditional chi-square tests for frequency tables using nominal categories' for the popular group of tests which do *not* require that the categories of the dependent variable be rankable (see section 9.8).

Degree of freedom (df)

When consulting tables of the chi-square distribution, the reader will notice that distributions are differentiated by a quantity known as degrees of freedom. He will also notice that the distribution which is suitable for use with a three-sample design is said to have only two degrees of freedom. Most beginning students in statistics resent this apparently gratuitous exercise in confusion. Although we could, if pressed, completely ignore the 'degrees of freedom' concept without prejudice to our ability to cope with rank sum analysis, it is instructive to pause briefly and examine the matter further.

In rank sum analysis we use H (or K) to indicate the degree of spread of the K rank sums. However, we can think of this spread as a compound measure based upon the distance of each individual rank mean from the mean value of the ranks. Thus we are interested in the probability of observing k such events (or better events) if the null hypothesis were true. However, the rank sums are not strictly independent. We can always work out the last rank sum if we already know the other $k - 1$ rank sums. This is because the rank sums must add up to $N(N + 1)/2$ which is the sum of the first N ranks. In this sense the last rank sum is fixed and not free. As a result, our problem has only $k - 1$ degrees of freedom, despite appearances to the contrary, and this hints at why we use the $k - 1$ df variant of the chi-square distribution.

In fact, we could rewrite our formulae for computing H and K in a manner which made no reference to the last rank sum. The only reason why we don't do this is that it would be computationally much less convenient. For example, for three samples we have

$$K = \frac{R_1^2}{n_1} + \frac{R_2^2}{n_2} + \left[\frac{\dfrac{N(N + 1)}{2} - R_1 - R_2}{N - n_1 - n_2} \right]^2$$

Notice that K is computed without reference to R_3.

One- and two-tailed tests

For the special case where we have only two samples, the specific (L) and non-specific (K) large-sample approximation procedures for evaluating the null hypothesis come very close to being the same test. The reason for this is that the chi-square distribution for one degree of freedom is in fact a disguised version of the normal distribution. Figure 3.5a shows that the normal distribution is made up of z scores lying between plus infinity and minus infinity. If we take the square of each of these scores and plot the resulting distribution as in figure 3.5b, we will note that all values are now positive and that values close to zero are most numerous. This new distribution is identical to the chi-square distribution for one degree of freedom shown in figure 3.4a.

At first sight this suggests that we could use the normal distribution to evaluate H by taking the square root and referring this to the normal distribution:

$$Z = \sqrt{H}$$

Unfortunately, it is not so straightforward because the right hand tail of the chi-square distribution is equivalent to *both* tails of the normal distribution. There-

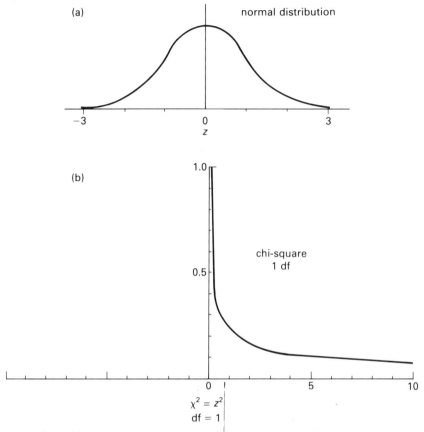

Figure 3.5   (a) Normal distribution and (b) chi-square distribution for one degree of freedom. The chi-square values are the squared values of z in the normal distribution

fore, if we wish to use the normal distribution to evaluate H we must find the area in *both* tails. This is why nonspecific tests have traditionally been known as 'two-tailed' tests. Fortunately, the normal distribution is perfectly symmetrical and we can find the area in both tails simply by doubling the appropriate area in the right hand tail of the normal distribution.

We can approach the relationship between the two tests from the point of view of the alternative hypothesis. When we have only two samples we can only have two specific alternatives to the null hypothesis: one that predicts that group A will have higher scores, and a second that predicts that group B will have higher scores. The nonspecific alternative embraces both possibilities whereas the specific test concentrates on only one. Thus, for any required degree of separation of the two groups, we are twice as likely to achieve this using the nonspecific procedure $(A > B \ or \ B > A)$ than using the specific test (only, say, $A > B$ acceptable).

Traditionally, the two approaches have been taught together even though conceptually they are far apart in the general case. This has caused considerable confusion in the minds of test users. Moreover, the one-tail/two-tail approach does nothing to prepare the beginner for the conceptual distinctions which need to be made for procedures dealing with more than two samples. No one-tail/ two-tail analogues exist with many-sample procedures, and procedures for testing specific and nonspecific alternative hypotheses do not grow naturally out of two-sample techniques. As a consequence, specific and nonspecific tests will be kept separate at all times even though this is a nontraditional approach.

Teaching experience has shown that the traditional approach to this business has given rise to much unnecessary confusion. Moreover, even a cursory examination of published research reports by experienced investigators using analysis of variance (ANOVA) shows that these confusions persist. In the case of ANOVA the distinction is between the F and t distributions. The F statistic should be used only for nonspecific alternative hypotheses and the t statistic should be used for specific alternative hypotheses (including trends, contrasts and predicted simple interactions). If these rules were observed, much nonsense would be avoided and a little more insight would creep into these difficult matters. With the unificiation of analysis of variance by *ranks* procedures, we have the opportunity to start again – this time, hopefully, with no confusion beween one- and two-tailed tests.

4 Types of statistical test

4.1 The problem

There are literally hundreds of different statistical tests which you might use to analyse your data from surveys or experiments. Although it is possible to argue that they constitute some of the most useful tests available, rank sum procedures form only a small proportion of all tests. Nevertheless, when you use them you are, by implication, preferring them to other tests. We must, therefore, pause to ask how they are different from other tests and why they are often chosen in preference to other techniques. Tests using ranks are not always the best tests and sometimes they are wholly inappropriate. This chapter will concentrate on the differences between rank sum tests and other tests and on the factors which help us decide which kind of test should be used. No attempt will be made to survey the whole knotty area of test selection in statistics, but the general issues surrounding tests using ranks will be introduced.

The tests in this book are all linked together by a simple family likeness. This is because they are all special cases of a single very general test. It is their membership of this family which distinguishes them as a group from other tests. There are many tests which use ranks and order information which are not included here. Some of these may look similar to the procedures described here and are, sometimes, equally useful. Failure to include them here implies no rejection of their value to the researcher. These tests (which include Kendall's rank correlation coefficient and Jonkheere's trend test) are not included because they do not fit into the analysis of variance by ranks scheme. Although these other tests are often just as sensitive and easy to use, there is no need to include them because adequate rank sum tests exist to test the same hypotheses. The discussion in the following is mainly intended to contrast rank sum techniques with tests which are very different in terms of both computational procedures and assumptions made about the quality of the data.

However, two tests have been included that are not part of the analysis of variance by ranks scheme. These are

(a) Wilcoxon's matched pairs, signed ranks test (sections 8.3 and 8.7)
(b) Chi-square (homogeneity) test for $k \times Q$ frequency table (section 9.8).

C

These are included for two reasons: firstly, to allow them to be contrasted with the tests which do belong to the scheme, and secondly, because their absence would weaken the book in its role as a source book for researchers. Rank sum analysis covers a very wide spectrum of applications but these two tests plug two significant gaps in the coverage.

4.2   Rank tests require rankable data

At a fundamental level, all of the tests in this book involve converting the data to ranks and then throwing the original data away. This may not always be obvious from a cursory inspection of the computational procedures of some of the tests (especially those using frequency tables) but it is so. Once the ranks have been found, the second step involves finding the sum of the ranks in each sample. The statistical test is then based on the rank sums for the samples. As a result the tests are called rank sum tests.

This means that rank sum tests can *only* be used if the data are rankable. Usually this presents no cause for confusion, but frequency tables may require care. Table 4.1 shows two frequency tables which look very similar at first glance but which require some thought before proceeding. In table 4.1a the categories of the dependent variable (exam performance) are clearly rankable. There are no problems here. In table 4.1b the situation looks very similar but is unsuitable for analysis using a rank sum test. This is because the three categories of the dependent variable (passed, failed, withdrew) are not obviously rankable. The reasons for withdrawing from a course are so variable (e.g. illness, boredom, lack of ability) that it is impossible to rank withdrawal as better or worse than failure. *One reason for not using analysis of variance by ranks occurs therefore when the dependent variable is not rankable.*

Table 4.1   Two frequency tables relating examination performance to year of the course entry. Note that in (b) the categories of the dependent variable (passed, failed and withdrew) are not truly rankable and therefore cannot be analysed in this form using rank sum analysis

(a)

	1980	1981	1982
61-100%	26	20	15
41-60%	14	16	15
Less than 40%	7	10	14

(b)

	1980	1981	1982
Passed	40	36	30
Failed	7	10	14
Withdrew	3	4	6

Table 4.2   Example of a frequency table which is suited to rank sum analysis even though the categories of the dependent variable (ice-cream flavour preference) and the independent variable (sex) are not apparently rankable. Two categories are always treated as rankable whatever their nature (see text)

	Boys	*Girls*
Chocolate	73	27
Vanilla	48	52

An interesting situation occurs when we have only two categories of the dependent variable (see table 4.2). It may seem that two categories are not rankable but, in fact, such frequency tables can *always* be analysed using rank sum techniques. The reason for this is a little obscure but two points should reassure the reader. Firstly, when we have only two categories, we achieve the same eventual result whichever way we rank the two categories (and there are only two ways). Secondly, the result using rank sum analysis is always the same as a result using the traditional chi-square (homogeneity) test (see section 9.8 and exercise 9.5, which does not assume that the categories are rankable). *This means that rank sum tests can be used whenever either the categories of the dependent variable are rankable or there are only two categories of the dependent variable.*

4.3   Information loss in rank tests

One of the problems of using ranks is that it may entail a loss of information. Remember that the original data are discarded once the conversion to ranks is completed. In certain circumstances the original values may contain crucial information with respect to the hypothesis being tested. Conversion to ranks may cause this information to be lost and a false research decision may be made (e.g. failure to reject the null hypothesis when it is false). Table 4.3 offers a contrived example designed to illustrate this possibility. In this example, each volunteer is tested twice – once after consumption of alcohol and once after consumption of an equal amount of water. Three of the volunteers (1, 3 and 4) were very much slower after alcohol whereas the remaining three (2, 5 and 6) were slightly faster. *On average* alcohol had a marked slowing effect (100 ms). Despite this large difference between the groups, the *rank sums* for the two groups are the same. Our ranking procedure is therefore insensitive to an effect which is readily seen using the original data alone. The problem occurs because our ranking technique is insensitive to the *size* of the difference between the scores in the two groups.

Table 4.3   A hypothetical example relating the effects of alcohol consumption to reaction time (measured in milliseconds). Notice that the control group (no alcohol) has a much faster mean reaction time even though the rank sum for both groups is the same

| Subject | Treatment group | | | |
| | Control (water) | | Alcohol | |
	Score (ms)	Rank	Score (ms)	Rank
1	374	*1*	593	*2*
2	384	*2*	380	*1*
3	310	*1*	440	*2*
4	275	*1*	520	*2*
5	380	*2*	375	*1*
6	420	*2*	410	*1*
mean score	357.2		453.0	
rank sum		*9*		*9*
median	377		425	

Sometimes the loss of information arising from the use of ranks can be serious, and then an accusation of *inefficiency* can reasonably be levelled at rank sum analysis. When the problem appears to be acute, the test user must switch to a more sensitive procedure. In this case a Student's *t* test for related scores or a Wilcoxon matched pairs, signed ranks test may be thought appropriate (see section 8.3). In doing so, the user must check that the data are suited to analysis by these other tests. Each test is valid only under certain conditions and there is no guarantee that an efficient test does exist to deal with any given set of data. An inefficient test is sometimes the only alternative.

The problem of inefficiency is not as serious as it might seem at first inspection. The example given in table 4.3 was specially constructed to magnify the effect. The two-sample matched pairs design is probably the least efficient of all the possible designs which can be analysed using rank sum tests. The data values in table 4.3 were also specially manipulated to highlight the possibility of information loss. In general, the effect is much less severe.

There is no information loss whatsoever in certain cases. These occur when the data are *collected in the form of ranks* (e.g. preference ratings) or in the form of ranked category frequencies (e.g. poor, fair, good). The problem of information loss only arises when the raw data begin life on a finely differentiated scale before they are transformed into ranks.

It is possible to become attached to information for its own sake, without pausing to ask whether it is relevant to the question being asked. Sometimes the ranking process may have the healthy effect of stripping the data of unreliable information. For example, the data in table 4.3 showed that there was a clear drop in performance following alcohol consumption. Nevertheless, it remains true that exactly half of the people in the experiment actually *improved* their reaction speed following alcohol, whereas half *showed a drop* in performance. Looked at this way, our results suggest that alcohol is just as likely to cause an improvement as a drop in performance. This seems like a paradox, but the matter can be understood in terms of the particular interests of the researcher. Does he wish to stress the typical *size* of the change in reaction time or does he wish to know merely the expected *direction* of the effect of alcohol? In the latter case, rank sum analysis is to be preferred. Otherwise, he should choose another test.

Sometimes it is not a matter of choice. Sometimes the information lost by using ranks is best lost by any criterion. It is difficult to supply a clear set of rules which define the situations where this is true, but a simple example will give the flavour of the problem. Consider two schoolboys: the first improved his score in a spelling test from 90 to 95 per cent correct whereas the second improved his score from 60 to 70 per cent correct. Which boy showed the greatest improvement? Clearly the second boy showed a 10 per cent improvement where the first showed only a 5 per cent improvement. However, you could argue that a 5 per cent improvement from 90 per cent was more difficult to achieve than a 10 per cent improvement from 60 per cent. The matter is not easily resolved simply in terms of arithmetic. Here, the information about the absolute size of the difference between the pairs of scores is not very useful and may even be misleading. Ranking may lose some of the detailed information present in the data, but that information which is preserved by ranking is nearly always certain and useful. In this case, the information preserved is that both boys improved.

In summary, we may say that tests using ranks can make less efficient use of the available data than some other tests. However, this is *only* true when the original scores lie on a finely differentiated scale and where the absolute values of scores can be meaningfully and consistently related to the hypothesis under test. The loss of information leads to inefficiencies which are only a matter of degree (to be discussed later). Sometimes the effect is severe but more frequently the effect is slight. In some circumstances, the loss of information may be beneficial in that it restricts the analysis to those aspects of the data which are both certain and useful.

4.4   Rank sum tests are distribution free

One of the advantages of using rank sum tests is that they do not involve any restrictive assumptions about the populations of scores from which the samples have been drawn. For example, some tests assume that the parent populations are normally distributed. Such tests form a category which is often loosely described as 'parametric'. Tests which do not make the assumption of normality are, equally loosely, called 'nonparametric', although the term 'distribution free' is more explicit and to be preferred. Rank sum tests are examples of distribution-free tests. Analysis of variance (ANOVA) and Student's *t* tests are examples of parametric tests.

Parametric tests have the capability of dealing directly with the original scores collected by the researcher. This avoids the loss of information associated with categorizing data or converting to ranks (see section 4.3). Such tests can also capitalize on the extra information implied by the knowledge that the parent populations are normally distributed. These two properties combine to make parametric tests into very sensitive analysis tools. However, they also limit the scope of the tests to situations where the crucial assumption of normality is valid. The advantage of rank sum tests is that they remain valid even when the normality assumption is violated. Rank sum tests may be slightly blunter tools than parametric tests but they are capable of dealing with a much wider range of types of data.

When the samples are drawn from normally distributed parent populations, either parametric or distribution-free tests can be used. However, when the samples are drawn from nonnormal parent distributions, distribution-free tests are necessary. When the parent populations are normally distributed, a parametric test always makes more efficient use of the available information in the data. The increase in efficiency varies in different situations and the matter is discussed more fully later. Although it may be inefficient to use a distribution-free test, it is certainly not an error or in any way illegal. When the population sampled is not normally distributed, then it is, strictly speaking, illegal to use a parametric test such as analysis of variance. The seriousness of this offence again varies from one data set to another in a complex manner. The availability of distribution-free tests means that this is a risk we need not take.

4.5   Rank sum tests concern population medians

All statistical tests attempt to permit broad generalizations based on a limited number of observations. We attempt to generalize from samples to the populations from which the samples were randomly drawn. One important distinction between most parametric tests and rank sum tests involves the nature of the inference made. Parametric tests make inferences about the *means* of populations whereas rank sum tests permit inferences about the *medians* of populations. Both means and medians are measures of 'central tendency' or 'middle point' of a collection of scores but they sometimes give different results. In table 4.3 the means and medians have been calculated for both groups. For neither group are the mean and median the same. In addition, there is a difference of 95.8 ms between the means of the two groups but only 48 ms between the medians. The divergence between means and medians is not always so great but the distinction must be borne in mind at all times. Although the same data may be analysed by both parametric and distribution-free tests, we must not casually assume that both tests are seeking answers to the same questions.

In fact, the matter goes a little deeper than this and concerns what statisticians call the 'model' which is used to conceptualize the problem. In parametric analysis the statistician assumes (rather simple-mindedly) that all the scores in a given group are increased or decreased by a constant which is characteristic of that group. His problem is to decide whether those constants are significantly different from zero. In the example given in table 4.3, we might say that alcohol slows down all reaction times by 95.8 ms. Although such a simple model is clearly unlikely to be the case, we assume it to be so as a convenience.

Rank sum tests have the virtue of making no such assumption. In this case the statistician asks (for the two-sample cases): 'If I were to choose a score (A) at random from one population and choose another score (B), also at random, from a second population, is it an equal chance that A will be greater than B?' In our example he is asking whether a randomly chosen normal reaction time is expected to be the same as (or less than) a randomly chosen reaction time influenced by alcohol. Parametric and rank sum tests are obviously asking different questions at this level – different questions which overlap considerably but sometimes give different answers.

4.6   Varieties of distribution-free tests

So far, the distinction between parametric and distribution-free tests has been emphasized. However, rank sum tests constitute only one category of distribution-free test. It is difficult to provide a simple map of this area of statistics because there are so many tests doing so many different jobs that no one has yet managed to sort them out into neat categories. It is the main purpose of this book to round up a number of these tests, show their fundamental similarity and then pack them neatly into a single framework for more convenient use. These tests have traditionally adopted their own names, notations, procedures and rationales (see chapter 5) but they can all be understood in terms of rank sum analysis. Even when this tidying-up operation is completed, we have only accounted for a proportion of the total number of published tests.

How then do rank sum tests differ from other distribution-free tests? In some cases the distinction is easily drawn in terms of the information used by the test. Rank sum methods require that the sample scores or categories be rankable whereas some other distribution-free tests are specially designed to cope with scores and categories which *cannot* be ranked (in so-called 'nominal scales'). They are obviously complementary to rank sum tests; they do a different kind of job.

Even so, there remain many kinds of distribution-free tests which use the same kind of order information as rank sum techniques and yet deliver slightly different answers to slightly different questions. It is the very existence of such tests which makes us acutely aware of the need to be conversant with the exact question asked by each and every test we use. No good can come of the common habit of thinking simply of a 'difference between two groups' or a 'trend across groups'. There are all kinds of differences and all kinds of trends; each test concerns itself with a different kind. One of the benefits of rank sum analysis comes from the fact that all of the tests in this framework are concerned with the same kind of difference and trend. Once the central principle has been grasped, it remains valid throughout the system. In contrast, other distribution-free procedures have a variety of different procedures and rationales.

It is not implied that any given rank sum test need be better suited to a given task than another, suitably chosen, distribution-free test. They are often very similar in terms of the efficiency with which they use the information available in the data. The main advantage of the rank sum system of tests presented in this book is that a wide range of tasks can be dealt with in a simple and consistent conceptual framework.

4.7   When rank sum tests should be used

Reasons for and against choosing a rank sum test in any given analysis can be broadly clsssified into the four areas: legality, logic, pragmatics and efficiency. *Legality* specifies the minimum conditions required before a test can be used. *Logic* is concerned with the nature of the conclusions we wish to draw. *Pragmatics* highlights the varying computational effort of using different methods as well as the issue of availability of a suitable test. *Efficiency* is concerned with the ability of tests to make use of the information available in the data. All of these topics are characterized by imprecision and controversy. The following sketch map merely highlights some of the more obvious landmarks in this treacherous territory.

Legality

Rank sum tests may *only* be used when the categories of the dependent variables are unambiguously rankable. If they are not rankable then a test for nominal categories must be used (e.g. chi-square homogeneity test). Even this simple restriction does not apply if there are only two categories of the dependent variable (see section 4.2). If a researcher wants to use a rank sum test even when the categories of the dependent variable are not rankable, he may look around for some meaningful way of combining categories so that he ends up with two categories. An example of this technique is given in table 4.4, where the categories 'failed' and 'withdrew' are combined into a 'did not graduate'

Table 4.4   An example of combining categories of the dependent variable to permit analysis by rank sum methods (see text and table 4.1)

	1980	1981	1982
Passed	40	36	30
Failed	7	10	14
Withdrew	3	4	6

	1980	1981	1982
Graduated	40	36	30
Did not graduate	10	14	20

Table 4.5   Sometimes a frequency table can be legitimately 'flipped' to permit analysis by rank sum methods

Independent variable

		Day duty	Night duty
	Present	41	34
Dependent variable	Sick	5	8
	Absented	4	8

Independent variable

		Present	Sick	Absented
Dependent variable	Day duty	41	5	4
	Night duty	34	8	8

category. This is an acceptable technique as long as the contrast between graduating and not graduating is relevant to the hypothesis under test. Another technique, illustrated in table 4.5, is to 'flip' the table on its side by swapping the dependent and independent variables. The user must take care that the analysis is still meaningful in terms of his original hypothesis.

The other 'legal' aspect of using a rank sum test is that it is often the test of choice when certain other eligible tests would themselves be illegal. For example, when the populations sampled are not normally distributed, a parametric test becomes, strictly speaking, illegal. It is not always a serious error to use a parametric test in such circumstances, but it is clearly wiser to use a distribution-free procedure as a conservative strategem. Sometimes it is possible to convert non-normal distributions to normal distributions by the simple expedient of transforming the scores by converting to logarithms or reciprocals. Although this may satisfy the legal requirements and permit the use of a parametric test, the score transformation can severely distort the hypothesis under test in a way which most users may not understand. For example, a difference in the mean logarithm of reaction times between two groups is not the same thing as a difference between mean reaction times. The problem can become acute when trends and contrasts are evaluated. A linear trend observed across scores on a logarithmic scale is not the same as a linear trend found using the original scores. Interaction effects found using raw scores may disappear following logarithmic transformation, and so on. If the user is not yet familiar with the implications of transforming scores, he is advised to leave the scores intact and settle for a distribution-free test such as rank sum analysis.

Logic

A rank sum test should be used when the hypothesis tested requires it. Rank sum tests permit inferences concerning the *ordering* of populations which have been sampled. Away from the physical sciences, a researcher is primarily concerned to know whether a given influence affects measurements on a particular variable. He merely wishes to show that when the influence is present the measurements show some increase (or decrease), and that this effect is loosely related to the strength of the influence. Rank sum tests are well suited to this kind of imprecise hypothesis. They are less well suited to situations where a researcher postulates a more precise relationship between variables (e.g. each additional cigarette smoked per day for 20 years subtracts one year from the smoker's life expectancy). Such precise hypotheses are best handled by parametric methods.

It is difficult to supply a simple set of principles to decide whether a given analysis is logically consistent with the aims of an investigation. Much depends upon the understanding of the investigator, who must bear in mind at all times the precise question which is being posed by a given statistical technique (see sections 2.10 and 4.5). Even a very simple understanding at an intuitive level is enough to avoid some of the commonest errors.

Pragmatics

For very simple experimental designs the researcher is typically faced with a bewildering array of alternative tests, but for complex designs with many independent variables there may be no test available at all. More often there is only one test, and the only sensible (pragmatic) approach is to use this test and then argue afterwards about the merits or demerits of doing so. Strictly speaking the test may be illegal, logically not quite appropriate, and very inefficient, but the results of the test will rarely be wholly misleading. One dodge is to adopt a conservative strategem and accept only significance levels of 1 or 0.1 per cent. Then you can argue that the true result will either be *more* significant, which is fine, or *less* significant, in which case it is hardly likely to be significant at less than the 5 per cent level - which, for many purposes, is also fine. This is not an ideal procedure but it is a lot better than simply abandoning your data.

Another pragmatic consideration is computational effort. Some tests are very easy to carry out in a few minutes with pencil and paper, whereas others can only realistically be carried out on a computer. There is a great deal to be said for choosing the least effortful test wherever there are at least two appropriate alternatives. Beginners often associate effort and computational complexity with respectability in statistics, but nothing could be more misguided. The most elegant method for answering a well expressed statistical question often involves the simplest of statistical tests. Simpler tests are easier to understand and result in less conceptual confusion. Their simple procedures are also less likely to give rise to computational error. Statistical significance achieved using elegant quick tests is just as sure a basis for inference as the result of the tedious and subtle manipulations of the more mathematically advanced techniques. Unfortunately, quick tests are often inefficient, and this is held against them by some purists. However, this is not serious, since the user is always free to have another attempt with a more efficient and laborious test if he fails to achieve significance first time round. This is not cheating - just effective utilization of the resources at his disposal.

The rank sum tests in this book are typically less computationally effortful than corresponding parametric (or some other distribution-free) alternatives. This is not always true when the problem involves large amounts of data which need to be ranked. In such cases the user is advised to recast his data in terms of a small number of categories (possibly only two) and attempt a quick analysis using a frequency table method. If significance is achieved using this (less efficient) method, there is no need to bother with the full ranking method. since this would merely be more significant still.

Efficiency

The most controversial aspect of rank sum tests involves their efficiency in comparison with other tests, in particular parametric tests such as analysis of variance. Research continues to discover the true relative efficiencies of these tests and the implications of this for statistical practice. It is a difficult topic for the beginner to digest and it is not clear, for reasons given later, that it matters all that much for our purposes. The interested reader will find that discussions abound in the literature, but a most digestible introduction is to be found in Marascuilo and McSweeney (1977, chapter 4).

Although the topic of relative efficiency is a most legitimate area of investigation for statisticians, an appreciation of the issues does little to improve the statistical practice of scientific researchers in their various far-flung disciplines. The reason for this is not clear at first sight since efficiency should be a prime concern for the hard-pressed investigator. If a more efficient test is a test which requires fewer observations to achieve a given level of significance in a particular *experiment*, then a more efficient test should be a time and money saver; so why is it not a pressing issue for the researcher?

The first reason lies in the triviality of the major conclusion of relative efficiency studies. This conclusion is that tests tend to be most efficient in the very situations for which they were designed. For example, when the data are sampled from normally distributed populations then parametric tests turn out to be most efficient. This is not surprising, since parametric tests make use of the information that the parent populations are normally distributed whereas rank sum tests do not make use of it. If the parent populations are not normally distributed, then things are different. In this case the parametric test is assuming something to be true which is not true, and this false assumption may put the parametric test at a disadvantage. Clearly, the researcher can avoid many problems by simply applying the maxim: 'Use a test in the situation for which it was designed.' Rank sum tests are designed for situations in which no confident statement can be made about the distribution of the populations which are the source of our data.

The second reason why efficiency studies do not help the researcher economize is that the test is not applied to the data until after they have been collected. Therefore, the use of the test comes too late to allow the investigator to economize on the number of observations. A more efficient test does not, in practice, lead to more efficient experiments but merely increases the likelihood of achieving significance with the available data.

One could imagine scenarios whereby informed calculations were made, in advance, as to how many observations would be necessary (taking into account the efficiency of the test to be used). Indeed, this is often recommended by advanced statisticians. However, it is no easy task since it assumes that the researcher has a much clearer idea of the required results than is typically the case. Moreover, what constitutes a 'reasonable' sample size in the eyes of the scientific community is typically at variance with the principles of statistics. An author is unlikely to get a result accepted for publication using a sample size of three – even if it is statistically highly significant! A wise scientist, therefore, chooses his sample sizes according to the general practice of this colleagues. As a result there is rarely any actual saving of time or money as a result of using an efficient test.

Tests applied in situations for which they were *not* designed are not, in fact, always greatly reduced in efficiency. In our case, a rank sum test used in a situation best suited to parametric analysis *may* be only marginally less efficient (95 per cent) than the more correct parametric test. This encourages us to use rank sum tests even when parametric tests are strictly appropriate. The only cost of doing so is a slight increase (5 per cent) in the required number of observations, and the main benefit is the economy of always using the same test. This kind of argument is one of the reasons why distribution-free tests have become so popular in the last two decades.

Of course, it is not quite so simple. Rank sum tests are not always so efficient. The user can usually spot the situations where rank sum analysis is inefficient. These occur where the data are reduced into a small number of categories. The solution to the efficiency problem, here, is simply not to reduce the data in this way. (If the data are categorized on arrival then you could not use a parametric test in any case, so the problem of choice does not arise.) The most inefficient design is the two-sample matched pairs arrangement illustrated in table 4.3. Here, each score in a pair is ranked 1 or 2 (i.e. is forced into a two-category system). This produces a dramatic loss of efficiency which is not easily remedied. If the rank sum test does not achieve significance the user must try the (parametric) paired t test or the (distribution-free) Wilcoxon matched pairs, signed ranks test.

4.8 Summary

In summary, there are many aspects to the problem of choosing the appropriate test for a given set of data. The first step is to narrow down the range of tests to those which are legally acceptable and those which are constructed to deal with your experimental design. If the sampled populations are normally distributed the main choice is between parametric and distribution-free tests. Parametric tests make more efficient use of the data but the savings are often small (except in the two-groups matched pairs design). If the parent population distribution is unknown, then the choice is between varieties of distribution-free tests. In this case the decision may simply rest upon computational simplicity. In general, tests which reduce the data to a small number of categories yield a reduction in computational effort at the expense of a loss of efficiency. Such tests are best when there is a great deal of data or when there are clear-cut differences between the samples. Rank sum tests offer a choice of fine-grain or frequency table methods but have the advantage of being unified into a single conceptual and computational framework. Although rank sum tests constitute only a fraction of all distribution-free tests, a rank sum equivalent exists for most other distribution-free tests.

5 Varieties of experimental design

5.1 Introduction

The statistical procedures in this book constitute a family of techniques for dealing with a wide range of different styles of experiment. Each test comes from the same stable but has been specially tailored for a particular kind of problem. In principle we need only one test and we could apply this 'omnibus' test to the whole range of designs. However, this is a very cumbersome test to work through by hand since multipurpose tools are always more complex and less manageable than streamlined special-purpose tools. We human calculators need procedures which are economically suited to particular situations, which minimize the amount of calculation needed and which allow us to keep a general grasp of what we are doing. These procedures are outlined in chapters 7–10. Although different they all follow a basically similar computational plan, and this is outlined in chapter 6 together·with detailed comments on particular numerical techniques which they have in common.

5.2 Describing different designs

Before attempting to carry out any of the tests we must be able to specify the design in simple terms, since the test we use depends upon the design of the experiment. The two basic concepts in describing a design are *samples* and *blocks* (see section 2.3). The simplest design of all is the control experiment consisting of two independent random samples. We used such an example in chapter 2 where a set of patients was randomly assigned to two groups; one group received ECT treatment, the other did not. In our terminology this was a *two-sample, single-block experiment*.

In that experiment we used only patients from one hospital, but a preferred procedure might have been to repeat the experiment in, say, three hospitals. Assuming that we are not specifically interested in the differences between hospitals because we only want to check on the generality of our results, we give the name 'blocks' to the data subsets formed by the different hospitals. Table 5.1

Table 5.1   Example of a two-sample, three-block experimental design. The scores are post-treatment depression ratings of ten patients from each of three hospitals in an (imaginary) investigation of the efficiency of electroconvulsive therapy (ECT) for endogenous depression

Blocks	Samples	
	ECT treatment	*Control (no treatment)*
Hospital A	16	18
	14	24
	15	26
	17	19
	21	20
Hospital B	17	16
	13	26
	18	23
	15	21
	23	18
Hospital C	19	24
	17	26
	18	26
	14	16
	21	20

illustrates the design. In our terminology, this design is a *two-sample, three-block design*. Table 2.2 gives another example of a two-sample, multiple block design.

We can carry our experiments with any number of samples. We might add a third sample to the design given in table 5.1 consisting of patients who receive two courses of ECT. It would then be a three-sample, three-block design. Likewise, we could vary the number of blocks by increasing or decreasing the number of hospitals sampled by our investigator. In the original study, the data were not subdivided into blocks because all patients came from the same hospital. In such a case we say that the design has only one block. It is a two-sample, one-block experiment.

The first step when choosing an appropriate test is to identify the design in terms of the number of samples and blocks. The major distinction between designs is whether they have two samples or more than two samples. Two-sample designs are covered in chapters 7 and 8 and many-sample designs are covered in

Table 5.2   Layout of chapters against designs dealt with

	Two samples	More than two samples *(many samples)*
Single block	chapter 7	chapter 9
Multiple blocks	chapter 8	chapter 10

chapters 9 and 10. The other major distinction between designs is whether they have a single block or more than one block. Single-block designs are covered in chapters 7 and 9 and many-block designs are covered in chapters 8 and 10. The system is summarized in table 5.2. We can see from table 5.2 that the two-sample, three-block design, illustrated in table 5.1, is catered for by procedures described in chapter 8. To keep our language simple we use the terms 'multiple blocks' and 'many samples' to represent 'more than one block' and 'more than two samples' respectively.

5.3   Samples and blocks

Usually it is a simple matter to decide which are the samples and which the blocks but it is important to consider how we know which are which. This will be useful later when the distinctions are sometimes blurred. One approach is to decide which variables are central to the hypothesis being studied. The levels of the *independent variable* will then identify the samples for us and help to distinguish them from the blocks.

Dependent and independent variables were discussed and defined in section 2.3. It should be clear that the key variables in the experiment referred to in table 5.1 are

(a) levels of treatment (treatment/no treatment)
(b) hospital (hospitals A, B and C)
(c) depression rating (on a scale from 0 to 30).

Our hypothesis specifies that we expect ECT treatment to reduce depression. We therefore identify treatment as our independent variable and depression rating as our dependent variable. The two levels of the independent variable define our samples (treatment/no treatment).

The hypothesis does not mention hospitals. They constitute *repetitions* of the basic experiment. We repeat the experiment in different hospitals in order to check that the effect (if any) of the ECT treatment is general to all hospitals and not just restricted to a single institution. Hospitals constitute the blocks variable and it is clear that *the different values of the blocks variable should ideally be selected at random*. To convince ourselves that the effect will work in any hospital we clearly need to sample the hospitals at random from the set of all appropriate institutions. Needless to say, this is an ideal which is only occasionally met in reality.

Notice that a researcher is not specifically interested in the differences between blocks. In our case it is almost certainly (and trivially) true that different hospitals will be dealing with different average levels of depression for all kinds of practical reasons. The purpose of separating patients into blocks is to preserve the principle of comparing like with like. Thus patients in hospital A should be compared only with patients in hospital A and not with patients in hospital B who may be different in many respects. If we did not segregate the scores into blocks, the effect of treatment would become confused with the effects of differences between hospitals. This confusion would reduce the sensitivity (power) of the test.

In summary, *samples are defined by levels of the independent variable whereas blocks are the result of repeating the basic experiment*.

5.4 Repeated measures

One of the most important applications of the blocks principle is that of repeated measures on the same subject. For example, the same animal or person may be treated with a number of different drugs to test a response of some kind. In this case the drugs represent the independent variable, the responses represent the dependent variable and the subjects (people, animals etc.) constitute the blocks. Since each subject experiences all the treatments he constitutes a mini-experiment. Multiple subjects are therefore repetitions of the basic experiments. *Subjects must therefore be a random sample of all possible subjects (i.e. blocks)*.

Despite the popularity of the repeated measures design, it is subject to many problems and it must be used only with great care (Poulton, 1973). When a person is exposed to a number of treatments one after the other, he is clearly subject to order effects, practice effects, sensitization effects etc. The experimenter is also likely to treat a person differently as he becomes more familiar with him. These effects can be minimized by randomizing the order of presentation or by using a balanced set of orderings across subjects, but there is no guarantee that these effects will be totally elimimated. These techniques certainly do reduce the contamination, but the effects are too subtle to be removed entirely by such simple ploys. Instead, the experimenter must agree to live with the uncertainty generated by the design in return for the economy and efficiency gained by using it. After all, if the results look promising, the experimenter can always repeat the experiment using a different design to confirm the preliminary results obtained using a repeated measures design.

5.5   Multiple independent variables

The greatest confusion between samples and blocks usually arises when the design is complicated by additional independent variables. Consider the design outlined in table 5.3. Here we have divided the patients in the study into males and females. At first sight it is tempting to class this as a two-sample, two-blocks design. This would be acceptable if sex were intended merely as a subdivision of the samples to test the generality of the results (that is, if the hypothesis to be tested did not concern itself with differences between the sexes).

If the experimenter is interested in the differences between males and females, the male/female distinction also becomes an independent variable, which complicates matters. The matter is simplified by the observation that we are obviously dealing with two separate hypotheses:

(1) that ECT treatment causes a reduction in depression ratings
(2) that males and females differ in the severity of their depression.

If this is the case, then the two hypotheses could be analysed separately. In the case of hypothesis 1, we can use treatment as the independent variable (samples) and consider the two sex groups as blocks. In the case of hypothesis 2 we can consider the two sex groups to be samples and use the treatment groups as blocks. This simply involves flipping the table on its side as in table 5.4 and proceeding normally. In this way, we are asking whether there are sex differences in depth of depression *irrespective of whether they are treated by ECT or not*.

More often, researchers wish to consider an even more complex hypothesis which involves both variables:

Table 5.3   Schema of an experimental design involving a possible second independent variable

	ECT treatment	*No treatment*
Male	*x*	*x*
	x	*x*
	x	*x*
	x	*x*
Female	*x*	*x*
	x	*x*
	x	*x*
		x

Table 5.4   Redesign of table 5.3 to test hypothesis concerning sex differences. Compare with table 5.3

Blocks	Samples	
	Male	Female
ECT treatment	x	x
	x	x
	x	x
	x	
No treatment	x	x
	x	x
	x	x
	x	x

Table 5.5   Redesign of tables 5.3 and 5.4 to test a hypothesis involving two independent variables

M_t Male, ECT treatment	F_t Female, ECT treatment	M_c Male, no treatment (control)	F_c Female, no treatment (control)
x	x	x	x
x	x	x	x
x	x	x	x
x		x	x

(3) that the female patients respond better than the male patients to ECT treatment in terms of depression scale ratings; *or*

(4) that male patients respond better to treatment.

Both variables are clearly involved in the hypothesis, and we have *two independent variables* (sex and treatment). Since we have two levels of each independent variable (male/female and treatment/no treatment), we end up with four samples as shown in table 5.5. This is now a four-sample, one-block design and can be analysed accordingly.

If we adopt hypothesis 3 (that females respond better to treatment), we must express our hypothesis clearly in terms of the rank ordering of the four expected sample rank means (see section 3.2). This has its own difficulties. For example, we may suggest that treated females (F_t) will show lower depression scores than treated males (M_t). They in turn will show lower scores than the control males (M_c) and control females (F_c) who should be similar. Expressed in shorthand, this might read

$$M_c = F_c > M_t > F_t$$

But the assumption that untreated (control) males and females are expected to have similar depression ratings is itself subject to alteration according to whether males or females are typically more depressed.

Parametric analysis of variance (ANOVA) has the same problem when dealing with interactions, but these are obscured from the test user by ritualistic application of computational formulae. The solution adopted by ANOVA in these circumstances is to hypothesize that treated females and untreated males when combined into a single group will have lower depression scores than treated males and untreated females combined into a single group. Expressed in shorthand it might look like this:

$$F_t + M_c > F_c + M_t$$

where + represents the union of two samples. It looks to be a curious solution, but it is sensitive to departures from a simple outcome where male and female patients are equally prone to depression and equally affected by the treatment. It is, of course, most sensitive to the situation where females benefit from the treatment but males get worse! If this approach is satisfactory to the user, then all of the ANOVA apparatus for handling interactions can be generalized with suitable care to rank sum analysis. This application is discussed further in chapter 12.

In conclusion, we must recognize that it is *the hypothesis which determines whether a variable is an independent, a dependent or merely a blocking variable*. In some situations, a number of variables can all stand as independent variables. This multiplies the number of samples and puts a special strain on the hypothesis to predict the relative ordering of all of them.

5.6   Frequency tables

So far we have only considered problems where the data have been explicitly presented score by score in the results table. A commonly used alternative method is the frequency table. Table 5.6 illustrates the use of a frequency table to present data from an experiment which is *structurally no different* from the

Table 5.6 Frequency table presentation of a two-sample, three-block design where each patient's improvement was rated on a three-point improvement scale

Blocks	Categories of the dependent variable	Sample	
		ECT	*No treatment*
Hospital A	improved	5	3
(block 1)	same	2	3
	worse	0	1
Hospital B	improved	7	4
(block 2)	same	1	3
	worse	0	0
Hospital C	improved	8	6
(block 3)	same	0	1
	worse	0	1

two-sample, three-block design illustrated in table 5.1. The important superficial difference is that the figures in the body of the table do not represent depression ratings of individual patients but specify the number of patients (frequency) being classed in a particular improvement category.

In table 5.6 the blocks are represented by *separate* frequency tables, and the rows of each table represent levels of the dependent variable (improved/same/worse). It is important not to confuse the two. This might be very easy since in table 5.1 the rows represented blocks whereas in table 5.6 the rows represent the dependent variable. This should not prove to be a problem, however, if the reader is relying on a proper intuition of the experimental design rather than superficial aspects of the data layout. Beginners are often caught out by the simple layout given in table 5.7 where the three categories of the dependent variable are likely to be confused with blocks. *The simple rule with frequency tables is 'one block per table'.* If we have only one table then we have only one block.

In this book we adopt the convention that samples (levels of the independent variable) are spread across the page from left to right, whereas blocks go from top to bottom of the page. For frequency tables, the levels of the dependent variable are also arranged down the side within the blocks. This is not

Table 5.7   Two-sample, one-block design presented in frequency table layout

	ECT treatment	*No treatment (control)*
Improved	5	3
Same	2	3
Worse	0	1

Table 5.8   Raw score tables derived from the frequency table 5.7

(a)

ECT	*Control*
better	better
better	better
better	better
better	same
better	same
same	same
same	worse

(b)

ECT	*Control*
1	1
1	1
1	1
1	0
1	0
0	0
0	-1

universal practice, however, and these superficial aspects of table layout should never be relied upon as a sole basis for deciding upon the design.

The use of frequency tables to present data is merely a stylistic feature and does not, it itself, represent a different kind of design. The data in table 5.7 could be presented in terms of the original raw scores where the assessment of every patient is presented explicitly. It can be seen that five people in the ECT group improved and two showed no change, whereas three people in the control group got better, three showed no change and one got worse (table 5.8a). If we assign a score of 1 for improvement, 0 for no change and −1 for worse, we can derive a more conventional table (table 5.8b). The frequency table (table 5.7) conveys exactly the same information as the raw score tables (table 5.8). *The difference is merely one of presentation.*

One consequence of this is that both frequency tables and raw score tables are analysed by the same procedures in this book where they have the same basic experimental design. The arithmetic used to compute the sample rank sums is, of course, different but the rest of the test is the same. This is confusing to some because many textbooks treat frequency tables in a quite different manner from raw score tables. The explanation lies with the nature of the categories of the dependent variable, which must always be rankable for rank sum analysis (see section 3.5), whereas most textbooks which deal with frequency tables are referring to situations where the categories are not rankable (i.e. nominal).

5.7 Rank sum analysis and traditional tests

The purpose of this book is to present a systematical approach to the analysis of data which are based on rank order scales. Many of the tests to be presented in later chapters are analogues of tests currently in use by researchers and statisticians. Table 5.9 attempts to organize some of these tests into a scheme which

Table 5.9 Summary of popular rank sum tests covered by rank sum analysis. There is no standard nomenclature for these tests

Samples	Blocks	Nonspecific H_1	Specific H_1
Many	Multiple	Benard and Van Elteren (1953)	Meddis (1980)
		Friedman* (1937)	Marascuilo and McSweeney*
		Cochran* (1950)	(1967)
		Spearman's* ρ	Page* (1963)
		Kendall's* W	
	Single	Kruskal-Wallis (Kruskal, 1952)	Marascuilo and McSweeney** (1967)
		$k \times q$ chi-square for frequency table	Dunn's test** (1964)
Two	Multiple	Wilcoxon stratified (1946, 1947)	One-tail equivalents of non-specific tests
		Sign test* (Dixon and Mood, 1946)	
		McNemar*	
	Single	Wilcoxon (1945), Mann-Whitney (1947)	
		$2 \times q$ chi-square for frequencies	
		Median test (Mood, 1950)	

*only one score per cell permitted
**equal sample frequencies required

relates them to the rank sum procedures described in chapters 7–11. Allowing for the minor caveats to be entered in the following, these tests are *mathematically equivalent* to rank sum analysis procedures. They give exactly the same numerical result. There is, therefore, no question as to which is the better test (the traditional test of the rank sum technique); they are the same test. Any differences are largely superficial and concern notation or arithmetical procedures. Rank sum procedures have a unified notation and a unified basic computational plan.

Some of the entries in table 5.9 are surprising even to experienced statisticians – for example, the inclusion of Spearman's two-variable and Kendall's many-variable correlation techniques. The mathematical justification of the inclusion of all of the tests in table 5.9 is to be found in chapter 14. It is also a considerable bonus that rank sum methods can successfully mimic certain traditional frequency table analysis techniques which were originally intended primarily for categories which were unrankable.

The major distinction in table 5.9 concerns the nature of the hypothesis under test, that is whether a specific alternative hypothesis (H_1) is being considered or whether the null hypothesis (H_0) is being tested against all possible hypotheses (nonspecific test). This distinction, which was discussed in detail in chapter 2, is very useful in discriminating pairs of tests (e.g. Page's test and Friedman's test) which deal with the same experimental design.

The distinction is somewhat blurred in the case of two-sample tests where there are only two possible alternatives to the null hypothesis – that either the first sample or the second sample is drawn from a population with a higher rank mean. The traditional practice has been simply to double the probability associated with the type I error (i.e. double the significance level) when converting the procedure from a test against a specific alternative to a test against both possible alternatives. This is the well known one-tail/two-tail conversion procedure which, despite its computational convenience, has caused confusion and suspicion in generation after generation of students. In this book the two procedures are kept quite separate at all times, notwithstanding tradition and notwithstanding the fact that they occasionally become procedurally very similar.

The relationship between the tests are discussed in some detail in chapter 14, where it is shown algebraically how the large-sample normal or chi-square approximations to the exact distributions of rank sum analysis are equivalent to the traditionally used formulae. However, it may be of some interest to review the scheme in outline as it appeared during the development of the systematization.

The basic idea is that some popular tests are merely special cases of other tests. For example, the two-tail version of the sign test – which deals with the two-sample, multiple block design – is a special case of Friedman's analysis of variance by ranks, which deals with many samples and multiple blocks. In other words both tests, when applied to a two-sample design, should give *exactly* the same result. In this sense, the sign test is completely redundant. However, the sign test is very quick and easy to do when compared with the Friedman test. For this reason we keep both tests alive, but it would seem to be an error to believe that they are two quite separate tests. The same argument applies to the two-tail version of the Wilcoxon/Mann-Whitney test (for two-sample, one-block designs) which is made redundant by the existence of the Kruskal-Wallis test (for many-sample, one-block designs). Once again they both give exactly the same result when applied to a two-sample, one-block design.

In their turn, both the Friedman test and the Kruskall-Wallis test are both special cases of a less well known test by Benard and Van Elteren (1953) which deals with many-sample, multiple blocks designs. The latter test makes all four other tests redundant in theory. This much is summarized in figure 5.1, which also includes other tests which spring up as special cases at various points in the system. All of the tests in this are to be viewed as tests of the null hypothesis against all possible alternatives (i.e. nonspecific tests).

Using similar logic we can create an analogous scheme in figure 5.2 for tests which evaluate a specific alternative to the null hypothesis. The tests in this scheme are relatively unfamiliar to researchers and largely recent in origin. Many of these tests involve special restrictions – for example, Page's test evaluates only trends and not comparisons – but the general scheme is similar to that of figure 5.1. It is important to note that the two-sample tests are here one-tailed (one-sided) versions of the comparable tests in figure 5.1.

Researchers who are familiar with traditional rank sum analysis may wish to analyse data twice, using the methods given in this book and the procedures which are being replaced. In each case, the same numerical result should be obtained. Exceptions occur with some frequency table procedures where the traditional technique may produce a smaller value of χ^2 because it ignores the information contained in the ordering of the categories of the dependent variable. Care should be taken always to use the correction factor for ties. This is particularly important for frequency tables, where the correction effect may be large. In some situations, results may differ by a factor of $N/(N-1)$, where N is the total sample size. This arises from minor differences in the way in which the chi-square approximations have been derived by different authors. For large N this difference factor is trivially small.

5.8 Novel aspects of rank sum analysis

So far, the relationship between rank sum analysis and existing tests has been stressed. It would be misleading, however, to suggest that analysis of variance by ranks did nothing more than systematize existing practice. On the contrary, the more general approach significantly expands the range of experimental designs which can now be analysed routinely by a researcher requiring a distribution-free approach. In the analysis of trends and contrasts, we can now analyse

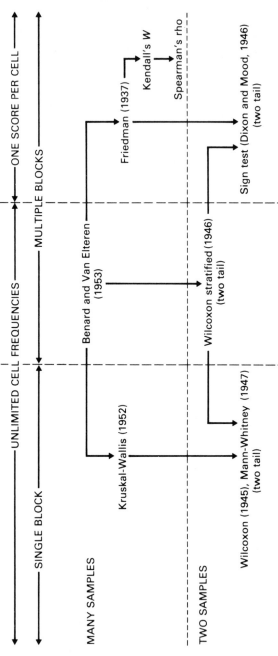

Figure 5.1 Traditional techniques for testing H_0 against all alternatives presented as special cases of each other (see text)

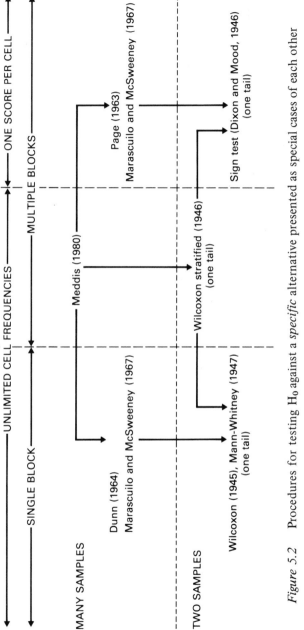

Figure 5.2 Procedures for testing H_0 against a *specific* alternative presented as special cases of each other

data sets with unequal sample sizes just as easily as the case with equal sample sizes. More usefully still, we can even analyse data sets with some empty cells.

More important is the new flexibility in handling data in the form of frequency tables. We can now cope easily with repeated frequency tables without recourse to such manoeuvres as combining them to make a single table. We also have the capability of testing for trends and contrasts across such tables. Moreover, we are able to perform such analyses within the rank sum framework, that is, without any new conceptual apparatus. This we can do as long as the categories of the dependent variable are either rankable or only two in number. When we use rank sum methods our analyses are either as sensitive or more sensitive than traditional frequency table techniques, which were usually based on the assumption that the categories were not necessarily rankable.

5.9   Correlation and concordance

Rank sum analysis of appropriate data can offer tests which are equivalent to Spearman's test of rank correlation or Kendall's coefficient of concordance. Both of these tests are measures of agreement between variables when a number of objects (or persons) are ranked with respect to their score on each variable. Spearman's test deals with the more common case where we wish to correlate only two variables. Kendall's test covers the case where we wish to assess the concordance between any number of variables. When Kendall's test is applied to only two variables it is equivalent to Spearman's test. Rank sum analysis can be used to give the same result as both of these popular tests.

Both tests can be construed as dealing with variants of Friedman's design (many samples, multiple blocks, one score per cell) where each block contains the data for a different variable. The samples represent the objects being ranked. An example is given in table 5.10, where the investigator wishes to know whether the retail price of hi-fi loudspeakers is correlated with the reported preferences of naïve listeners in standardized listening tests. Here, the data form

Table 5.10   Preferences and cost of seven pairs of loudspeakers in standardized listening tests

Blocks	Samples: loudspeaker pairs						
	A	B	C	D	E	F	G
Preference (1 = most preferred)	2	4	1	7	3	6	5
Cost (£)	150	275	840	35	80	60	45

seven samples (i.e. seven sets of loudspeakers) and two blocks (preference rating and price) with one score per cell. Procedures for analysing the data are given in chapter 11.

It may seem strange to mix data on different variables in the same table but this does not do injury to any of the principles of rank sum analysis. This is because data in different blocks are never mixed in any way. They are ranked quite separately. In multiple block designs, the method of rank sum analysis works by measuring the extent of agreement between the pattern of ranks in the blocks. In most applications, blocks represent repeated mini-experiments and we are interested to know whether the same result emerges from each. In the case of correlation or concordance analysis we simply take advantage of this property to ask whether the same pattern of ranks occurs in the different blocks. In this latter case, however, the blocks represent quite different variables. The principle remains the same in that we are measuring agreement between patterns of rankings across blocks in both cases.

5.10 Summary

Analysis of variance by ranks can be used to assess data from a wide range of experimental designs. To be able to use the system effectively the user must be able to distinguish and describe different designs. The key descriptors are the numbers of samples and blocks. Samples are defined by levels of the independent variable. Blocks represent independent subsets of the data which constitute repeated mini-experiments. The popular repeated measures design is an example of a multiple blocks design. Results presented in the form of frequency tables do not indicate a special design, and such results are to be analysed using similar techniques to those used for raw scores. Many of the tests within the analysis of variance by ranks framework are directly equivalent to currently popular distribution-free tests. However, the rank sum system significantly extends the range of designs that can be analysed. It has the additional benefit of arranging all of the tests within a common conceptual and computational framework.

Part II

6 Computational procedure

6.1 Introduction

All of the computational procedures in chapters 7 to 13 are based on the same five-step plan. This chapter introduces the plan by explaining and elaborating on each of the steps. Many details are presented once only here to avoid endless repetition later. This allows a more economical presentation of individual worked examples and supplies a resource of more detailed discussion which the reader may use whenever he becomes confused while working through later chapters. The beginner, therefore, should use chapter 6 in parallel with chapters 7 to 10 when familiarizing himself with new techniques.

6.2 Notation

Every effort has been made to keep the notation consistent throughout the book. Table 6.1 explains briefly the usage of each symbol and, where possible, refers to the section where some further explanation might be found. This table is repeated on the endpapers of the book for easy reference.

Tables 6.2 and 6.3 illustrate the deployment of some of the symbols used.

6.3 Step 1 – formulate the problem

It is a common mistake among beginners to launch directly into the arithmetic of a test without pausing to make explicit the ideas that are being assessed. There are many different experimental designs which look deceptively similar when the results are written down on paper. It is essential to pause to consider precisely what is going on before selecting the procedures to be applied. The simple act of writing down the hypotheses and naming the variables can promote insights which will save a whole wasted morning spent on the mechanics of a test which will later prove to have been the wrong one. Even if you do go wrong, it will be so much easier for an adviser to put you back on the right track if he has an explicit statement of what the problem is and what the key variables are. Much statistical practice among students and young researchers is simply concerned with the ritualistic application of 'tests' without which their results will never gain social approval. In fact, the tests should be a routine and highly practical tool used to formulate simple answers to simple questions about the results which a researcher has collected. The tests are only meaningful if the researcher genuinely wants to know the answer to the question. If he does want the answer, it is likely that he will have little difficulty in framing the question, especially if he has been in the habit of making his hypotheses explicit from the very start. It is a question of whether we want to baptize our results with statistics or subject them to a statistical analysis. Step 1 is intended for those who choose the latter path.

Specify the hypotheses

Null hypothesis (H_0). Every test has a null hypothesis at its heart and the main purpose of the test is to decide whether or not we have enough evidence to reject it (see section 2.2). In essence, H_0 specifies no reliable association between the independent and dependent variables. It does not refer directly to the results which are about to be analysed but to the world from which these results have been sampled.

Specific alternative hypothesis (H_1). Only some tests have a specific alternative. For convenience we call these specific tests. By proposing H_1, the researcher is declaring a viewpoint which he wishes to adopt at the expense of H_0. Thus if H_0 is rejected, the researcher wishes to construe this rejection as support for his own theoretical orientation. A specific test is, therefore, designed so that H_0 will *only* be rejected if the results are, at least approximately, in line with H_1. When we have only two samples, there are only two possible forms for H_1 – that group A has higher scores than group B or *vice versa*. If there are many samples, then the number of possible specific alternative hypotheses increases considerably (see section 2.10).

General alternative hypothesis. In a nonspecific test, the researcher merely wishes to know if there are any reliable effects in the results without specifying them in advance. In this case, the general alternative hypothesis is simply that H_0 is untrue. The presence of any substantial nonrandom effect, in whatever direction, can be used as evidence for rejecting H_0. The disadvantage of using nonspecific tests is that the nonrandom effects need to be much larger than in specific tests to be considered as equally convincing evidence for rejecting H_0.

D

Table 6.1   List of symbols

A	Number of cases in the upper left hand cell of a 2×2 frequency matrix $(f_{1,2})$. See section 7.5
B	Number of cases in the upper right hand cell of a 2×2 frequency matrix $(f_{1,2})$. See section 7.5
b_i	Total number of cases in the ith block. See section 5.3
C	Number of cases in the lower left hand cell of a 2×2 frequency matrix $(f_{2,1})$. See section 7.5
d_i	Difference between a pair of ranks, used by Spearman in his method for computing ρ. See section 11.2
D	Number of cases in the lower right hand cell of a 2×2 frequency matrix $(f_{2,2})$. See section 7.5
E, E_{ij}	Expected number of cases in the cell in the ith row and the jth column of a frequency table when H_0 true. See section 9.8
$E(L)$	Expected value of L when H_0 true. The average of L over all permutations of the ranks. See section 3.2
f_{ijq}	Number of cases in the qth category of the dependent variable in the jth sample in the ith block. See sections 5.6 and 10.7
H	Statistic measuring the dispersion of the sample rank means; approximately distributed as chi-square. See section 3.3
H'	H when corrected for ties. See section 3.3
H_0	Null hypothesis. See section 2.9
H_1	Specific alternative hypothesis. See section 2.10
i	Index normally reserved for the block number (e.g. b_i)
j	Index normally reserved for the sample number (e.g. R_j)
k	Number of samples. Note that k is lower case
K	Statistic reflecting differences between the rank means. See section 3.3. Note that K is upper case
L	Statistic measuring agreement between data and a specific hypothesis. See section 3.2.
M	Median; a value exceeded by half of the scores in the data set. See section 2.9
m	Number of blocks. See section 5.3
m'	Number of blocks which are not completely tied. See section 8.2
n_{ij}	Number of cases in the cell in the ith block of the jth sample
n_j	Number of cases in the jth sample $\left(\text{i.e. } n_j = \sum_{i}^{m} n_{ij} \right)$
n	Number of cases in each cell when cell frequencies are all equal
N	Total number of cases $\left(\text{i.e. } N = \sum_{j}^{k} n_j \right)$
O, O_{ij}	Observed (as opposed to expected) number of cases in the cell in the ith row and jth column of a frequency table. See section 9.8

Table 6.1 continued

---

p	Probability; usually the probability of obtaining the results under discussion, or similarly extreme results, when H_0 is true
q	Index normally reserved for the number of a category of the dependent variable. See section 8.9
Q	Number of categories of the dependent variable. See section 8.9
r	The rank of a particular score
$r_q\ (r_{iq})$	The rank shared by all scores in the qth category of the dependent variable (in the ith block). See section 6.4
R_j	The sum of all ranks in the jth sample. See section 3.2
$\bar{R}_j$	The mean rank in the jth sample. See section 3.2
R_+	The sum of the ranks of positively signed differences; used in the Wilcoxon matched pairs test. See section 8.3
R_-	The sum of the ranks of negatively signed differences; used in the Wilcoxon matched pairs test. See section 8.3
$S_i.$	The number of 1s in the ith block when the data consist of only 0 and 1 (dichotomy). See section 10.4
$S._j$	The number of 1s in the jth sample when the data consist of only 0 and 1 (dichotomy). See section 10.4
t, t_q	Number of cases in the qth category of the dependent variable (or the qth row or column total). See section 5.6
Σt_q	Cumulative total of cases in the qth *and lower* categories of the dependent variable. See section 6.4
$T, (T_i)$	Correction factor for ties (in the ith block). See section 3.2
$\text{var}(L)$	Variance or dispersion of all possible values of L when H_0 true (i.e. overall permutations of the ranks). See section 3.2
W	Kendall's coefficient of concordance. See section 11.4
Z	Statistic for evaluating L. Z is approximately normally distributed. See section 3.2
Z'	Z corrected for ties. See section 3.2
z	Critical value of the normal distribution
χ^2	(pronounced kigh square) Critical value of the chi-square distribution. See section 3.3
λ_j	(pronounced lambda) Coefficients which represent numerically the specific alternative hypothesis. See section 3.2
ρ	(pronounced roe) Spearman's coefficient of rank correlation. See section 11.2
$\bar{\rho}$	Average value of ρ over all possible pairs of variables in a concordance analysis. See section 11.4

Table 6.1   continued

$\displaystyle\sum_{i}^{m}$	Instruction to add all the values to the right of the symbol by systematically changing the index i from a starting point of 1 to finish at m
σ_L	Standard deviation of L; the square root of var(L). See section 3.2
$X > Y$	X is greater than Y
$X \geqslant Y$	X is greater than or equal to Y
$X < Y$	X is less than Y
$X \leqslant Y$	X is less than or equal to Y

Table 6.2   Illustration of deployment of some of the symbols used in table 6.1. Layout is for individual scores

Blocks (matched sets)	Samples					Block frequency
	1	2	... j	...	k	
1	n_{11} (cell frequency)	n_{12}	...	n_{ij}	... n_{ik}	b_1
. . .	. . .	. . .	. . .	. . .	. . .	. . .
i	n_{i1}	n_{i2}		n_{ij}	n_{ik}	b_i
. . .	. . .	. . .	. . .	. . .	. . .	. . .
m	n_{m1}	n_{m2}		n_{mj}	n_{mk}	b_m
rank sums	R_1	R_2	...	R_j	... R_k	N (total
sample frequency	n_1	n_2	...	n_j	... n_k	frequency)
rank means	$\bar{R}_1$	$\bar{R}_2$	...	$\bar{R}_j$	... $\bar{R}_k$	

Table 6.3 Illustration of deployment of some of the symbols used in table 6.1. Layout is for repeated frequency tables

Block	Category	Sample					Category frequency	Block frequency
		1	...	j	...	k		
1	1	f_{111}	...	f_{1j1}	...	f_{1k1}	t_{11}	
	$\vdots$	$\vdots$		$\vdots$		$\vdots$	$\vdots$	
	q	f_{11q}	...	f_{1jq}	...	f_{1kq}	t_{1q}	b_1
	$\vdots$	$\vdots$		$\vdots$		$\vdots$	$\vdots$	
	Q	f_{11Q}		f_{1jQ}		f_{1kQ}	t_{1Q}	
		n_{11}		n_{1j}		n_{1k}		
		(cell frequency)						
$\vdots$								$\vdots$
i	1	f_{i11}	...	f_{ij1}	...	f_{ik1}	t_{i1}	
	$\vdots$	$\vdots$		$\vdots$		$\vdots$	$\vdots$	
	q	f_{i1q}		f_{ijq}		f_{ikq}	t_{iq}	b_i
	$\vdots$	$\vdots$		$\vdots$		$\vdots$	$\vdots$	
	Q	f_{i1Q}		f_{ijQ}		f_{ikQ}	t_{iQ}	
		n_{i1}	...	n_{ij}	...	n_{ik}		
$\vdots$								$\vdots$
m	1	f_{m11}	...	f_{mj1}	...	f_{mk1}	t_{m1}	
	$\vdots$	$\vdots$		$\vdots$		$\vdots$	$\vdots$	
	q	f_{m1q}		f_{mjq}		f_{mkq}	t_{mq}	b_m
	$\vdots$	$\vdots$		$\vdots$		$\vdots$	$\vdots$	
	Q	f_{m1Q}		f_{mjq}		f_{mkQ}	t_{mQ}	
		n_{m1}	...	n_{mj}	...	n_{mk}		N (total frequency)
rank sums		R_1	...	R_j	...	R_k		
sample frequencies		n_1	...	n_j	...	n_k		
rank means		$\bar{R}_1$	...	$\bar{R}_j$	...	$\bar{R}_k$		

Specify the variables

Variables are discussed at length in section 2.3. The two main variables (usually independent and dependent) must be identified at this point together with any blocking variable which may be present. This is a convenient place to specify the nature of the scales on which each variable is assessed (e.g. number of groups, rankable scale, set of rankable categories). Clearly the main variables identified in this section must agree with those identified in the null hypothesis. This is an opportunity to supply more details of the variables, such as how they were measured in the particular investigation.

Order the samples or specify the coefficients

Every specific test is accompanied by a set of coefficients (one for each sample) which identify the expected ordering of the sample rank means. Any numbers can be used as coefficients as long as the *pattern* of coefficients mirrors the expected pattern of rank means. This freedom to choose coefficients can be used to reduce computational labour while obtaining the same final result. For example the following two sets of coefficients both represent a monotonic increasing trend:

λ_j	1	2	3	4	5
λ_j	-2	-1	0	1	2

However, the second set will be easier from a computational point of view – unless you become easily confused by the use of negative numbers! Two different sets of coefficients will always give the same result if they form a straight line when corresponding pairs of coefficients are plotted on a graph (see section 3.2).

With two-group tests we can avoid the explicit use of coefficients by the simple expedient of ordering the groups in accordance with our specific alternative hypothesis. Take the sample with the predicted higher rank mean and make this the first group of scores. The rest of the test is conducted on the assumption that the coefficients are $\lambda_j = 1, 0$ (see section 3.2). All of the computational formulae for that test have been rewritten on that assumption, and so there is no need to treat the coefficients as variable or even to mention them at all.

For nonspecific tests we occasionally need to order the groups in the same way. This is particularly necessary for exact tests where tables have to be prepared (by the author) with critical values for either unusually large values of rank sums or unusually small values. In this book, the tables are arranged for assessing unusually large values of rank sums and it is, therefore, less confusing to adopt the convention of *always assigning the sample with the highest observed rank mean to be group 1*. Classroom experience shows that this avoids a number of complications. The catch is that, at this stage, the researcher may not know which sample has the higher rank mean. He must, of course, simply make an informed guess and be prepared to change it later if necessary.

6.4 Step 2 - find the sample rank sums and rank means

In this section we move the problem from the parametric to the nonparametric domain. We convert each score to its rank equivalent and, from this point on, we see the problem entirely in terms of ranks, sample rank sums and sample rank means. The original scales on which the measurements are made are lost. We need the rank means to simplify our interpretation of the differences between samples. We need the rank sums for our evaluation of the statistical significance of these differences in step 4 (section 6.6).

Prepare your computational layout

Although it sounds too obvious to repeat, it does bear repetition that a clear and consistent layout makes for fewer mistakes, easier correction of the mistakes which do occur, and better advice from a second person who might be invited to advise on a problem. Whenever possible, the following rules are applied in this book to make the presentation of quite varied problems consistent:

(a) The *samples form the columns* of the layout and the *blocks form the rows*. This applies even to frequency table problems.
(b) Whenever practicable, the scores within a block are rearranged to show their rank order and to make ranking easier.
(c) The ranks are written alongside the original scores for easy checking. If this would make the presentation too cluttered, then the ranks are put into a separate table which maps on to the first.
(d) The sample rank sums (R_j), the sample frequencies (n_j) and the sample rank means $(\bar{R}_j)$ are all placed at the foot of the sample columns – in that order.

Rank the scores

Ranking takes two major forms - the assignment of individual ranks to individual scores in the raw score case, and the assignment of shared ranks to groups of scores in the same category of the dependent variable in the case of frequency tables. They are basically the same process since shared ranks occur even when ranking individual scores, but they look superficially different and will be discussed separately in the following.

Both systems require that *low scores are assigned low ranks*. Any deviation from this rule causes confusion. 'Low score' in this context means low numerically and has nothing to do with the value or worth of the score. Thus if a winning runner's performance is measured in terms of time taken he will be assigned the lowest rank (i.e. shortest time), but if it is measured in terms of speed then he will be assigned the highest rank (i.e. greatest speed). In some situations, however, the categories of the dependent variable do not have an obvious numerical equivalent. For example, the scale (a) happy, (b) neither happy nor sad, and (c) sad is clearly formed of ordered categories, but there is no unambiguous way of deciding which should have high or low ranks. In these cases an arbitrary decision must be made, and the researcher must keep clearly in mind the decision he has made if he is to interpret the rank means correctly.

Raw scores. Table 6.4 shows one way of assigning individual ranks to a moderately large group of scores. This illustration is taken from section 7.7. The

Table 6.4   Layout for ranking individual scores and evaluating tie lengths. Data taken from the example in section 7.7

Sample	Score	Order	Rank	Tie length
1	23	1	1	
2	28⎫	2	2.5⎫	
2	28⎭	3	2.5⎭	2
2	29	4	4	
1	30	5	5	
1	32	6	6	
2	36⎫	7	7.5⎫	
2	36⎭	8	7.5⎭	2
2	38	9	9	
1	40	10	10	
1	42	11	11	
2	43	12	12	
1	44⎫	13	15 ⎫	
1	44⎪	14	15 ⎪	
1	44⎬	15	15 ⎬	5
2	44⎪	16	15 ⎪	
2	44⎭	17	15 ⎭	
2	45	18	18	
1	46⎫	19	20 ⎫	
2	46⎬	20	20 ⎬	3
2	46⎭	21	20 ⎭	
1	48⎫	22	23 ⎫	
2	48⎬	23	23 ⎬	3
2	48⎭	24	23 ⎭	

Table 6.4 continued

Sample	Score	Order	Rank	Tie length
1	50 ⎫	25	26 ⎫	
1	50 ⎬	26	26 ⎬	3
2	50 ⎭	27	26 ⎭	
1	52 ⎫	28	29 ⎫	
1	52 ⎬	29	29 ⎬	3
1	52 ⎭	30	29 ⎭	
2	53	31	31	
1	54 ⎫	32	34.5 ⎫	
1	54 ⎪	33	34.5 ⎪	
1	54 ⎪	34	34.5 ⎪	6
2	54 ⎪	35	34.5 ⎬	
2	54 ⎪	36	34.5 ⎪	
2	54 ⎭	37	34.5 ⎭	
1	56 ⎫	38	41 ⎫	
1	56 ⎪	39	41 ⎪	
1	56 ⎪	40	41 ⎪	
1	56 ⎬	41	41 ⎬	7
2	56 ⎪	42	41 ⎪	
2	56 ⎪	43	41 ⎪	
2	56 ⎭	44	41 ⎭	
1	58 ⎫	45	46 ⎫	
1	58 ⎬	46	46 ⎬	3
1	58 ⎭	47	46 ⎭	
2	60 ⎫	48	48.5 ⎫	2
2	60 ⎭	49	48.5 ⎭	
1	62 ⎫	50	50.5 ⎫	2
2	62 ⎭	51	50.5 ⎭	
2	64	52	52	
2	66 ⎫	53	53.5 ⎫	2
2	66 ⎭	54	53.5 ⎭	
2	67	55	55	
1	68	56	56	
1	70	57	57	
1	72 ⎫	58	58.5 ⎫	2
1	72 ⎭	59	58.5 ⎭	
2	78	60	60	
1	82	61	61	
1	100	62	62	

procedure is as follows:

(a) Write down all the scores (within a given block and irrespective of sample) in ascending order.

(b) Write the sample number of the score to the left of the score.

(c) Number each score from one upwards in the column marked 'order'.

(d) Bracket the scores having the same value (ties).

(e) If a score is untied, then its rank is the value in the 'order' column and it is said to belong to a tie of length one ($t = 1$).

If a set of scores is tied then the rank for each score in the tie is the average of all the 'order' values in the tie. For example, five scores have a value of 44 and their 'order' scores stretch from 13 to 17. We therefore calculate the shared rank as

$$r = \frac{13 + 14 + 15 + 16 + 17}{5} = 15$$

We assign the shared rank 15 to each score and note that the tie length is five. Alternatively, we might subtract half of the tie length from the highest 'order' score and then add a half to find the shared rank thus:

$$r = 17 - \frac{5}{2} + \frac{1}{2} = 15$$

The latter method is useful when there are many scores in a tie. It is the same method as that used for frequency tables.

NB   The term 'order score' is not used in the technical statistical sense but merely to distinguish a preliminary from a final ranking value.

(f) Write the tie lengths in the end column for later use, if required, in the correction for ties procedure. Use only one length for each tie; omit tie lengths for untied scores.

Frequency tables. Table 6.5 shows the same data after assignment to a reduced set of categories being ranked by the frequency table method. Notice that the row totals (t_q) are, in fact, tie lengths since they represent the number of scores in the same category. The adjacent column gives the cumulative frequency (Σt_q), which is simply the number of scores in this and all previous categories (i.e. superior rows). The computation of shared ranks is now effected very economically using the expression

$$r_2 = \Sigma t_q - \frac{t_q}{2} + 0.5$$

where q is the category number, as is illustrated in table 6.5.

Table 6.5 Computation of shared ranks for a frequency table with ordered categories. Data are taken from table 6.4

Category	Sample 1	Sample 2	Row total t_q	Cumulative total Σt_q	Shared rank r_q
0–44	8	9	17	17	9
45–53	7	7	14	31	24.5
54–59	10	6	16	47	39.5
60–100	7	8	15	62	55
				N 62	

$r_1 = 17 - \frac{17}{2} + 0.5 = 9$

$r_2 = 31 - \frac{14}{2} + 0.5 = 24.5$

$r_3 = 47 - \frac{16}{2} + 0.5 = 39.5$

$r_4 = 62 - \frac{15}{2} + 0.5 = 55$

When creating arbitrary categories as we have done in this example it is a good idea to arrange to have, as nearly as possible, the same number of scores in each category as the data structure will allow. Obviously a category boundary cannot go through the middle of a tie, but it is worth fiddling a little to get *the most even distribution of scores possible*. This has two advantages. Firstly, it can be shown that, for a fixed set of categories, the information lost by the categorization procedure is least if the scores are approximately evenly spread. Secondly, different final results can be achieved by changing the category boundaries, and it seems intuitively wrong that we should fiddle the boundaries solely to optimize the results with respect to our prior hypotheses. An objective rule such as that proposed, if invariably applied, should avoid the temptation (conscious or unconscious) to influence the outcome of the analysis.

Find the sample rank sums (R_j)

In the case of individually ranked scores, the sample rank sum is simply the sum of the ranks in the sample. For frequency tables we must take the number of scores in a given category and multiply by the shared rank and repeat for each category. Thus for sample 1 in table 6.5 we find the rank sum thus:

$$R_1 = 8(9) + 7(24.5) + 10(39.5) + 7(55) = 1023.5$$

and for sample 2:

$$R_2 = 9(9) + 7(24.5) + 6(39.5) + 8(55) = 929.5$$

Because there are often different numbers of scores in each sample, the rank sums should not be routinely used as a basis for interpreting the differences between the samples. The sample rank means (see later) should be used for this purpose.

Check that the ranking has been carried out correctly

This is a most important step and should never be omitted even by experienced researchers. The checks proposed in each chapter are different but all are variations on a common theme. Any set of ranks (from 1 to N) must have the following sum of ranks:

$$\Sigma r = N(N + 1)/2$$

In our example (tables 6.4 and 6.5) we have 62 ranks; therefore

$$\Sigma r = 62(63)/2 = 1953$$

This value corresponds with our sum of ranks based on the sample rank sums:

$$R_1 + R_2 = 1953$$

This rule applies even when there are heavy ties. In the most general case we have m blocks of scores and b_i separate ranks in each group. Since each block is ranked separately we can use the general check sum

$$\sum_i^k R_j = \frac{1}{2} \sum_i^m b_i(b_i + 1)$$

for all purposes. It is cumbersome, however, and has been simplified for special cases wherever possible.

If the check sum does not work out, then there is no purpose served by proceeding. You must go back and get the ranking right. Usually ranking errors result from a failure to follow the computational methods outlined above. Attempts to rank quickly, 'by eye', may or may not succeed. Methodical ranking almost always succeeds and, after a little practice, can be as quick.

Find the sample rank means ($\bar{R}_j$)

The sample rank mean is simply the sample rank total divided by the number of scores in that sample:

$$\bar{R}_j = R_j/n_j$$

The main use is for interpreting the differences between the samples. This is only an approximation to the true sample rank mean, which involves finding the cell rank mean for each sample/block cell and then averaging over samples. However, the approximation is very good almost all of the time. If the pattern of cell frequencies (n_{ij}) fluctuates wildly from block to block it might, however, be useful to find the sample rank mean by the longer route through the cell rank means.

6.5 Step 3 – interpret the rank means

This is probably the most important step in the computations and yet is also the most commonly omitted. Its inclusion is crucial in distinguishing between statistics as a ritual baptism of data and statistics as an analytical tool. There is no need for this section to be long winded. It need only include a brief statement of what the differences in the numerical values of the rank means signify in common sense terms. In the case of a specific test, it should be made clear whether or not the pattern of differences corresponds, at least approximately, with the pattern predicted on the basis of the specific alternative hypothesis. If it does not then there is little point in proceeding further, since the test will not yield a significant result. If the difference is large but in the wrong direction, then now is the time to change tack and attempt a nonspecific test in step 4 (section 6.6).

6.6 Step 4 – compute and evaluate our key statistics

In steps 1–3 the procedure has been very similar, whatever the test, but in step 4 considerable variation is introduced. The procedure differs for specific and non-specific tests, and there is further variation between exact and approximate tests (see chapter 3). To an extent we may view these differences as superficial since they share the same two-stage process:

compute the key statistic
evaluate the key statistic.

The key statistic is a number whicn reflects the disagreement between our data and the null hypothesis (H_0). *The larger the value of this statistic, the less likely it is that* H_0 *can be true.* The valuation stage is simply an attempt to assess just how likely it is that H_0 is true given our key statistic. Since chapter 3 reviews these points at length, they will not be pursued here. The following notes refer only to specific computational issues in the second stage.

Evaluate the key statistic

For exact tests the key statistic can be evaluated by simply consulting the appropriate exact table in the appendix. For approximate tests the key statistic needs to be expressed as a variable which is approximately normally distributed (Z) or approximately distributed as chi-square (H). Note that we use lower case z and χ^2 for variates distributed truly as normal or chi-square, and upper case Z and H to represent approximations. In all cases, we have to consult a table where the relevant row of critical values looks like this:

	5%	2.5%	1%	0.1%
$x \geqslant$				

The tables in this book have all been arranged so that our key statistic must be *equal to or larger* than a critical value to be significant at the corresponding significance level. This is to avoid confusion, but it is not universally true of statistical tables, and the symbol $\geqslant$ is included as a reminder. (Tables I and J and extracts from them are exceptions; in these the key statistic must be equal to or less than ($\leqslant$) a critical value.)

Clearly, a result which is significant at the 1 per cent level will also be significant at 2.5 and 5 per cent levels. It is the researcher's job to find the most significant possibility for his data. In some exact tables, blanks will be found most often in the 0.1 per cent column. These blanks indicate that it is simply not possible to achieve significance at this level since there is no permutation of ranks which is so rare.

The percentage notation for significance is used throughout the book because it is most readily grasped by students with little mathematical ability. However, professionals prefer to use probability notation and this is normally required for publication in learned journals. The conversion simply requires moving the decimal point two places to the right:

Significance level (%)	*Probability notation*
5	$p < 0.05$
2.5	$p < 0.25$
1	$p < 0.01$
0.1	$p < 0.001$

We use the 'less than' symbol ($<$) to remind the reader that our result was larger than the critical value. Of course, when it is exactly the same we should use the equal sign.

Consulting tables of chi-square involves the use of 'degrees of freedom'. This idea is briefly described in section 3.5. Usually the concept is redundant for most practical purposes since the number of degrees of freedom is equal to the number of samples less one. However, we cannot omit the concept altogether because it is needed for the chi-square homogeneity test (section 9.8) and for *post hoc* tests (chapter 12). This is a pity because it creates unnecessary confusion and despondency in beginning students. In the early stages, it is best to apply the degrees of freedom expressions quite mechanically. Their 'meaning' can only be explained fully in more advanced courses.

When using *exact tables* the significance levels quoted do not always correspond closely with the exact probability associated with the key statistic. This point is discussed in sections 3.2 and 3.3. In this case, however, it is always true to say that the exact probability is 'equal to or less than' the significance level quoted in the table.

A negative value for Z normally means that the specific alternative hypothesis is contradicted by the evidence. Sometimes, however, it simply means that the test user either (a) has his samples the wrong way round (two-sample case), or (b) has his λ coefficients in inverse order (this is particularly easy with frequency tables). If either of these two possibilities are the case, the solution is simply to ignore the negative sign.

Correct for ties

Approximate tests can be performed without correcting for ties. This omission is justified by the fact that ties in the data render the tests more conservative. However, correction for ties is always appropriate if the best possible estimate is required, for example, for publication of results. The correction factors (T_i) have been designed so that they are optional and can be easily included or omitted.

Each T_i is computed using the expression

$$T_i = 1 - \frac{\sum_q (t_q^3 - t_q)}{b_i^3 - b_i}$$

where t_q is the qth tie in this block and b_i is the total number of scores in the block.

If $t = 1$ then $t^3 - t = 0$ and contributes nothing to the calculations. Values of $x^3 - x$ for $x = 1$ to 400 can be found in table D in the appendix. These can be used to find values of $(t^3 - t)$ as well as $(b_i^3 - b_i)$. When there is only one block, $b_i = N$ and

$$T = 1 - \frac{\sum(t^3 - t)}{N^3 - N}$$

We can illustrate the use of this formula with the data in table 6.5. The tie lengths in this case are the row totals 17, 14, 16 and 15.

$$T = 1 - \frac{(17^3 - 17) + (14^3 - 14) + (16^3 - 16) + (15^3 - 15)}{62^3 - 62}$$

$$= 1 - \frac{4896 + 2730 + 4080 + 3360}{238\,266}$$

$$= 1 - \frac{15\,066}{238\,266} = 0.937$$

The effect of ties is very much smaller, as a rule, when tie lengths are short relative to the total number of scores. The ties in table 6.4 are 2, 2, 2, 2, 2, 2, 3, 3, 3, 3, 3, 5, 6, and 7.

$$T = 1 - \frac{6(2^3 - 2) + 5(3^3 - 3) + (5^3 - 5) + (6^3 - 6) + (7^3 - 7)}{62^3 - 62}$$

$$= 1 - \frac{6(6) + 5(24) + 120 + 210 + 336}{238\ 266}$$

$$= 1 - \frac{822}{238\ 266} = 1 - 0.0034 = 0.997$$

When there are no ties in a block, $T_i = 1$. When all the scores are tied, $T_i = 0$. T_i is always between 0 and 1. Any other value signifies a computing error.

Note that tie correction procedures are different for the correlation methods given in chapter 11, and many of the earlier comments do not apply there.

6.7   Step 5 – draw conclusions

Drawing conclusions following a statistical analysis is often difficult because the methods of hypothesis testing statistics do not always follow everyday logic. Statistical analysis helps us to decide whether or not the null hypothesis (H_0) is true. It does not say anything directly about any other hypothesis (see chapter 2). The following procedure should lead to useful conclusions and avoid logical blunders:

(1) If the results are statistically significant, reject H_0. If the results are not statistically significant do not reject H_0. It is an error to infer that H_0 is true simply on the grounds of a nonsignificant result. This error is called 'proving the null hypothesis'. You may accept H_0 to be true following a nonsignificant result, just as you may assume a man to be guilty until proven innocent. What you can't do, using the methods in this book, is logically infer it to be true.

(2) If you reject H_0, you now have the job of replacing it with some alternative. In the case of a specific test, you may say that the data support the specific alternative hypothesis (H_1). The analysis does not prove it to be true in any logical sense but it does offer a degree of support. All the analysis does is to attach a likelihood to the possibility of making a mistake by inferring that H_0 is false. The rest is informed conjecture.

In the case of a *nonspecific* test, there are many problems attached to the business of adopting an alternative hypothesis. Rejecting H_0 merely implies that there is some nonrandom influence at work causing the sample means to be more widely spread than we would expect if H_0 were true. It is certainly unacceptable to assume that the observed detailed pattern of differences between the sample means or any pair of means is reliable. Drawing specific conclusions calls for specific tests, and the *post hoc* methods of chapter 12 may need to be invoked before this can be done with any certainty. The only exception occurs when we are dealing with two-sample designs. Then, the position is akin to that of the specific test. When H_0 is untrue, the only logical alternative is that the observed difference between the two groups is statistically reliable.

(3) Be careful when wording your conclusion not to infer too much on the basis of a statistically significant result. It is acceptable to expect that a similar experiment would yield a similar result if the same population of subjects, the same investigator, the same methods and the same measures were used. A simple change to one of the details can and sometimes does lead to a quite different result. An awareness of this fact can avoid over-statement when drawing conclusions.

(4) Whenever possible, round off your conclusions by a simple statement in plain English of what you believe the analysis to have shown. Remember that your readers will often not understand the logic of hypothesis testing but might be prepared to trust your conclusions if you express them in nontechnical language. The omission of this last vital step is the commonest example of poor style to be encountered in the reporting of scientific findings.

7 Two independent samples

7.1 Introduction

This chapter is concerned exclusively with the simplest and least complicated of all experimental designs, that of two independent samples. It may not be the most sophisticated, the most impressive, or the most efficient of designs, but it has much to commend it. It is simple to implement, simple to analyse and simple to interpret. After many years of trying to unravel the spaghetti designs of my colleagues and students. I have come at last to see the true worth of this humble workhorse. Its utilization is often the hallmark of the clear thinker, the person who knows what he wishes to prove and who knows the quickest and most certain way to do so. Not all problems are suited to this design, of course, but it is worth thinking long and hard before opting for a more complex approach which will often bring with it ambiguities of interpretation, difficulties of analysis, and increased opportunities for computational error.

Two samples are said to be independent if the individual scores in each sample are not related to each other in any systematic way. The simplest way of obtaining two independent samples is to draw them at random from two different populations, or alternatively to take a collection of individuals (animals, objects, etc.) and divide them using some random process into two groups. In the latter case we are then in a position to treat one group differently from the other with a view to deciding whether the treatment has any noticeable effect. Because the same individual cannot be used in both groups, an independent samples design is relatively expensive to implement compared with a repeated measures arrangement (see chapter 8).

For many people, a truly controlled situation arises only when each individual is used twice in both the treatment and the control conditions (i.e. a repeated measures design). Only then is the comparison obviously valid. However, repeated measures on the same individual raise a number of special problems concerning such effects as practice and fatigue. The independent samples design avoids these problems, although it does so at the expense of using a comparison between two different samples. Acceptance of the validity of the design using two independent samples depends crucially on a faith in the ability of random processes to generate two genuinely comparable groups of individuals. Such a faith can only be strengthened by experience but it is helpful to remember that the overall similarity of two random samples increases as the sample size gets bigger. Statistical hypothesis testing takes into account the possible initial dissimilarity between the samples by making full use of a knowledge of the sample sizes.

Experienced observers will recognize that much of this chapter is a variation on a theme by Wilcoxon (1945) and Mann and Whitney (1947). This applies even to the quick tests for 2 x 2 frequency tables. It will surprise many, however, that I have omitted a description of Fisher's exact test for 2 x 2 frequency tables, which is a popular test in the armoury of researchers in the life sciences. Strictly speaking, Fisher's test is not a special case of the analysis of variance by ranks procedure, but it might have merited inclusion in this chapter as a supplementary test by virtue of its popularity and its ease of use. For technical reasons, which cannot be raised here, there is no exact distribution on which to base a test for the 2 x 2 table within the analysis of variance by ranks scheme. For this reason alone, Fisher's exact test might have served the purpose of filling a gap. Why then has it been omitted?

The problem is that Fisher's test assumes that the marginal totals in the 2 x 2 frequency table are constrained (or 'fixed') by the experimental situation. For many years this assumption has worried statisticians who have applied the test to problems in the life sciences where these totals are not necessarily fixed at all. The average researcher, however, has ignored these qualms. He was only too happy to enjoy the benefits of a very quick and easy test, assured moreover by the possibly misleading adjective 'exact' in the test's title. Theory, however, has a habit of catching up and overtaking practice.

Recent evaluations of the test have been less generous. Liddell (1976) suggests that 'when at least one margin is free, the conditional exact test is highly conservative and of low power and, indeed, irrelevant'. Upton (1982), in a very thorough analysis of the problem, agrees that 'the exact test is inappropriate for the 2 x 2 comparative trial' and recommends a chi-square approximation which is directly equivalent to the analysis of variance by ranks procedure given in sections 7.5 and 7.9. The matter remains controversial but it is becoming increasingly clear that Fisher's exact test should, in future, be restricted to a rather limited class of problems which are of minority interest only in the life sciences. As a consequence of these and other considerations, Fisher's test has been omitted from this book.

This chapter is devoted to eight examples of procedure for analysing two-sample designs. Four tests for evaluating specific hypotheses are followed by four analogous nonspecific tests. The reasons for keeping these similar tests quite separate are given in section 2.10. All tests in this book follow a basic five-step plan which is discussed in some detail in chapter 6. To avoid repetition the basic computational steps only are presented in this chapter, and the reader should consult chapter 6 for further explanation. Detailed accounts of such concepts as significance levels, exact distributions, hypothesis, variables, etc. are given in chapters 1-5. Justification for individual computational formulae will be found in chapter 14.

A   Tests against a specific alternative hypothesis

7.2   Exact test (specific)

When sample sizes are small, it is possible to prepare exact tables of critical values of L based upon every possible permutation of the ranks. These tables have been worked out for sample sizes of 20 and under and appear in table E in the appendix. Exact tables have two main advantages. Firstly, they give a more accurate statement of the likelihood of obtaining an observed value of L when the null hypothesis is true (see section 3.2). Secondly, they avoid a whole computational step of finding Z (or H), which represent approximations to the normal (or chi-square) distribution. Exact tests have the additional advantage of being easier to present to beginners since a smaller number of concepts is required for their understanding.

Unfortunately, exact tables must be devised on the assumption that there are no ties in the data - otherwise we would need a new set of exact tables for each configuration of ties, which would be impossible. We can still use the exact tables when ties are present as an approximate guide if the number of ties is small. This is at the cost of a small (but undefined) decrease in the power of the test (i.e. its ability to reject the null hypothesis when it is false). If the number of ties is very large, then the approximate test given in section 7.3 may be more accurate than the use of exact tables because it has provision for correcting for ties.

Table E (see appendix) extends as far as sample sizes of 20 but this cutoff point is quite arbitrary. The table could just as easily have been larger or smaller. The approximate method given in section 7.3 does not suddenly become valid around sample sizes of 20. On the contrary, the approximation is quite useful at the smallest sample sizes. It merely becomes more accurate as the sample sizes increase. As a rule of thumb, we may say that the approximation is always useful near the 2.5 per cent significance region and only slowly acquires accuracy in the extreme 0.1 per cent region. Exact tables are certainly more accurate (when there are no ties) but their main benefit is probably ease of use. When publishing experimental results, analysis using exact tables is always to be preferred, however.

Example

Hunt (1979) investigated the theory that hypnosis could be explained in terms of 'obedience behaviour'. As part of a larger experiment, she compared the hypnotizability of two groups of volunteers. In one group (control) each subject was hypnotized in isolation. In the other group (experimental) each subject was hypnotized in the presence of an obedient stooge. Both groups were tested for depth of hypnotic trance using the Stanford hypnotic susceptibility scale which is composed of 12 different tests. In the experimental group the obedient stooge faked a positive response on all 12 tests, and it was predicted that this would cause the volunteer subjects who were hypnotized in the presence of the stooge to give higher scores on the depth of hypnosis measure. Depth of hypnosis was measured as the number of tests yielding a positive result. Table 7.1 gives the number of positive tests for ten subjects in the control group and nine subjects in the experimental group. (One score has been dropped from Hunt's original experimental group to illustrate the procedure for unequal sample sizes.) Do the data support the thesis that the presence of a conforming stooge leads to a deeper measured hypnotic trance in a hypnotized subject?

Table 7.1 Depth of hypnosis scores for example in section 7.2

Control group (subject alone)	12 8 9 8 10 7 12 12 8 5
Experimental group (subject with obedient stooge)	12 12 9 11 10 11 12 8 12

Step 1 - formulate the problem
(a) Specify the hypotheses:
   Null hypothesis (H_0): the depth of hypnosis for subjects hypnotized alone will be no different from that for subjects hypnotized in the presence of a conforming stooge.
   Specific alternative hypotheses (H_1): subjects hypnotized in the presence of a conforming stooge will show greater hypnosis depth scores.
(b) Specify the variables:
   Independent variables: presence/absence of conforming stooge.
   Dependent variable: hypnosis depth score, range 0-12. The greater the score the deeper the trance.
(c) Order the samples:
   Sample 1 must be the sample with the highest expected mean rank under the specific alternative hypothesis. In our case, we expect the experimental group (stooge present) to have higher scores (and hence a higher mean rank). We therefore arrange it in table 7.2 that the experimental group is sample 1 and the control group is sample 2.

Table 7.2 Computational layout for example in section 7.2

Sample 1 Experimental (obedient stooge)		Sample 2 Control (no stooge)	
Raw score	*Rank*	*Raw score*	*Rank*
8	*4.5*	5	*1*
9	*7.5*	7	*2*
10	*9.5*	8	*4.5*
11	*11.5*	8	*4.5*
11	*11.5*	8	*4.5*
12	*16*	9	*7.5*
12	*16*	10	*9.5*
12	*16*	12	*16*
12	*16*	12	*16*
		12	*16*
rank sum	R_1 108.5	R_2	81.5
sample size	n_1 9	n_2	10
rank mean	$\bar{R}_1$ 12.06	$\bar{R}_2$	8.15

Step 2 – find the sample rank sums and means

(a) Arrange the scores in columns as in table 7.2. Put the scores in ascending order of size for easy ranking.

(b) Rank the scores *irrespective of sample*, remembering to give low scores low ranks. Write the ranks alongside the raw scores.

(c) Find the two sample rank sums (R_j) by totalling the ranks in the two columns. In our case $R_1 = 108.5$ and $R_2 = 81.5$.

(d) Check that the ranking has been carried out correctly. The total sum of ranks should equal $N(N + 1)/2$, where N is the total number of scores ranked:

$$N(N + 1)/2 = R_1 + R_2$$
$$(19)(20)/2 = 108.5 + 81.5$$
$$190 = 190$$

(e) Note the two sample sizes ($n_1 = 9$ and $n_2 = 10$) and compute the sample rank means ($\bar{R}_j = R_j/n_j$). In our case $\bar{R}_1 = 12.06$ and $\bar{R}_2 = 8.15$.

Step 3 - interpret the rank means

When testing a specific hypothesis we must check that the rank means are ordered in accordance with our expectations. If they are not so ordered, it is not meaningful to proceed any further. In our case we predicted that hypnosis in the presence of a conforming stooge would generate higher depth scores $(\bar{R}_1 > \bar{R}_2)$, which is in agreement with our computations so far. We may therefore proceed to step 4.

Step 4 - compute and evaluate our key statistic (L)

(a) Find L, which in the two-sample case is simply the sum of the ranks in the sample with the expected higher rank mean:

$$L = R_1 = 108.5$$

(b) Evaluate L by consulting table E (in the appendix) for $L = 108.5$, $n_1 = 9$ and $n_2 = 10$. The relevant row for our sample sizes is

n_1	n_2	5%	2.5%	1%	0.1%
9	10	$L \geqslant 121$	125	129	137

To be significant at any of the four levels, L must be equal to or greater than the corresponding critical value. To be significant at the lowest level (5 per cent), our value of L (108.5) must be greater than or equal to 121, which it is not. Our result is, therefore, not significant ($p > 0.05$).

(c) Correct for ties:

No correction for ties is possible in the exact test. In our case this is unfortunate, since the correction would make our result slightly more significant, possibly bringing it even closer to the critical value for the 5 per cent significance level.

Step 5 - draw conclusions

Because our result is not significant, we may not reject H_0 in favour of H_1 (see step 1). In simple terms, the data do not warrant the conclusion that a conforming stooge facilitates the hypnosis of a second person.

7.3   Two-sample approximate test (specific)

The approximate test is normally used only when an appropriate exact table is not available. This will occur if at least one of the sample sizes is larger than 20. The test is sometimes called a 'large-sample test', although this is somewhat misleading since the approximation remains useful down to quite small sample sizes. This is fortunate because computers are more easily programmed to perform an approximate test because it does not require the storage of bulky sets of exact tables. In addition the approximate method may also be used for smaller sample sizes if there are many ties in the data, because it permits a correction for ties whereas the exact test does not. Despite the differences, both tests will usually arrive at the same conclusion in terms of rejecting or accepting H_0 (see exercise 7.1 for an example).

Example

Amlaner and McFarland (1981) investigated the relationship between the posture of herring gulls and their response threshold to mild electric shock. Two of the postures they studied were

(a) eyes open, rest posture – standing with head erect and eyes open
(b) eyes blinking, rest/sleep posture – standing with head drawn into the breast and bill slightly lowered.

They expected that the rest/sleep posture would be associated with higher response thresholds. These thresholds were measured by passing a gentle electric current through a metal grid which had been placed in the bird's nest. The current intensity was regularly increased through ten possible intensities (from lowest to highest) until the bird responded by changing its posture or by looking around. Their results are given in table 7.3 (except for ten scores deleted, at random, to permit a more manageable presentation). A low score indicates a low threshold. Do the data support their contention that the rest/sleep posture is associated with higher thresholds?

Table 7.3   Response threshold measures for two groups of gulls. Low numbers indicate low electric current intensities on an arbitrary scale

Eyes open, rest posture	3 2 4 4 2 2 3 2 4 2 3 4 3 2 1 4 2 2 2 2 4 1 3
Eyes blinking, rest/sleep posture	5 4 4 5 5 10 8 10 4 10 5 10 8 8

Step 1 – formulate the problem
(a) Specify the hypotheses:
    Null hypothesis (H_0): response thresholds are the same for the two postures.
    Specific alternative hypothesis (H_1): the eyes blinking, rest/sleep posture is associated with higher response thresholds.
(b) Specify the variables:
    Independent variable: variation in posture, in this case two different postures (i) eyes open, rest posture and (ii) eyes blinking, rest/sleep posture.
    Dependent variable: response threshold expressed in terms of ten different

levels of intensity of electric current. The lowest intensity is numbered 1 and the highest is numbered 10.

(c) Order the samples:
Sample 1 must be the sample with the highest expected rank mean. In our example, we expect the rest/sleep posture to be associated with the highest thresholds and hence the highest rank mean. The rest/sleep group is accordingly chosen as sample 1 (table 7.4).

Table 7.4   Computational layout for example in section 7.3

Sample 1 Eyes blinking, rest/sleep posture		Sample 2 Eyes open, rest posture	
Raw score	Rank	Raw score	Rank
4	22	1	1.5
4	22	1	1.5
4	22	2	7.5
5	28.5	2	7.5
5	28.5	2	7.5
5	28.5	2	7.5
5	28.5	2	7.5
8	32	2	7.5
8	32	2	7.5
8	32	2	7.5
10	35.5	2	7.5
10	35.5	2	7.5
10	35.5	3	15
10	35.5	3	15
		3	15
		3	15
		3	15
		4	22
		4	22
		4	22
		4	22
		4	22
		4	22
rank sum R_1 418		R_2 285	
sample size n_1 14		n_2 23	
rank mean $\bar{R}_1$ 29.86		$\bar{R}_2$ 12.39	

Step 2 - find the sample rank sums and means
(a) Arrange the scores in columns as in table 7.4. Put the scores in ascending order of size for easy ranking.
(b) Rank the scores irrespective of sample, remembering to give low ranks to low scores. Write the ranks alongside the raw scores.
(c) Find the two sample rank sums (R_j) by totalling the ranks in the two columns. In our example $R_1 = 418$ and $R_2 = 285$.
(d) Check that the ranking has been carried out correctly. The total sum of ranks should equal $N(N + 1)/2$ where N is the total number of scores ranked:

$$N(N + 1)/2 = R_1 + R_2$$
$$(37)(38)/2 = 418 + 285$$
$$703 = 703$$

(e) Note the two sample sizes ($n_1 = 14$ and $n_2 = 23$) and compute the two sample rank means ($\bar{R}_j = R_j/n_j$). In our example, $\bar{R}_1 = 29.86$ and $\bar{R}_2 = 12.39$.

Step 3 - interpret the rank means
We predicted that birds in the rest/sleep posture would show higher response thresholds ($\bar{R}_1 > \bar{R}_2$), which is in agreement with our results. We may, therefore, proceed to step 4.

Step 4 - compute and evaluate our key statistics (L and Z)
(a) Find L which, in the two-sample cases, is simply the rank sum of sample 1:

$$L = R_1 = 418$$

Find Z:

$$Z = \frac{L - \frac{1}{2}(N + 1)\,n_1}{\sqrt{[\frac{1}{12}(N + 1)\,n_1 n_2]}} = \frac{418 - \frac{1}{2}(38)(14)}{\sqrt{[\frac{1}{12}(38)(14)(23)]}}$$
$$= 4.76$$

(b) Evaluate Z by consulting tables of the normal distribution (table A in the appendix). The critical values are reproduced here:

5%	2.5%	1%	0.1%
$z \geq 1.64$	1.96	2.33	3.10

To be significant at any of the four levels, Z must be equal to or greater than the corresponding critical value. Our value of Z (4.76) is greater than the highest critical value and is therefore significant at the 0.1 per cent significance level ($p < 0.001$).

(c) Correct for ties:

The tie correction procedure will increase the value of Z, making it more significant. Since in our example Z is already very significant, tie correction serves little purpose, but the computations are given here for the purposes of illustration. First of all find T the correction factor:

$$T = 1 - \frac{\Sigma(t^3 - t)}{N^3 - N}$$

where $N = n_1 + n_2$, the total number of observations. Table 7.5 identifies seven ties and their tie lengths (t) and computes ($t^3 - t$) for each. $\Sigma(t^3 - t) = 2040$, so we have

$$T = 1 - \frac{2040}{37^3 - 37} = 1 - \frac{2040}{50\,616} = 0.960$$

Table D in the appendix gives values for ($x^3 - x$).

We may now apply the correction

$$Z' = Z/\sqrt{T} = 4.76/\sqrt{0.960} = 4.86$$

Our new corrected value of Z is only very slightly larger than the old value (4.76) and continues to be significant at better than the 0.1 per cent level.

Table 7.5 Tie correction procedure for example in section 7.3

Raw score	Tie length (t)	$(t^3 - t)$
1	2	6
2	10	990
3	5	120
4	9	720
5	5	120
8	3	24
10	4	60
	$\Sigma(t^3 - t)$	2040

Step 5 - draw conclusions

Because our result is statistically significant ($p < 0.001$), we may reject H_0 in favour of H_1 (see step 1). In simple terms, the data support the idea that gulls in an eyes blinking, rest/sleep posture show higher response thresholds than gulls in an eyes open, rest posture.

7.4 Two-sample test (specific) for frequency tables

This test for analysing a two-sample frequency table differs from the corresponding approximate test for raw scores only in the method for computing the rank sums. Its main use is for analysing data involving large sample sizes. By reducing the scale of the dependent variable into a small number of categories, considerable computational savings can often be achieved at the cost of a small reduction in the sensitivity of the test. In practice, significant results based on large samples are usually very significant indeed. A small loss of sensitivity is, therefore, often of little consequence. The other major application of this method is for data which arrive already in the form of frequency tables. In such cases, the test user should make sure that the categories of the dependent variable are indeed rankable: *the test is quite meaningless if this condition is not met.*

Example

The following data were collected as part of a much larger investigation (in which the author was involved) of patients' reactions to aspects of the British National Health Service. Following consultations with their doctor; a larger number of patients were asked to rate their satisfaction with the consultation on a three-point scale (dissatisfied, satisfied or very satisfied). The investigators anticipated that obtaining a prescription for medication from the doctor would act as a contributing factor to the overall satisfaction of the patient. A note was therefore made of whether or not a prescription had been issued for that particular interview. The data are given in table 7.6. Do they support the expec-

Table 7.6 Satisfaction ratings of 521 patients following consultation with a doctor (see example in section 7.4)

	Given prescription	Not given prescription
Very satisfied	188	82
Satisfied	140	75
Dissatisfied	17	19

tation that prescriptions for medication increase the patients' satisfaction with the consultation?

Step 1 - formulate the problem
(a) Specify the hypotheses:
  Null hypothesis (H_0): obtaining a prescription from the doctor does not influence a patient's satisfaction with the interview.
  Specific alternative hypotheses (H_1): obtaining a prescription makes a patient *more* satisfied with the interview.
(b) Specify the variables:
  Independent variable: obtaining/not obtaining a prescription.
  Dependent variable: satisfaction following the interview. Three levels of satisfaction were used (very satisfied, satisfied, dissatisfied).
(c) Order the samples:
  Sample 1 must be the sample with the highest expected rank mean. In a frequency table, categories near the bottom of the table have higher shared ranks. According to H_1, patients who are not given a prescription are expected to be dissatisfied (i.e. near the bottom of the table). Therefore, they are expected to have the higher rank and should be chosen as sample 1 (see table 7.7).

Step 2 - find the sample rank sums and means
(a) Draw up the frequency table as in table 7.7. Find the category totals (t_q) and the cumulative category totals (Σt_q).
(b) Find the shared ranks for each category using the expression

$$r_q = \Sigma t_q - t_q/2 + 0.5$$

Table 7.7 Computational layout for example in section 7.4

	Sample 1 No prescription	Sample 2 Given a prescription	Total t_q	Cumulative total Σt_q	Shared rank r_q
Very satisfied	82	188	270	270	135.5
Satisfied	75	140	215	485	378
Dissatisfied	19	17	36	521	503.5
rank sum	R_1 49 027.5	R_2 86 953.5			
sample size	n_1 176	n_2 345	N 521		
rank mean	$\bar{R}_1$ 278.56	$\bar{R}_2$ 252.04			

In our example we have

$$r_1 = 270 - 270/2 + 0.5 = 135.5$$
$$r_2 = 485 - 215/2 + 0.5 = 378$$
$$r_3 = 521 - 36/2 + 0.5 = 503.5$$

(c) Find the sample rank sums for each sample by multiplying the shared rank for each category by the number of scores in that category and adding:

$$R_1 = 82\,(135.5) + 75\,(378) + 19\,(503.5) = 49\,027.5$$
$$R_2 = 188\,(135.5) + 140\,(378) + 17\,(503.5) = 86\,953.5$$

(d) Check that the ranking has been carried out correctly. The total sum of ranks should equal $N(N + 1)/2$:

$$N(N + 1)/2 = R_1 + R_2$$
$$(521)(522)/2 = 49\,027.5 + 86\,953.5$$
$$135\,981 = 135\,981$$

(e) Note that the two sample sizes ($n_1 = 176$ and $n_2 = 345$) and compute the two sample rank means, $\bar{R}_j = R_j/n_j$. In our example, $\bar{R}_1 = 278.56$ and $\bar{R}_2 = 252.04$.

Step 3 - interpret the rank means
We predicted that patients who did not receive a prescription would show greater dissatisfaction (i.e. higher mean rank),which is in agreement with our results. We may, therefore, proceed to step 4.

Step 4 - compute and evaluate our key statistics (L and Z)
(a) Find L, which in this case is simply

$$L = R_1 = 49\,027.5$$

Find Z:

$$Z = \frac{L - \frac{1}{2}(N + 1)\,n_1}{\sqrt{[\frac{1}{12}(N + 1)\,n_1 n_2]}} = \frac{49\,027.5 - \frac{1}{2}(522)(176)}{\sqrt{[\frac{1}{12}(522)(176)(345)]}}$$
$$= 1.90$$

(b) Evaluate Z by consulting tables of the normal distribution (table A in the appendix). The critical values are reproduced here:

5%	2.5%	1%	0.1%
$z \geqslant 1.64$	1.96	2.33	3.10

To be significant at any of the four levels, Z must be equal to or greater than the corresponding critical value. Our value of Z (1.90) is only greater than the critical value at the 5 per cent level of significance ($p < 0.05$).

(c) Correct for ties:
The tie correction procedure will make our value of Z larger and even more significant. The tie lengths can be found in the category total (t_q) column of table 7.7. Table D in the appendix will be found useful for finding values for ($t^3 - t$):

$$\Sigma(t^3 - t) = (270^3 - 270) + (215^3 - 215) + (36^3 - 36)$$

$$= 19\,682\,730 + 9\,938\,160 + 46\,620$$

$$= 29\,667\,510$$

Find T:

$$T = 1 - \frac{\Sigma(t^3 - t)}{N^3 - N}$$

$$= 1 - \frac{29\,667\,510}{521^3 - 521} = 1 - \frac{29\,667\,510}{141\,420\,240} = 0.79$$

We can now correct Z:

$$Z' = Z/\sqrt{T} = 1.90/\sqrt{0.79} = 2.14$$

Our new value of Z' (2.14) is now greater than the critical value at the 2.5 per cent significance level ($p < 0.025$).

Step 5 - draw conclusions
Because our result is significant ($p < 0.025$) we may reject H_0 in favour of H_1 (see step 1). In simple terms, the data support a conclusion that patients who are given a prescription during a consultation with a doctor will subsequently express greater satisfaction with that consultation.

7.5   2 × 2 frequency table quick test (specific)

When we have only two categories of the dependent variable, the two-sample frequency table test procedure can be greatly simplified. It becomes so simple that it begins to look like a different test. However, the test user will do well to remember that it is basically the same test as the two approximate tests which have preceded it (sections 7.3 and 7.4). The enormous simplification has been achieved because there are only four scores involved in the analysis. It means that we can immediately compute Z' (i.e. Z corrected for ties) with a single formula. As a consequence, the 2 × 2 frequency table is very useful as a quick test, and experienced researchers use it a great deal in the preliminary analysis of data in order to get the 'feel' of their results. One way of doing this is to recast all values of the dependent variable into two categories, such as above and below the median score or some similar scheme. The increase in speed is offset by a reduction in the sensitivity of the test. However, when the sample size is large, this need not be a serious problem. Notice that a result which is significant using the 2 × 2 quick test will always be more significant if a full test is used, so that a subsequent full test is often not necessary if a satisfactory level of significance has already been achieved using the quick test.

To avoid confusion, the experienced researcher should note that the 2 × 2 quick test given here is not the same as the normal approximation to Fisher's exact test. It is similar but is different in two respects. Firstly, the denominator of our main computational formula contains the expression $(N - 1)$ where the approximation to Fisher's exact test uses N. Secondly, our formulation has no 'correction for continuity', nor does it admit of any (see Upton (1982) for a thorough discussion). The two formulae do often given similar results but their underlying rationale is quite different.

No exact tables can be drawn up for the 2 × 2 quick test (unlike Fisher's exact test) because of difficulties in deciding what the dependent variable marginal totals should be. This is the general problem of finding exact tables when there are ties in the data. As a result we do not know for sure how good our approximate test really is. We do know that it gradually deteriorates as the sample sizes get smaller. Various considerations, however, lead us to believe that the approximation is never so poor as to be useless. If the user merely wants to know if the results are significant at approximately the 5 per cent level, he should bear in mind that an error of ±2 per cent may be involved in the tail region in the estimate. Therefore, he should be prepared to insist on a critical value appropriate to the 2.5 per cent level in order to be sure of being significant at the 5 per cent level. This rule of thumb is only necessary for very small sample sizes such as $N < 20$, when extreme levels of significance are, in any case, unlikely to be achieved.

The distinction between the dependent and independent variable can become blurred in the 2×2 frequency test. The same result is, however, obtained whichever variable is chosen as the independent variable. This is a simple consequence of the symmetry of the 2×2 table. Indeed, the test is often used as a test of association between two variables neither of which need be directly influenced by the other (see the following example). Another consequence of having only two categories of both variables is that neither is required to belong to a rankable scale. As a result the test works equally well with nominal categories (e.g. Englishman, Irishman) and rankable categories (e.g. high, low). Both considerations have contributed to the versatility and popularity of the test.

Example

When individuals are shown certain kinds of striped patterns (e.g. black/white horizontal stripes, three stripes per centimetre), they often report sensations such as colour, shimmer, dazzle, glare, bending of the lines, fading and flickering. They may also report such adverse effects as eyeache, tiredness, headache or dizziness, according to Wilkins et al. (1984), who decided to look for a link between these two phenomena. They showed a horizontal striped pattern, which was divided into a left and right half, to 50 women and asked them to decide whether the visual effects were either equally on both sides of the display or predominantly on only one side, either left or right.These two responses were classified as 'bilateral' and 'unilateral'. The same women also completed a questionnaire about their recent headaches with a view to deciding whether they typically affected both sides of the head (bilateral), mainly one side of the head, either left or right (unilateral), or sometimes one side of the head and sometimes the other (inconsistent). Twenty-four of the women reported inconsistent lateralization and their scores were deleted from the data set. The results for the remaining 26 women are shown in table 7.8. Do these results support the contention of Wilkins et al. (1984) that lateralization of the visual effects on viewing stripes is associated with lateralization of headaches?

Table 7.8 Data for example in section 7.5 (see text)

| | *Illusions in visual hemifields* | |
	Bilateral	*Unilateral*
Bilateral headaches	$A = 13$	$B = 1$
Unilateral headaches	$C = 6$	$D = 6$

E

Step 1 – formulate the problem

(a) Specify the hypotheses:

Null hypothesis (H_0): lateralization of headaches is not related to lateralization of special visual effects (illusions) on viewing horizontal stripes (visual gratings).

Specific alternative hypothesis (H_1): lateralization of headaches is positively associated with lateralization of the visual effects (i.e. when the illusions are unilateral the headaches are unilateral too).

(b) Specify the variables:

In this example it is difficult to decide which variable is dependent and which independent, since both phenomena (headaches, visual effects) are presumed to be dependent upon a common neurological process. We shall therefore avoid the issue by calling them variables 1 and 2:

variable 1: lateralization of visual effects (bilateral/unilateral)
variable 2: lateralization of headaches (bilateral/unilateral).

(c) Order the samples'

Identify the left and right frequencies in the top row of the table as A and B respectively and the frequencies in the bottom row as C and D. Swap the columns, if necessary, so that the ratio $A/(A + C)$ is expected to be larger than the ratio $B/(B + D)$ if the specific alternative hypothesis be true. In our case we expect bilateral illusions to be associated more with bilateral headaches (score A) than unilateral headaches (score C). Bilateral illusions are therefore assigned to the first column and we do not need to rearrange table 7.8.

Step 2 – find the sample rank sums and means

This step is omitted from the quick test.

Step 3 – interpret the rank means

In the absence of any sample rank means, we observe that our specific hypothesis implies that

$$A/(A + C) > B/(B + D)$$

(i.e. that the proportion of bilateral headaches among persons with bilateral illusions will be greater than the proportion of bilateral headaches among persons with unilateral illusions), and this should now be checked:

$$13/(13 + 6) > 1(1 + 6)$$
$$0.68 > 0.14$$

Our results are therefore in line with the specific hypothesis under test and we may proceed to step 4.

Step 4 - compute and evaluate our key statistic (Z')
(a) Compute Z' (which is automatically corrected for ties):

$$Z' = \frac{AD - BC}{\sqrt{\dfrac{(A+B)(C+D)(A+C)(B+D)}{N-1}}} = \frac{(13)(6) - (1)(6)}{\sqrt{\dfrac{(14)(12)(19)(7)}{25}}}$$

$$= 2.41$$

(b) Evaluate Z' by consulting tables of the normal distribution (table A in the appendix). The critical values are reproduced here:

5%	2.5%	1%	0.1%
$z \geqslant 1.64$	1.96	2.33	3.10

To be significant at any of the four levels, Z' must be equal to or greater than the corresponding critical value. Our value of Z' (2.41) is therefore significant at the 1 per cent level of significance ($p < 0.01$).

Step 5 - draw conclusions
Because out result is significant at the 1 per cent level we may reject H_0 in favour of H_1 (see step 1). In simple terms, the data support the view that lateralization of headaches is associated with lateralization of visual effects attendant upon viewing narrow black/white striped patterns.

B   Tests against a nonspecific alternative hypothesis

Whereas specific tests predict the direction of the difference between two treatments (e.g. treatment A > treatment B), nonspecific tests make no prediction; they allow either direction of difference as evidence against H_0 being true. Traditionally, in the two-sample case, specific and nonspecific tests have been treated as minor variants ('one tailed' and 'two tailed' respectively) of the same test. The decision to deal with them quite separately here has been defended at length in section 2.10. The distinction between specific and nonspecific tests is quite fundamental to hypothesis testing. Any economy achieved by blurring the distinction in the two-sample case is purchased at the expense of conceptual clarity.

    If a design involves two independent samples and no prediction was made (or could reasonably have been made), prior to the collection of the data, of the direction of the difference between the two treatments, then a nonspecific test is appropriate. If a prediction was made but failed, then a nonspecific test may be used to evaluate the statistical significance of the difference that was found. This latter case is an example of *post hoc* analysis, and the procedure is justified in chapter 12 and illustrated in section 7.6.

7.6   Exact test (nonspecific)

The only important practical difference between the specific and nonspecific exact test is that they use different tables of the exact distribution of L. The following example was chosen to illustrate the situation where the difference between the two samples was not in the expected direction but may yet be statistically significant in the reverse direction.

Remember that further detailed advice on each computational step can be found in chapter 6.

Example

This example follows from the example in section 7.2. It is taken from an experiment by Hunt (1979) which investigated the possibility that hypnotic phenomena could be explained in terms of 'obedience behaviour'. Here we compare her control group, where each subject was hypnotized in isolation, with a second experimental group, where each subject was hypnotized alongside a stooge who was behaving according to a prearranged plan. This plan involved simulating positive responses to only the first six tests of the depth of hypnotic trance administered to both subjects. Following the sixth test the stooge declared the whole thing to be 'ridiculous' and refused to respond to the remaining six tests but did agree to remain quietly in the room until testing was complete. The depth of hypnotic trance (given as the number of tests yielding a positive response) for each subject is given in table 7.9. On this occasion, we might have expected the stooge's behaviour to reduce the depth of trance of the subjects in the experimental group compared with those in the control group. Since this appears not to be the case, a test of a specific hypothesis is not in order and a nonspecific test is required.

Table 7.9   Depth of hypnosis scores for example in section 7.6

Control group (subject alone)	12 8 9 8 10 7 12 12 8 5
Experimental group (disruptive stooge)	12 8 8 8 12 9 11 8 12 12

Step 1 - formulate the problem
(a) Specify the hypotheses:
Null hypothesis (H_0): the depth of hypnosis for subjects hypnotized alone will be unaffected by the presence of the experimenter's stooge.
There is no specific alternative hypothesis. The general alternative hypothesis is simply that H_0 is untrue.
(b) Specify the variables:
Independent variable: presence/absence of disruptive stooge.
Dependent variable: hypnosis depth score, range 0-12. The greater the score, the deeper the trance.
(c) Order the samples:
Sample 1 is the sample with the higher rank mean. Strictly speaking we do not know which this is until we are part way through out computations. Usually, however, we can tell this by inspection even at this stage. In our case, the experimental group appears to have higher scores and will be named sample 1. If this initial guess proves wrong, it is easy to rename the groups later.

Table 7.10   Computational layout for example in section 7.6

Sample 1 Experimental group (disruptive stooge)		Sample 2 Control group (no stooge)	
Raw score	Rank	Raw score	Rank
8	6	5	1
8	6	7	2
8	6	8	6
8	6	8	6
9	10.5	8	6
11	13	9	10.5
12	17	10	12
12	17	12	17
12	17	12	17
12	17	12	17
rank sum R_1 115.5		R_2 94.5	
sample size n_1 10		n_2 10	
rank mean $\bar{R}_1$ 11.55		$\bar{R}_2$ 9.45	

Step 2 – find the sample rank sums and means
(a) Arrange the scores in columns as in table 7.10. Put the scores in ascending order of size for easy ranking.
(b) Rank the scores, irrespective of sample, remembering to give low scores low ranks.
(c) Find the two sample rank sums (R_j) by totalling the ranks in the two columns. In our case $R_1 = 115.5$ and $R_2 = 94.5$.
(d) Check that that the ranking has been carried out correctly. The total sum of ranks should equal $N(N + 1)/2$ where N is the total sample size $(n_1 + n_2)$:

$$N(N + 1)/2 = R_1 + R_2$$
$$(20)(21)/2 = 115.5 + 94.5$$
$$210 = 210$$

(e) Note the the sample sizes $(n_1 = 10$ and $n_2 = 10)$ and compute the sample rank means $(\bar{R}_j = R_j/n_j)$. In our case, $\bar{R}_1 = 11.55$ and $\bar{R}_2 = 9.45$.

Step 3 – interpret the rank means
The rank means suggest that the disruptive stooge is increasing the depth of trance in those hypnotized alongside the stooge compared with unaccompanied subjects in the control group. At this stage we should confirm that sample 1 does, in fact, have the higher rank mean (see step 1(c)). In our case it does.

Step 4 – compute and evaluate our key statistic (K)
(a) Find K, which in the two-sample case is simply the rank sum of sample 1 (the sample with the higher expected sample rank mean):

$$K = R_1 = 115.5$$

(b) Evaluate K by consulting table F (in the appendix) for $K = 115.5, n_1 = 10,$ $n_2 = 10$. The relevant row for our purpose is

n_1	n_2	5%	2.5%	1%	0.1%
10	10	$K \geqslant 132$	135	139	147

To be significant at any of the four levels, K must be equal to or greater than the corresponding critical value. Our value of K is not significant at any of these levels $(p > 0.05)$.
(c) Correct for ties:
No correction for ties is possible in the exact test. When ties exist this renders the exact test conservative.

Step 5 – draw conclusions

Because our result is not significant we may not reject H_0 (see step 1). In simple terms, the data do not warrant the conclusion that the disruptive stooge affects the depth of trance in a hypnotized subject.

7.7 Approximate test (nonspecific)

From a purely computational point of view, the nonspecific approximate test is the same as its specific counterpart (section 7.3) except for the use of the chi-square approximation in place of the normal approximation in step 4. It should be used in preference to the specific test only when no prediction has been made concerning the direction of the difference between the two groups or when the actual direction is contrary to that predicted.

A nonspecific test can be used to check against unwanted bias when creating groups which should be similar at the beginning of an investigation. If a statistical test shows that a control group is significantly different from an experimental group *even before the treatment is applied*, this clearly undermines the logic of the experiment from the outset. If the subjects were assigned at random to the groups, we have no reasonable basis for predicting the direction of any difference between them, and a nonspecific test is clearly indicated. Unfortunately, a failure to achieve a significant result does not prove that the selection was unbiased, but a significant result is a useful warning that something may have gone wrong at this early stage.

Example

Johnstone et al. (1980) studied 62 psychiatric patients suffering from endogenous depression who were randomly allocated to either (a) a course of eight *real* electroconvulsive therapy (ECT) sessions or (b) a course of eight *simulated* ECT sessions. All patients were assessed on the Hamilton depression rating scale *before any treatment* and their ratings are given in table 7.11. A higher rating

Table 7.11 Hamilton depression ratings for two groups of patients (example in section 7.7)

Simulated ECT (control group)

28	45	66	38	44	43	50	48	54	48	
62	56	29	78	46	54	44	36	58	54	
60	56	36	67	66	28	60	64	46	56	53

Real ECT (experimental group)

62	50	44	30	58	56	52	54	48	40	
42	23	52	32	72	82	72	56	54	46	
58	52	44	56	70	54	50	44	68	56	100

indicates a more severe depression. Is there any indication that the two groups were different before the experiment began?

Step 1 - formulate the problem
(a) Specify the hypotheses:
Null hypothesis (H_0): no bias has affected the selection of the two groups and the depth of depression is expected to be the same for both groups. There is no specific alternative hypothesis. The general alternative hypothesis is that H_0 is untrue.
(b) Specify the variables:
Independent variable: group assignment; control/experimental.
Dependent variable: depression rating; high scores, more depression.
(c) Order the samples:
The order of the samples is irrelevant.

Step 2 - find the sample rank sums and means
(a) Arrange the scores in columns as in table 7.12. Put the scores in ascending order of size for easier ranking.
(b) Rank the scores irrespective of sample, remembering to give low scores low ranks. When there are many scores (as in this case), take particular care over the ranking; it is easy to make time-consuming mistakes (see section 6.4 for advice). Write the ranks next to the raw scores.
(c) Find the two sample rank sums (R_j) by totalling the ranks in the two columns. In our example $R_1 = 1023$ and $R_2 = 930$.
(d) Check that the ranking has been carried out correctly:

$$N(N + 1)/2 = R_1 + R_2$$

$$(62)(63)/2 = 1023 + 930$$

$$1953 = 1953$$

(e) Note the two sample sizes ($n_1 = 31$ and $n_2 = 31$) and compute the two sample rank means ($\bar{R}_j = R_j/n_j$). In our case $\bar{R}_1 = 33$ and $\bar{R}_2 = 30$.

Step 3 - interpret the rank means
The experimental group shows slightly higher depression ratings than the control group.

Step 4 - compute and evaluate our key statistic (H)
(a) Find H:

$$H = \frac{12}{N(N + 1)} \left(\frac{R_1^2}{n_1} + \frac{R_2^2}{n_2} \right) - 3(N + 1)$$

Table 7.12   Computational layout for example in section 7.7

Sample 1 (real ECT)		Sample 2 (simulated ECT)	
Score	Rank	Score	Rank
23	1	28	2.5
30	5	28	2.5
32	6	29	4
40	10	36	7.5
42	11	36	7.5
44	15	38	9
44	15	43	12
44	15	44	15
46	20	44	15
48	23	45	18
50	26	46	20
50	26	46	20
52	29	48	23
52	29	48	23
52	29	50	26
54	34.5	53	31
54	34.5	54	34.5
54	34.5	54	34.5
56	41	54	34.5
56	41	56	41
56	41	56	41
56	41	56	41
58	46	58	46
58	46	60	48.5
62	50.5	60	48.5
68	56	62	50.5
70	57	64	52
72	58.5	66	53.5
72	58.5	66	53.5
82	61	67	55
100	62	78	60

sample rank sum	R_1	1023		R_2	930
sample size	n_1	31		n_2	31
sample rank mean	$\bar{R}_1$	33			30

$$= \frac{12}{(62)(63)}\left(\frac{1023^2}{31} + \frac{930^2}{31}\right) - 3(63)$$

$$= 0.428$$

(b) Evaluate H by consulting tables of chi-square (table B in the appendix) for $H = 0.428$. In two-sample problems we have only one degree of freedom (df = 1). The appropriate critical values are, therefore:

Degrees of freedom	5%	2.5%	1%	0.1%
1	$\chi^2 \geqslant 3.8$	5.0	6.6	10.8

To be significant our value of H must be greater than or equal to the critical χ^2 value at the appropriate level. Our value of H (0.428) is clearly too small and is therefore not significant at any level ($p > 0.05$).

(c) Correct for ties:
Using Table D (in the appendix) set up a tie length table as in table 7.13.

Table 7.13   Computational layout for finding T, correction for ties (example in section 7.7)

Value	Tie length (t)	$(t^3 - t)$
28	2	6
36	2	6
44	5	120
46	3	24
48	3	24
50	3	24
52	3	24
54	6	210
56	7	336
58	3	24
60	2	6
62	2	6
66	2	6
72	2	6
	$\Sigma(t^3 - t)$	822

Note that only those scores which are tied need to be included.

$$T = 1 - \frac{\Sigma(t^3 - t)}{N^3 - N}$$

$$= 1 - \frac{822}{238\ 266} = 0.997$$

We can now find H', the corrected value of H:

$$H' = H/T = (0.428)/0.997$$

$$= 0.429$$

Our value of H' is clearly still not significant.

Step 5 - draw conclusions

Because our result is not significant, we cannot reject H_0. This means that the data do not support any suggestion of bias in selecting the patients for the two groups. This does not prove, of course, that no bias is present. However, since the experimental condition has the more depressed patients, any demonstration that the treatment produces better results than the control condition will obviously be all the more impressive.

7.8 Approximate test (nonspecific) for frequency tables

This test is basically the same as that for individual scores (section 7.7) and is similar to its specific counterpart (section 7.4) except for the use of the chi-square approximation in place of the normal approximation. The preliminary notes for these sections should therefore be consulted. This nonspecific test should only be used in preference to its specific equivalent either when there is no predicted direction of difference between the groups or when this prediction does not agree with the observed results.

Example

The Eysenck personality inventory (EPI) measure of introversion–extroversion was used (among other measures) by Parkes (1982) in an investigation of the causes of job dissatisfaction among student nurses. No advance prediction was made concerning the direction of the relationship between extroversion and job satisfaction since this was simply an exploratory measure. A nonspecific test is therefore indicated. Nurses were asked to rate their satisfaction as (a) happy with the job or (b) would rather change jobs. Some nurses had already changed

Table 7.14   Extroversion and job satisfaction among student nurses (see example in section 7.8)

	High extroversion	Low extroversion
Happy with the job	31	76
Would rather change jobs	11	18
Already changed jobs	9	9

jobs by the time the assessment was made. The data are given in table 7.14. Is the personality measure related to job dissatisfaction?

Step 1 – formulate the problem
(a) Specify the hypotheses:
Null hypothesis (H_0): there is no relationship between the EPI measure of extroversion/introversion and job satisfaction among student nurses.
There is no specific alternative hypothesis. The general alternative hypothesis is that H_0 is false.
(b) Specify the variables:
Independent variable: extroversion. This is normally measured on a many-point scale. To simplify the problem this has been reduced to two samples of nurses, (i) high and (ii) low on extroversion.
Dependent variable: job satisfaction rated on a three-point scale. For the purposes of this analysis, we assume that these three points are legitimately rankable. However, nurses may leave their job for many reasons other than dissatisfaction. This problem is tackled in section 7.9.
(c) Order the samples:
The order of the samples is irrelevant.

Step 2 – find the sample rank sums and means
(a) Draw up the frequency table as in table 7.15. Find the category totals (t_q) and the cumulative category totals (Σt_q).
(b) Find the shared ranks for each category using the expression

$$r_q = \Sigma t_q - t_q/2 + 0.5$$

In our example we have

$$r_1 = 107 - 107/2 + 0.5 = 54$$
$$r_2 = 136 - 29/2 + 0.5 = 122$$
$$r_3 = 154 - 18/2 + 0.5 = 145.5$$

Table 7.15 Computational layout for example in section 7.8

	Extroversion		Category total t_q	Cumulative total Σt_q	Shared rank r_q	$t_q{}^3 - t_q$
	High	Low				
Happy with job	31	76	107	107	54	1 224 936
Rather change jobs	11	18	29	136	122	24 360
Already changed jobs	9	9	18	154	145.5	5 814
rank sum	R_1 4325.5	R_2 7609.5				Σ 1 255 110
sample size	n_1 51	n_2 103				
mean rank	$\bar{R}_1$ 84.8	$\bar{R}_2$ 73.9				

(c) Find the sample rank sums for each sample by multiplying the shared rank for each category by the number of scores in that category and adding:

$$R_1 = 31(54) + 11(122) + 9(145.5) = 4325.5$$
$$R_2 = 76(54) + 18(122) + 9(145.5) = 7609.5$$

(d) Check that the ranking has been carried out correctly:

$$N(N + 1)/2 = R_1 + R_2$$
$$(154)(155)/2 = 4325.5 + 7609.5$$
$$11\,935 = 11\,935$$

(e) Note the two sample sizes ($n_1 = 51$ and $n_2 = 103$) and compute the two sample rank means ($\bar{R}_j = R_j/n_j$). In our example, they are $\bar{R}_1 = 84.8$ and $\bar{R}_2 = 73.9$.

Step 3 - interpret the rank means
Sample 1, the high extroversion nurses, has a higher rank mean. Since higher ranks in this case reflect great job dissatisfaction, the data indicate, therefore, less job satisfaction in our sample of extrovert nurses.

Step 4 - compute and evaluate our key statistic (H)
(a) Find H:

$$H = \frac{12}{N(N + 1)} \left(\frac{R_1{}^2}{n_1} + \frac{R_2{}^2}{n_2} \right) - 3(N + 1)$$

$$= \frac{12}{(154)(155)} \left(\frac{4325.5^2}{51} + \frac{7609.5^2}{103} \right) - 3(155)$$

$$= 2.05$$

(b) Evaluate H by consulting tables of the chi-square distribution (table B in the appendix) for $H = 2.05$. In two-sample problems we have only one degree of freedom (df $= 1$). The appropriate critical values are therefore

Degrees of freedom	5%	2.5%	1%	0.1%
1	$\chi^2 \geqslant 3.8$	5.0	6.6	10.8

To be significant, our value of H must be equal to or greater than the value of χ^2 for the appropriate level. Our value of H (2.05) is clearly too small and is therefore not significant at any level ($p > 0.05$).

(c) Correct for ties:

A correction for ties usually has a large effect in frequency table examples where ties are often extensive. Use table D in the appendix to find values of $(t^3 - t)$ and enter them in table 7.15:

$$\Sigma(t_q^3 - t) = (107^3 - 107) + (29^3 - 29) + (18^3 - 18)$$

$$= 1\,224\,936 + 24\,360 + 5814$$

$$= 1\,225\,110$$

Find T, the correction factor:

$$T = 1 - \frac{\Sigma(t^3 - t)}{N^3 - N} = 1 - \frac{1\,255\,110}{154^3 - 154}$$

$$= 0.656$$

Compute H':

$$H' = H/T = 2.05/0.656 = 3.12$$

Our value of H' is much larger but remains less than the critical value of χ^2 at the 5 per cent level of significance ($p > 0.05$).

Step 5 - draw conclusions

Because our result is not significant, we may not reject H_0. In simple terms, our data do not support the idea of any relationship between this measure of introversion-extroversion and job satisfaction among student nurses.

7.9   2 × 2 frequency table quick test (nonspecific)

See section 7.5 for comments on this design. The nonspecific version of this test should only be used in preference to its specific counterpart if no prediction has been made concerning the direction of the difference between the samples, or if this prediction is contradicted by the results.

Example

In the previous example (section 7.8), student nurses were rated as high or low on an extroversion measure. Their job satisfaction was also rated on a three-point scale: (a) happy with their job, (b) would rather change jobs and (c) have already changed jobs. The last category is problematic since there are many reasons for changing a job other than job dissatisfaction. To meet this objection we shall reanalyse the data in table 7.14, but this time the nurses who have already changed jobs will be omitted from the computations. This leaves us with the 2×2 frequency table shown in table 7.16. Given this new data arrangement, is there any support for the idea that extroversion is linked either with job satisfaction or with job dissatisfaction?

Table 7.16   Data for example in section 7.9

	Extroversion		
	High	Low	
Happy with job	$A = 31$	$B = 76$	
Would rather change jobs	$C = 11$	$D = 18$	$N = 136$

Step 1 - formulate the problem
(a) Specify the hypotheses:
Null hypothesis (H_0): there is no relationship between the EPI measure of extroversion/introversion and job satisfaction among student nurses.
There is no specific alternative hypothesis. The general alternative hypothesis is that H_0 is false.
(b) Specify the variables:
Independent variable: extroversion. This measure has been reduced to two categories (a) high extroversion and (b) low extroversion.
Dependent variable: job satisfaction rated on a two-category scale.
(c) Order the samples:
The order of the samples is irrelevant.

Step 2 - find the sample rank sums and means
This step is omitted from the quick test.

Step 3 - interpret the rank means
By examining the ratios of high/low satisfaction for both samples we see that the

nurses in the high extroversion sample show less satisfaction with their jobs:

high extroversion nurses ratio: $A/(A + C) = 31/42 = 0.74$

low extroversion nurses ratio: $B/(B + D) = 76/94 = 0.81$

Step 4 - compute and evaluate our key statistic (H)
(a) Find H:

$$H = \frac{(N - 1)(AD - BC)^2}{(A + B)(C + D)(A + C)(B + D)}$$

$$= \frac{(135)[31(18) - 76(11)]^2}{(107)(29)(42)(94)}$$

$$= 0.85$$

(b) Evaluate H by consulting tables of the chi-square distribution (table B in the appendix) for $H = 0.85$. In two-sample problems we have only one degree of freedom (df $= 1$). The appropriate critical values are therefore

Degrees of freedom	5%	2.5%	1%	0.1%
1	$\chi^2 \geqslant 3.8$	5.0	6.6	10.8

To be significant our value of H must be equal to or greater than the critical value for χ^2 at the appropriate level. Our value of H is clearly not significant at any level ($p > 0.05$).
(c) Correct for ties:
The above procedure automatically corrects for ties.

Step 5 - draw conclusions
Because our result is not significant, we may not reject H_0. In simple terms, our data do not support the idea of any relationship between this measure of introversion–extroversion and job satisfaction among student nurses.

Exercises

7.1 Repeat the example in section 7.2 using the approximate test, with and without correction for ties. Compare both results with those from section 7.2.
7.2 Repeat the example in section 7.3 using a test for frequency tables using seven categories of the dependent variable.

Table 7.17   Depression ratings for exercise in section 7.6

Simulated ECT			Real ECT		
Subject	Before	After	Subject	Before	After
1	28	3	1	62	0
2	45	7	2	50	0
3	66	7	3	44	2
4	38	8	4	30	2
5	44	10	5	58	2
6	43	10	6	56	2
7	50	10	7	52	2
8	48	12	8	54	4
9	54	14	9	48	4
10	48	16	10	40	6
11	62	16	11	42	6
12	56	18	12	23	7
13	29	18	13	52	8
14	78	20	14	32	8
15	46	21	15	72	8
16	54	22	16	82	8
17	44	22	17	72	8
18	36	23	18	56	8
19	58	23	19	54	12
20	54	25	20	46	14
21	60	28	21	58	20
22	56	28	22	52	20
23	36	28	23	44	23
24	67	29	24	56	30
25	66	30	25	70	30
26	28	34	26	54	33
27	60	40	27	50	36
28	64	42	28	44	38
29	46	44	29	68	38
30	56	48	30	56	40
31	53	48	31	100	76

7.3 Repeat the example in section 7.4 combining the categories 'very satisfied' and 'satisfied' into a single category and use a 2×2 test. Compare your results with those from section 7.4.

7.4 Oakley (1979) removed the cerebral cortex from two rats and, after they had recovered from the surgical procedures, he compared their performance with six untreated rats on a simple learning task. The number of trials required to master the task are given below for each animal. Do the data justify the conclusion that decortication impairs learning on this task?

Decorticates	9	47				
Normals	0	3	14	2	2	8

Attempt both an exact and an approximate test and compare the results.

7.5 Howard, Miller and Calver (1981) investigated the theory that phsyical contact between two people encourages more honest interactions. To do this they left a coin on the counter in a public telephone call box. When the next user of the telephone came out of the box, one of the experimenters approached him and asked if he had found the coin. Half of the people were touched on the sleeve while the question was being asked, where the remainder were not touched at all. Of 20 people studied, all picked up the coin but only eight returned it to the investigator. The results are summarized as follows. Is the investigators' theory supported by the data?

	Touched	*Not touched*
Returned coin	8	0
Kept coin	2	10

7.6 The example in section 7.7 looked at the Hamilton depression ratings of 72 psychiatric patients before they received *either* real *or* simulated ECT. Table 7.17 shows the ratings both before and after the treatment. Since the two samples are not statistically significantly different before treatment we can simply compare the two samples of depression ratings after treatment. Attempt this analysis. Compare your result with that obtained after the analysis of the same data suggested in exercise 8.5.

7.7 Repeat the example in section 7.7 using frequency table methods. Most of the work has already been done in sections 6.4 and 6.6. How does your result differ from that of section 7.7?

8 Two related samples

8.1 Introduction

The expression 'two related samples' is used whenever the individuals in one sample are systematically related to the individuals in the second sample. This is most obviously the case in a repeated measures design where a single person or animal may contribute a score from two conditions – say before and after some treatment. It is equally the case when two different people are matched on certain variables such as age, sex, height etc., but one receives the treatment and the other does not. In both situations we are dealing with a matched pair of scores. When we have a number of matched pairs we say that we have two related samples. We are not, of course, restricted to *pairs* of scores. For example, we may wish to study small groups of young athletes from a number of schools where within each school some do and some do not receive a special kind of coaching. In this example, each school forms a 'matched set' and the statistical analysis will need to keep them separate.

Considerable efficiency may be introduced by using matched sets of scores because the process of comparing like with like reduces the confusing inter-individual variation which occurs when two randomly chosen individuals are used for comparison purposes. The more careful the matching process, the greater the gain in efficiency. This matching must be based on the most relevant variables for optimum effect. If we are studying the effect of a drug on performance, then matching must be based upon such variables as age and weight since both factors affect a person's response to a drug. Little efficiency would be achieved by matching on the basis of socioeconomic status. Matching, however, takes time and involves additional data gathering. This is the price of the extra efficiency, and each researcher must weigh the costs and benefits in his own particular situation before choosing a related samples design.

The most perfect variety of matching is to use the same individual as his own control. This repeated measures process has the additional advantage of reducing by half the number of subjects in an experiment. Since it is a lot easier and cheaper to use the same individual twice than to find and investigate two different people, we can understand the popularity of the repeated measures technique. However, this economy of effort is bought at a high risk of ambiguity because the very business of taking measurements repeatedly is itself a complex factor which is easily confounded with the main effect under study. There are a number of counter-balancing strategies which seek to minimize these confounding effects, but it is likely that none of these is wholly successful. As a result, each researcher should carefully weigh the advantages and disadvantages of using repeated measures before beginning his investigation. The question he must ask is whether or not the confounding aspects of practice, sensitization, boredom, familiarity etc. will render ambiguous the interpretation of his results.

Analysis of variance by ranks is at its least sensitive when dealing with a matched pairs design (see sections 8.2 and 8.6). This is not a problem when statistical significance is achieved, but presents difficulties when results fall just short of significance. In this case, we might reasonably feel that a more sensitive test would have given a more useful answer. One solution is to use a popular test proposed by Wilcoxon (1945) which is known by the ungainly title of the 'Wilcoxon matched pairs, signed ranks test'. The test is useful for our purposes because it is more sensitive and it is still a nonparametric test. Although this test is not, strictly speaking, a member of the analysis of variance by ranks family, it is fully described in sections 8.3 and 8.7 as a method of plugging the sensitivity gap. The test is very popular with researchers in the life sciences because it is quick to apply and appears to involve few assumptions concerning the data. There are, however, a number of largely unacknowledged difficulties with the test which are discussed in section 8.3; these limit its application somewhat, and must be borne in mind when selecting an appropriate test.

The format of chapter 8 is basically the same as that of chapter 7, with a clear division between tests of a specific alternative hypothesis (section A) and nonspecific tests (section B). To avoid too long a presentation, the exact and approximate tests have been combined into a single section for the important matched pairs designs. General guidance on computational procedures, together with explanations, is to be found in chapter 6, which should be consulted whenever amplification is required. In the related samples design the matched pairs or matched sets correspond to blocks in the analysis of variance by ranks framework. The symbol m, therefore, refers to the number of matched pairs or matched sets or blocks. In the case of multiple frequency tables, each table is a block, and on this occasion m will be the number of frequency tables.

A   Tests against a specific alternative hypothesis

8.2   Matched pairs: exact and approximate (specific)

The matched pairs design is the simplest, and by far the most popular, example of the configuration of two related samples. The analysis of variance by ranks approach to its evaluation has been with us for a long time in the form of the 'sign test' (Dixon and Mood, 1946). This test simply counts the number of pairs in the sample which give results in favour of the hypothesis, and refers this number to an exact table. The only difference between the sign test and the test described in the following lies in the method of computation. This is slightly more long-winded in rank sum analysis but has the benefit of preserving the basic pattern of ranking and summing which unites all of the tests in the book. Otherwise, they are basically the same test. They share the advantage of computational speed and the disadvantage of relatively low sensitivity.

Example

Dixon and Spitz (1980) presented subjects with a film whose soundtrack was not exactly synchronous with the picture. Their apparatus allowed them to slowly increase the degree of synchronism between the two until subjects noticed that they were 'out of sync'. They discovered that people were less sensitive to the asynchronism when the sound was delayed compared with when the picture was delayed. They explained this tolerance to auditory delay as a consequence of the fact that sound travels slower than light and, as a result, people would often find themselves in a situation where the sound arrived later than the visual component. They preferred an explanation in terms of perceptual learning rather than some innate compensating mechanism. In support of this hypothesis they draw attention to the fact that greater auditory delays are tolerated for films of people speaking than for a film of a hammer hitting a peg. They claim that speaking represents a 'more frequently experienced relationship between seeing and hearing'. Table 8.1 gives data for 18 subjects who each experienced both 'hammer' and 'speech' conditions. The figures are the maximum auditory delays in milliseconds tolerated by subjects before they reported the delays 'out of sync'. Do the data support Dixon and Spitz's hypothesis that the effect is a result of perceptual learning? Their hypothesis leads to the prediction that greater delays will be tolerated for the 'speech' condition.

Step 1 - formulate the problem
(a) Specify the hypotheses:
   Null hypothesis (H_0): there is no difference in the amount of auditory delay tolerated by subjects when presented with 'hammer' and 'speech' films.
   Specific alternative hypothesis (H_1): greater auditory delay will be tolerated for 'speech' than for 'hammer' films.
(b) Specify the variables:
   Independent variable: type of film (i) 'speech' and (ii) 'hammer'.
   Dependent variable: amount of auditory delay specified in milliseconds tolerated by subjects before reporting the film 'out of sync'.
   Blocks: subjects. Each subject yielded one score for each condition.

Table 8.1 Degree of auditory delay (ms) tolerated by subjects watching a film with asynchronous sound track in example of section 8.2

Subject	Hammer	Speech
1	142	380
2	243	342
3	182	182
4	101	124
5	197	267
6	168	224
7	159	198
8	106	180
9	188	254
10	409	389
11	150	259
12	175	179
13	114	184
14	175	270
15	158	358
16	188	234
17	207	366
18	117	210

(c) Order the samples:
Sample 1 must be the sample with the highest expected mean rank. Greater auditory delay is expected for the 'speech' film. We therefore make the 'speech' sample into sample 1 in table 8.2.

Step 2 – find the sample rank sums and means
(a) Arrange the scores in columns as in table 8.2.
(b) Rank the scores in each matched pair by assigning the ranks 1 and 2 to the two scores. *If the pair is tied, delete it from the calculations.*
(c) Find the two sample rank sums (R_j) by totalling ranks for the two samples. In our case, $R_1 = 33$ and $R_2 = 18$.
(d) Check that the ranking has been carried out correctly:

$$R_1 + R_2 = 3m'$$
$$33 + 18 = 3(17)$$
$$51 = 51$$

where m' is the number of untied matched pairs (blocks).

Table 8.2 Computational layout for example in section 8.2

Subject	Sample 1 Speech film		Sample 2 Hammer film	
	Score	Rank	Score	Rank
1	380	2	142	1
2	342	2	243	1
3	182	deleted	182	deleted
4	124	2	101	1
5	267	2	197	1
6	224	2	168	1
7	198	2	159	1
8	180	2	106	1
9	254	2	188	1
10	389	1	409	2
11	259	2	150	1
12	179	2	175	1
13	184	2	114	1
14	270	2	175	1
15	358	2	158	1
16	234	2	188	1
17	366	2	207	1
18	210	2	117	1
rank sum	R_1	33	R_2	18
sample size	n_1	17	n_2	17
rank mean	$\bar{R}_1$	1.94	$\bar{R}_2$	1.06

(e) Compute the sample rank means ($\bar{R}_j = R_j/n_j$). In our case, $\bar{R}_1 = 1.94$ and $\bar{R}_2 = 1.06$.

Step 3 – interpret the rank means
We predicted that greater auditory delay would be tolerated by the 'speech' sample ($\bar{R}_1 > \bar{R}_2$), which is in agreement with our computations.

Step 4 – compute and evaluate our key statistic (L)
Exact method
(a) Find L, which is simply

$$L = R_1 = 33$$

(b) Evaluate L by consulting table G (in the appendix) for $L = 33$ and $m' = 17$. The relevant row for our problem is

m'	5%	2.5%	1%	0.1%
17	$L \geqslant 30$	30	31	33

To be significant, L must be equal to or greater than the critical value at the appropriate level. Our value of L (33) is, therefore, significant at better than the 0.1 per cent level ($p < 0.001$).

Approximate method

(a) Find L:

$$L = R_1 = 33$$

Find Z:

$$Z = \frac{2L - 3m'}{\sqrt{m'}} = \frac{2(33) - 3(17)}{\sqrt{17}} = 3.64 \tag{8.1}$$

where m' is the number of *untied* matched pairs and Z is already corrected for ties.

(b) Evaluate Z by consulting tables of the normal distribution (table A in the appendix). The critical values are

5%	2.5%	1%	0.1%
$z \geqslant 1.64$	1.96	2.33	3.10

To be significant at any of the four levels, Z must be greater than or equal to the corresponding critical value. Our value of Z (3.64) is greater than the highest critical value and is therefore significant at the 0.1 per cent level ($p < 0.001$).

Step 5 - draw conclusions

Because our result is significant at better than the 0.1 per cent level, we may reject H_0 (see step 1) in favour of H_1. In simple terms, the data support a conclusion that greater auditory delay is tolerated for asynchronous sound on speech films.

8.3   Wilcoxon's matched pairs test (specific)

A rank sum test is undoubtedly insensitive when dealing with matched pairs of scores. In the worst cast, we may need up to 37 per cent more data in order to achieve the same level of statistical significance as another more sensitive test. Of course, if the data in hand already produce a significant result, then this is not a problem; a more sensitive test will only make our result more significant still, and the final decision (to reject H_0) will be the same. However, if the data only just fail to achieve significance, then we can be fairly sure that a more sensitive test would result in a different decision. There are two popular alternative tests – parametric analysis of variance (in the form of Student's related *t* test), and the nonparametric Wilcoxon's matched pairs test. If our samples have been drawn from normally distributed populations, then Student's *t* test is the most appropriate procedure. If not, then Wilcoxon's test should be considered.

This test is based on the difference scores (for each matched pair) obtained by subtracting the score for sample 2 from its paired score in sample 1. If the alternative hypothesis is true, we expect more positive than negative differences. However, we also expect the positive differences to be on average larger than the negative differences. To assess this, the differences are ranked. In order to use the test legitimately, we must assume *that these differences are meaningfully rankable*. This assumption is crucial, though almost always ignored in popular presentations of the test. For most data the assumption is true but occasionally it is not, and failure to spot the fact can result in a quite meaningless application of the test.

A simple example should be enough to illustrate the fact that difference scores are not always meaningfully rankable. Consider a rifle marksmanship competition in which three competitors shoot at 20 moving targets in conditions of (a) high and (b) low illumination. Their score is simply the number of direct hits. The question is whether high illumination improves their performance. Scores for three marksmen are as follows:

Marksman	High illumination	Low illumination	Difference
1	17	6	11
2	20	10	10
3	20	18	2

Which marksman showed the greatest improvement? Marksman 1 showed the largest difference, but the other two marksmen were prevented from exceeding his difference by the fact that 20 was the maximum score. No matter how much they improved it would be impossible to increase their score by 11 points. This 'ceiling' effect renders meaningless any attempt to rank the differences.

Many rating scales have 'floor effects' too, especially when scores hover around a minimum score of zero. Furthermore, some scales assign different significance. to changes at different points on the scale. For example a small increase in salary may mean a great deal to a poor man but very little to a rich man. The situation is difficult to characterize precisely, but the test user should always carefully look at the nature of the differences he is about to rank before committing himself to a Wilcoxon matched pairs test. The following example is taken from a paper in a reputable journal which uses Wilcoxon's test in a marginal case, suitable for class discussion. The implied comment does not, of course, detract from the author's published conclusions since significance would have been achieved even if the less sensitive analysis of variance by ranks technique had been used.

Wilcoxon's matched pairs test is not part of the analysis of variance by ranks scheme. Its inclusion here is an attempt to plug a gap caused by the low power of the rank sum equivalent test (section 8.2). It is possibly the most popular and the most useful nonparametric test which is not part of the rank sum scheme, and that in itself is an argument for its inclusion.

Example

Taub et al. (1971) examined the commonly held belief that too much sleep can impair intellectual performance on the following day. They compared the performance of 12 subjects on the mornings following (A) two normal nights' sleep and (b) two nights of 'extended sleep'. Sleep was extended by the simple expedient of causing the subjects to retire to bed three hours earlier than the control group. This led to an average of two hours extra sleep on the extended sleep night. In the morning they were given a number of tests of ability to think quickly and clearly. Among these tests was a 'vigilance test' which involves listening over a period of 45 minutes for a small number of quiet bleeps against a noisy background in headphones. Subjects scored one point for every bleep they failed to detect and one point for every 'false alarm' when they reported a bleep which had not, in fact, occurred. The scores given in table 8.3 represent an average of two nights. Do the data support the view that too much sleep can be bad for you?

Step 1 - formulate the problem
(a) Specify the hypotheses:
     Null hypothesis (H_0): extra sleep does not affect performance on a subsequent vigilance task.
     Specific alternative hypothesis (H_0): extra sleep impairs performance on a subsequent vigilance task.
(b) Specify the variables:
     Independent variable: presence/absence of extra sleep.
     Dependent variable: performance on a vigilance task. Performance is measured as the sum of errors of omission and commission. The figures in table 8.3 represent an average over two nights.

Table 8.3 Combined errors of commission and omission on a vigilance task following nights of normal and extended sleep (Taub et al., 1971). Each value in the table is the mean of two sessions following on two different days

Subject	Normal sleep	Extended sleep
1	7.5	8
2	9	8.5
3	14	15
4	4	1.5
5	11.5	21
6	10.5	15.5
7	3	9
8	25.5	38
9	2.5	10
10	10.5	10.5
11	9.5	16
12	20.5	41

Table 8.4 Computational layout for Wilcoxon's matched pairs test (example in section 8.3)

1 Subject	2 Sample 1 Extended sleep	3 Sample 2 Control	4 Difference	5 Ranks	6
				Positive	Negative
1	8	7.5	+0.5	1.5	
2	8.5	9	−0.5		1.5
3	15	14	+1.0	3	
4	1.5	4	−2.5		4
5	21	11.5	+9.5	9	
6	15.5	10.5	+5.0	5	
7	9	3	+6.0	6	
8	38	25.5	+12.5	10	
9	10	2.5	+7.5	8	
10	10.5	10.5	deleted		
11	16	9.5	+6.5	7	
12	41	20.5	+20.5	11	

m', number of untied pairs 11 $\quad\quad R_+$ 60.5 $\quad R_-$ 5.5

(c) Order the samples:
Sample 1 must be the sample with the highest expected mean score. In our example, we expect subjects with the extra sleep to have more errors. We therefore make them sample 1 in table 8.4.

Step 2 - find the rank sums (R_+ and R_-) of the positive and negative differences

(a) Arrange the scores in columns as in table 8.4. Put the difference between each pair of scores in column 4. The difference is found by subtracting the sample 2 score from the sample 1 score. Delete any pair with zero difference from the calculations.

(b) Rank the differences irrespective of sign (i.e. ignore the fact that a score is positive or negative when ranking). Put the ranks associated with a positive difference in column 5 and put the ranks associated with a negative difference in column 6.

(c) Find the sum of the ranks in column 5 (R_+) and the sum of the ranks in column 6 (R_-). In our case $R_+ = 60.5$ and $R_- = 5.5$.

(d) Check that the ranking has been carried out correctly:

$$m'(m' + 1)/2 = R_+ + R_-$$
$$11(12)/2 = 60.5 + 5.5$$
$$66 = 66$$

where m' is the number of untied matched pairs.

Step 3 - interpret the rank sums
When testing a specific hypothesis we must check that the frequency and size of the differences in favour of the hypothesis are greater than those against. In other words we expect $R_+ > R_-$. In our example, this is the case. The extended sleep group performed worse than the control group.

Step 4 - evaluate our key statistics (R_- and Z)

Exact method
(a) Evaluate R_- by consulting table I (in the appendix) for $m' = 11$ and $R_- = 5.5$. The critical row for our purpose is

m'	5%	2.5%	1%	0.1%
11	$R_- \leqslant 13$	10	7	1

To be significant at a given level, R_- must be *less than or equal to* the corresponding critical value. Our value of R_- (5.5) is significant at better than the 1 per cent level ($p < 0.01$).

Note   In this book it is unusual to have tables which require a statistic to be *less* than a critical value. Such a table is used here to permit rapid calculation. When the sample size is large, the skilled test user may wish to perform the ranking process only up to the point where he can definitely specify R_-. since R_- may only involve one or two of the lower ranks, this procedure can save a considerable amount of computational labour. When there are no negative differences, $R_- = 0$ always and no ranking is necessary at all.

Approximate method
(a) Find Z:

$$Z = \frac{m'(m' + 1)/4 - R_-}{\sqrt{[m'(m' + 1)(2m' + 1)/24]}}$$

$$= \frac{(11)(12)/4 - 5.5}{\sqrt{[(11)(12)(23)/24]}} = 2.44 \tag{8.2}$$

note that m' is the number of matched pairs *after* deletion of tied pairs.
(b) Evaluate Z by consulting table A (in the appendix), which is reproduced here:

5%	2.5%	1%	1.0%
$z \geqslant 1.64$	1.96	2.33	3.10

To be significant at any of the four levels Z must be equal to or greater than the corresponding critical value. Our value of Z is therefore significant at better than the 1 per cent level ($p < 0.01$). Notice that the exact and approximate results agree quite well.

Step 5 - draw conclusions
Because our result is significant we may reject H_0 (see step 1) in favour of H_1. In simple terms, the data support the conclusion that extended sleep impairs performance on vigilance tasks.

8.4   Two-sample matched sets test (specific)

The matched pairs analysis given in section 8.2 is a simplified version of a more general test which allows any number of scores in each matched set. Although this design is potentially very useful, it has attracted very little attention since its introduction by Wilcoxon (1946, 1947), and despite extensions by Bradley (1968) and Meddis (1975). Researchers typically condense the data in matched sets to matched pairs by averaging and then using some tests such as those given in sections 8.2 and 8.3. This is, in fact, the technique adapted by Taub et al. (1971) whose paper formed the basis of the example in section 8.3. Although this is a perfectly legitimate procedure, it does result in some loss of information which would be preserved using the procedure outlined in the following example.

Example

Bradley and Meddis (1974) investigated the well known phenomenon whereby sleepers may incorporate the sound of their morning alarm into the content of a dream. and, by transforming the significance of the sound, fail to wake up in response to it. In a sleep laboratory, Bradley and Meddis played white noise through a loudspeaker near to the bed when the sleeper was dreaming (as indicated by rapid eye movements and other electroencephalographic criteria). They gradually increased the intensity of the noise until the subject woke up. On waking the researchers asked the subjects to recount the dream they were having immediately prior to waking. Later, independent judges who read the dream transcript decided whether or not the white noise had been incorporated into the dream (for example as a rustling of leaves in a tree). The arousal thresholds (measured as sound intensity at the time of arousal) for 39 awakenings are given in table 8.5. Only eight subjects were used, and scores for each subject are treated as a separate matched set. Do the data show that stimuli incorporated into dreams are less potent arousers?

Step 1 – formulate the problem
(a) Specify the hypotheses:
     Null hypothesis (H_0): incorporation of a stimulus into a dream will not affect the arousal threshold to that stimulus.
     Specific alternative hypothesis (H_1): the arousal threshold to a stimulus will be increased if the stimulus is incorporated into a dream.
(b) Specify the variables:
     Independent variable: incorporation/nonincorporation of the stimulus into the dream.
     Dependent variable: sound intensity of the stimulus required to wake the subject. This is measured in decibels (dBA); the more decibels, the louder the stimulus.
     Blocks: subjects.
(c) Order the samples:
     Sample 1 must be the same with the highest expected rank mean. In our case, this is the sample where the stimulus is incorporated into the dream.

Step 2 – find the sample rank sums and means
(a) Put the scores within each cell into ascending order for easy ranking. This has already been done in table 8.5.

Table 8.5 Arousal thresholds (sound intensity, dBA) from dreaming sleep for eight subjects each woken a variable number of times (Bradley and Meddis, 1974)

Subject	Sample 1 Stimulus incorporated	Sample 2 Stimulus not incorporated
1	59	54 54 72
2	63	52 54 56
3	53 66 76	48 52 57
4	70	50 51 51 66
5	62 64 86	66
6	70 82	68
7	68 77	62 64 64 77
8	69 73 73 77	56 64 79

(b) Rank the scores within each matched set separately and note the ranks in a new table (see table 8.6); it would be confusing to mix raw scores and ranks for this design.

(c) Find the two sample rank sums by totalling the ranks in columns 2 and 3 in table 8.6. In our case $R_1 = 65.5$ and $R_2 = 55.5$.

(d) Check that the ranking has been carried out correctly. The sum of the ranks in each row should equal $b_i(b_i + 1)/2$, where b_i is the number of cases in each block:

Block	Ranks	Sum	Expected
1	$3 + 1.5 + 1.5 + 4$	10	$4(5)/2 = 10$
2	$4 + 1 + 2 + 3$	10	$4(5)/2 = 10$
3	$3 + 5 + 6 + 1 + 2 + 4$	21	$6(7)/2 = 21$
4	$5 + 1 + 2.5 + 2.5 + 4$	15	$5(6)/2 = 15$
5	$1 + 2 + 4 + 3$	10	$4(5)/2 = 10$
6	$2 + 3 + 1$	6	$3(4)/2 = 6$
7	$4 + 5.5 + 1 + 2.5 + 2.5 + 5.5$	21	$6(7)/2 = 21$
8	$3 + 4.5 + 4.5 + 6 + 1 + 2 + 7$	28	$7(8)/2 = 28$

(e) Note the cell frequencies (n_{i1} and n_{i2}) in columns 4 and 5 in table 8.6. Then by totalling these columns find the sample sizes ($n_1 = 17$ and $n_2 = 22$). Find the sample rank means ($\bar{R}_j = R_j/n_j$). In our case, $\bar{R}_1 = 3.85$ and $\bar{R}_2 = 2.52$.

Step 3 – interpret the rank means

We predicted that arousal thresholds would be higher when the stimuli were incorporated into the dream ($\bar{R}_1 > \bar{R}_2$), and this is in fact the case.

Table 8.6 Computational layout for example in section 8.4

1 Subject	2 Stimulus Sample 1 Incorporated (ranks)	3 Sample 2 Not incorporated (ranks)	4 n_{i1}	5 n_{i2}	6 $(n_{i1} + n_{i2} + 1)$	7 $n_{i1}(n_{i1} + n_{i2} + 1)$	8 $n_{i1}n_{i2}(n_{i1} + n_{i2} + 1)$
1	3	1.5 1.5 4	1	3	5	5	15
2	4	1 2 3	1	3	5	5	15
3	3 5 6	1 2 4	3	3	7	21	63
4	5	1 2.5 2.5 4	1	4	6	6	24
5	1 2 4	3	3	1	5	15	15
6	2 3	1	2	1	4	8	8
7	4 5.5	1 2.5 2.5 5.5	2	4	7	14	56
8	3 4.5 4.5 6	1 2 7	4	3	8	32	96
	R_1 65.5	R_2 55.5	n_1 17	n_2 22		Σ 106	Σ 292
	n_1 17	n_2 22					
	R_1 3.85	R_2 2.52					

Step 4 – compute and evaluate our key statistics (L and Z)
(a) Find L, which in our case is simply the rank sum of sample 1:

$$L = R_1 = 65.5$$

For each matched set compute the values $(n_{i1} + n_{i2} + 1)$, $n_{i1}(n_{i1} + n_{i2} + 1)$ and $n_{i1}n_{i2}(n_{i1} + n_{i2} + 1)$. Enter these values in table 8.6 in columns 6, 7 and 8 respectively. Find the totals (Σ) of columns 7 and 8 and enter these at the foot of the respective columns. Now find Z:

$$Z = \frac{2L - \Sigma n_{i1}(n_{i1} + n_{i2} + 1)}{\sqrt{[\Sigma n_{i1}n_{i2}(n_{i1} + n_{i2} + 1)/3]}}$$

$$= \frac{2L - (\text{sum column 7})}{\sqrt{[(\text{sum column 8})/3]}}$$

$$= \frac{2(65.5) - 106}{\sqrt{[292/3]}} = 2.53$$

(b) Evaluate Z by consulting tables of the normal distribution (table A in the appendix), whose critical values are reproduced here:

5%	2.5%	1%	0.1%
$z \geqslant 1.64$	1.96	2.33	3.10

To be significant at any of the four levels, Z must be equal to or greater than the appropriate critical value. Our value of Z (2.53) is therefore significant at better than the 1 per cent level ($p < 0.01$).

(c) Correct for ties:
The tie correction procedure is very time consuming for this design since it involves computing a tie correction factor for each matched set and then recomputing column 8 of table 8.6 using this expression:

$$n_{i1}n_{i2}(n_{i1} + n_{i2} + 1) T_i$$

where

$$T_i = 1 - \frac{\Sigma(t_{iq}^3 - t_{iq})}{b_i^3 - b_i}$$

Here t_{iq} is the length of the qth tie in the ith block (matched set) and b_i is the total number of scores in the ith block (matched set). Unless the ties are long relative to the block frequencies (b_i), this correction is best omitted.

F

For the purposes of illustration, however, the recalculation of

$$\Sigma n_{i1} n_{i2} (n_{i1} + n_{i2} + 1) T_i$$

is shown as follows:

Sub-ject	t_{iq}	T_i	$n_{i1} n_{i2} (n_{i1} + n_{i2} + 1)$	Corrected
1	2	$1 - (2^3 - 2)/(4^3 - 4) = 0.9$	15	13.5
2	−	1	15	15
3	−	1	63	63
4	2	$1 - (2^3 - 2)/(5^3 - 5) = 0.95$	24	22.8
5	−	1	15	15
6	−	1	8	8
7	2,2	$1 - [(2^3 - 2) + (2^3 - 2)]/(6^3 - 6)$ $= 0.94$	56	52.64
8	2	$1 - (2^3 - 2)/(6^3 - 6) = 0.98$	96	94.08

$$\Sigma n_{i1} n_{i2} (n_{i1} + 1) T_i \quad 284.02$$

We may now find the corrected value of Z:

$$Z' = \frac{2L - \Sigma n_{i1}(n_{i1} + n_{i2} + 1)}{\sqrt{[\Sigma n_{i1} n_{i2}(n_{i1} + n_{i2} + 1) T_i/3]}}$$

$$= \frac{2(65.5) - 106}{\sqrt{[284.02/3]}} = 2.57$$

This is a small increase, but one hardly worth the computational effort in this case. Our result remains statistically significant at the 1 per cent level.

Step 5 - draw conclusions
Because our result is significant ($p < 0.01$), we may reject H_0 (see step 1) in favour of H_1. In simple terms, the data support the conclusion that stimuli are less potent arousers if they become incorporated into the theme of the sleeper's dream.

8.5 Two-sample test for multiple frequency tables (specific)

A single frequency table is treated in rank sum analysis as a single block and dealt with accordingly (see sections 7.5 and 9.4). When we have more than one table, we treat each table as a separate block (related set of scores). Multiple frequency tables occur most often when data are collected in batches at

Table 8.7   Data for example in section 8.5

| *Block 1: women volunteers* | *Illusions in visual hemifields* | |
	Sample 1 *Bilateral*	*Sample 2* *Unilateral*
Bilateral headaches	13	1
Unilateral headaches	6	6

Block 2: neurology patients

Bilateral headaches	13	3
Unilateral headaches	8	14

different times, in different places or from different types of individuals. One way of coping with multiple tables is simply to add them together to form a single table and then work on the resulting table. This method has the disadvantage of diluting the significance of the difference between the two samples as a result of mixing in the differences between the tables. This section provides a means of analysing multiple tables while keeping them separate.

Example

Wilkins et al. (1984), using volunteer women, found that lateralization of headaches was associated with lateralization of special visual effects when viewing patterns comprised of narrow horizontal stripes (see section 7.5 for more details). With a view to increasing their sample size, they repeated this experiment with 38 neurology patients whose primary complaint was headache. The results are given in table 8.7. They analysed the tables individually and reported two separate results, but these could have been combined using the following procedure.

Step 1 – formulate the problem
(a) Specify the hypotheses:
Null hypothesis (H_0): lateralization of headaches is not related to lateralization of special visual effects (illusions) on viewing horizontal stripes (visual gratings).
Specific alternative hypothesis (H_1): lateralization of headaches is positively associated with lateralization of the visual effects (i.e. when the illusions are unilateral the headaches will tend to be unilateral).

(b) Specify the variables:
In this example it is difficult to decide which variable is dependent and which independent, since both phenomena (headaches and visual effects) are presumed to be dependent upon a common neurological process. We shall therefore avoid the issue by calling them variables 1 and 2:

variable 1: lateralization of visual effects (bilateral/unilateral)
variable 2: lateralization of headaches (bilateral/unilateral).

Blocks: (1) normal women (2) neurology patients.

(c) Order the samples:
Sample 1 must be the sample with the higher expected mean rank. Remember that the category at the bottom of each frequency table (unilateral headaches) has the highest rank. Since we expect unilateral illusions to be associated with more unilateral headaches, this group should have the higher mean rank. We must therefore make the unilateral illusions group into sample 1 (see table 8.8).

Table 8.8   Computational layout for example in section 8.5

	Illusions		Total t_q	Cumulative total Σt_q	Shared rank r_q
	Sample 1 Unilateral	Sample 2 Bilateral			
Block 1: women volunteers					
Bilateral headaches	1	13	14	14	7.5
Unilateral headaches	6	6	12	26	20.5
	$n_{1,1}$ 7	$n_{1,2}$ 19		b_1 26	
Block 2: neurology patients					
Bilateral headaches	3	13	16	16	8.5
Unilateral headaches	14	8	22	38	27.5
	$n_{2,1}$ 17	$n_{2,2}$ 21		b_2 38	
rank sum	R_1 541	R_2 551			
sample size	n_1 24	n_2 40			
rank mean	$\bar{R}_1$ 22.54	$\bar{R}_2$ 13.77			

Step 2 - find the sample rank sums and means
(a) Draw up the frequency tables as in table 8.8. Find the category totals (t_q) and the cumulative category totals (Σt_q).
(b) Find the shared ranks for each category. Do this separately for each table using the expression

$$r_q = \Sigma t_q - t_q/2 + 0.5$$

For the first table (women volunteers) we have

$$r_1 = 14 - 14/2 + 0.5 = 7.5$$

$$r_2 = 26 - 12/2 + 0.5 = 20.5$$

For the second table (neurology patients) we have

$$r_1 = 16 - 16/2 + 0.5 = 8.5$$

$$r_2 = 38 - 22/2 + 0.5 = 27.5$$

(c) Find the rank sum for each sample by multiplying the shared rank for each category by the number of scores in that category and adding:

$$R_1 = (1)(7.5) + (6)(20.5) + (3)(8.5) + (14)(27.5) = 541$$
$$R_2 = (13)(7.5) + (6)(20.5) + (13)(8.5) + (8)(27.5) = 551$$

(d) Check that the ranking has been carried out correctly. The total sum of ranks should equal the sum of the ranks in each table:

$$R_1 + R_2 = \sum_i b_i(b_i + 1)/2$$

$$541 + 551 = (26)(27)/2 + (38)(39)/2$$
$$1092 = 1092$$

where b_i is the number of scores in the ith block (frequency tables).
(e) Note the sample sizes ($n_1 = 24$ and $n_2 = 40$) and compute the sample rank means ($\bar{R}_j$). In our case $\bar{R}_1 = 22.54$ and $\bar{R}_2 = 13.77$.

Step 3 - interpret the rank means
We predicted that unilateral headaches would be more likely to be associated with unilateral illusions. This is the case for our two samples.

Step 4 - compute and evaluate our key statistics (L and Z)

(a) Find L:

$$L = R_1 = 541$$

Find Z:

$$Z = \frac{2L - \Sigma n_{i1}(b_i + 1)}{\sqrt{[\Sigma n_{i1}n_{i2}(b_i + 1)/3]}}$$

but first find

$$\sum_i^m n_{i1}(b_i + 1) = 7(27) + 17(39) = 852$$

$$\sum_i^m n_{i1}n_{i2}(b_i + 1) = (7)(19)(27) + (17)(21)(39) = 17\,514$$

Now

$$Z = \frac{2(541) - 852}{\sqrt{(17\,514/3)}} = 3.01$$

(b) Evaluate Z by consulting tables of the normal distribution (table A in the appendix). The critical values are

	5%	2.5%	1%	0.1%
$z \geqslant$	1.64	1.96	2.33	3.10

To be significant at any of the four levels, Z must be equal to or greater than the appropriate critical value. Our value of Z (3.01) is significant at better than the 1 per cent level ($p < 0.01$).

(c) Correct for ties:
Tie correction usually has a large effect on the value of Z in frequency table analyses and is therefore worth doing even though the computations can be cumbersome. We need a tie correction factor for each table:

$$T_i = 1 - \frac{\Sigma(t_q^3 - t_q)}{b_i^3 - b_i}$$

For table 1,

$$T_1 = 1 - \frac{(14^3 - 14) + (12^3 - 12)}{26^3 - 26}$$

$$= 1 - \frac{2730 + 1716}{17\,550} = 0.747$$

For table 2,

$$T_2 = 1 - \frac{(16^3 - 16) + (22^3 - 22)}{(38^3 - 38)}$$

$$= 1 - \frac{4080 + 10\,626}{54\,834} = 0.732$$

These two corrections are then applied to the denominator in the following manner:

$$\Sigma n_{i1} n_{i2}(b_i + 1) T_i = (7)(19)(27)(0.747) + (17)(21)(39)(0.732)$$

$$= 3591(0.747) + (13\,923)(0.732)$$

$$= 12\,874$$

We can now find the corrected value of Z:

$$Z' = \frac{2L - \Sigma n_{i1}(b_i + 1)}{\sqrt{[\Sigma n_{i1} n_{i2}(b_i + 1) T_i/3]}}$$

$$= \frac{2(541) - 852}{\sqrt{(12\,874/3)}} = 3.51$$

Our new value of Z' (3.51) is substantially greater than our previous value of Z (3.01) and is now significant at the 0.1 per cent level ($p < 0.001$).

Step C - draw conclusions
Because our result is statistically significant at the 0.1 per cent level we may reject H_0 (see step 1) in favour of H_1. In simple terms, the data support a conclusion that people who have unilateral headaches have a stronger tendency to report unilateral illusions when narrow horizontal stripes are viewed.

B   Tests against a nonspecific alternative hypothesis

The four tests in this section run parallel with the four tests in section A of this chapter. The important difference is that, in this section, the tests are not carried out with the aim of rejecting the null hypothesis in favour of a specific alternative suggested by a theory or previous experience. Instead they merely ask whether or not some influence is at work causing a difference between the two samples *in either direction*. When we have two samples, there is only one difference to consider and this difference can have only two directions (positive and negative). As a consequence, tests of a nonspecific hypothesis for two-sample designs have traditionally been called 'two-tail' tests (see section 2.10), and this terminology will be found to be widely used in the research literature. For specific tests we use approximations to the normal distribution, whereas for non-specific tests we use approximations to the chi-square distribution.

The notes on the individual tests in section A are often applicable to the corresponding tests in section B but will not be repeated here. The interested reader is therefore advised to consult section A as necessary.

8.6   Matched pairs: exact and approximate tests (nonspecific)

The major difference between the specific and nonspecific versions of this test is that the user has no prior basis for predicting which sample will have the higher sample rank mean. Consult section 8.2 for further notes on this design.

Example

Dixon and Spitz (1980) have shown that asynchronism between a film and its soundtrack is tolerated to a greater degree for speech material than for a hammer hitting a peg when the sound trails behind the film (see section 8.2 for more details). They predicted this relationship on the grounds that

(a) sound is normally slightly behind vision; and
(b) speech is a more frequently experienced phenomenon.

As a consequence people will learn to tolerate the delay more in situations where the delay is more frequently experienced. They also studied a control condition where the sound was slightly *in advance* of the film. The theory proposed by Dixon and Spitz does not predict any difference between 'hammer' and 'speech' conditions for auditory advancement since auditory advancement is highly unusual in real life and little relevant perceptual learning should occur. It is, therefore, useful to look at this condition to see whether any difference between 'hammer' and 'speech' conditions occurs despite an absence of a prediction. The results given in table 8.9 represent the maximum auditory advancement

Table 8.9 Degree of auditory advancement (ms) tolerated by subjects watching a film with asynchronous soundtrack

Subject	Hammer	Speech
1	93	169
2	81	262
3	85	59
4	49	79
5	113	103
6	101	101
7	74	83
8	40	112
9	39	41
10	111	144
11	47	123
12	58	55
13	58	112
14	75	98
15	39	235
16	54	120
17	117	141
18	40	132

(measured in milliseconds) before each subject reported the film to be 'out of sync'.

Step 1 – formulate the problem
(a) Specify the hypotheses:
   Null hypothesis (H_0): there is no difference in the amount of auditory advancement tolerated by subjects when presented with asynchronous 'hammer' and 'speech' films.
   There is no specific alternative hypothesis. The general alternative hypothesis is simply that H_0 is 'untrue.
(b) Specify the variables:
   Independent variable: type of film (i) 'speech' and (ii) 'hammer'.
   Dependent variable: amount of auditory advancement (in milliseconds) tolerated by subjects before reporting the film 'out of sync'.
   Blocks: subjects. Each subject yielded one score for each condition.
(c) Order the samples:
   Sample 1 must be the condition with the higher rank mean. Inspection of table 8.9 reveals that the figures are typically higher in the 'speech' row.

We must, therefore, assign the speech scores to sample 1. If this turns out later not to have the higher rank mean, the samples will need to be switched.

Step 2 - find the sample rank sums and means
(a) Arrange the scores in columns as in table 8.10.
(b) Rank the scores in each matched pair by assigning the ranks 1 and 2 to the two scores. If a pair is tied, delete them from the calculations.
(c) Find the two sample rank sums (R_j) by totalling the ranks in each column. In our case $R_1 = 31$ and $R_2 = 20$.

Table 8.10   Computational layout for example in section 8.6

Subject	Sample 1 Speech film		Sample 2 Hammer film	
	Score	Rank	Score	Rank
1	169	2	93	1
2	262	2	81	1
3	59	1	85	2
4	79	2	49	1
5	103	1	113	2
6	101	deleted	101	deleted
7	83	2	74	1
8	112	2	40	1
9	41	2	39	1
10	144	2	111	1
11	123	2	47	1
12	55	1	58	2
13	112	2	58	1
14	98	2	75	1
15	235	2	39	1
16	120	2	54	1
17	141	2	117	1
18	132	2	40	1
rank sum	R_1	31	R_2	20
sample size	n_1	17	n_2	17 $(m' = 17)$
rank mean	$\bar{R}_1$	1.82	$\bar{R}_2$	1.18

(d) Check that the ranking has been carried out correctly:

$$R_1 + R_2 = 3m'$$
$$31 + 20 = 3(17)$$
$$51 = 51$$

where m' is the number of *untied* matched pairs.

(e) Note the two sample sizes ($n_1 = 17$ and $n_2 = 17$) and compute the sample rank means ($\bar{R}_j = R_j/n_j$). In our case $\bar{R}_1 = 1.82$ and $\bar{R}_2 = 1.18$.

Step 3 - interpret the rank means

The rank means suggest that greater auditory advancement is tolerated in the 'speech' sample. At this stage we should also note that $\bar{R}_1$ is, in fact, greater than $\bar{R}_2$ and that it is not necessary to change the samples round as explained in step 1(c).

Step 4 - compute and evaluate our key statistics (K and H)

Exact method

(a) Find K, which is simply

$$K = R_1 = 31$$

(b) Evaluate K by consulting table H (in the appendix) for $K = 31$ and $m' = 17$. The relevant row for our problem is

m'	5%	2.5%	1%	0.1%
17	$K \geqslant 30$	31	32	33

where m' is the number of untied pairs.

To be significant at any of the four levels, K must be equal to or greater than the corresponding critical value. Our value of K is, therefore, significant at better than the 2.5 per cent level ($p < 0.025$).

Approximate method

(a) Find K:

$$K = R_1 = 31$$

Find H (corrected for ties):

$$H = (2K - 3m')^2/m'$$
$$= [2(31) - 3(17)]^2/17 = 7.12 \qquad (8.4)$$

where m' is the number of *untied* matched pairs.

(b) Evaluate H by consulting chi-square table B in the appendix for $H = 7.12$. In two-sample problems we have only one degree of freedom (df $= 1$). The appropriate critical values are, therefore:

Degrees of freedom	5%	2.5%	1%	0.1%
1	$\chi^2 \geqslant 3.8$	5.0	6.6	10.8

To be significant our value of H must be equal to or greater than the corresponding critical value. Our value of H (7.12) is, therefore, significant at the 1 per cent level ($p < 0.01$).

Step 5 - draw conclusions

Because our result is significant ($p < 0.025$) we may reject H_0 (see step 1). In simple terms, the data support a conclusion that people tolerate greater auditory advancement for speech material than for a hammer hitting a peg.

This result is in the same direction for auditory delay (see section 8.2). The position adopted by Dixon and Spitz would imply an effect for auditory delay but not for auditory advancement. In fact, the results are very much more significant ($p < 0.001$) for auditory delay than advancement, which is in agreement with their position. Unfortunately we cannot compare significance levels, and a more direct test of this difference is required. The matter is taken a step further in exercise 8.3.

8.7   Wilcoxon's matched pairs test (nonspecific)

The nonspecific version of Wilcoxon's test is used when there is no prior basis for predicting the observed direction of the difference between the two samples. See section 8.3 for further explanatory notes.

Example

Langford, Meddis and Pearson (1973) studied the speed of reaction of subjects following spontaneous arousals from sleep during the night. To do this they waited all night until their sleeping subjects woke up of their own accord. Then, instead of allowing them to return to sleep, as they would normally have done, the experimenters switched on an alarm which the subjects knew had to be switched off quickly using a switch by the bed. The researchers measured how quickly they were able to do this. They were interested in the state of mind of individuals following spontaneous arousals from different kinds of sleep. The results given in table 8.11 are the mean of two reaction times following an arousal from active ('dreaming') sleep and the mean of two reaction times following an arousal from light quiet ('nondreaming') sleep for eight subjects.

Table 8.11   Reaction time (seconds) to an alarm *following* spontaneous arousal from two kinds of sleep. Each value is the mean of two reaction times (i.e. following two separate awakenings)

Subject	Active sleep	Light quiet sleep
1	1.625	4.25
2	1.75	2.5
3	2	2.75
4	2.875	2.375
5	4	5.5
6	15	17
7	3	4.75
8	3	3.5

Previous animal studies had suggested that active sleep was the deepest kind, but these results go counter to this expectation. A nonspecific test is therefore required. Is the difference between the two samples great enough to warrant the conclusion that human reactions are quicker following spontaneous arousals from active sleep?

Step 1 - formulate the problem
(a) Specify the hypotheses:
Null hypothesis (H_0): reaction time following a spontaneous arousal from sleep is unaffected by the kind of sleep preceding the arousal. There is no specific alternative hypothesis. The general alternative hypothesis is that H_0 is false.
(b) Specify the variables:
Independent variable: kind of sleep preceding the arousal: (i) active (or dreaming) sleep and (ii) light quiet (or nondreaming) sleep. This variable was measured using polygraphic indicators.
Dependent variable: reaction time (seconds) to switch off an alarm placed near the bed.
(c) Order the samples:
This is not necessary for this test.

Step 2 - find the rank sums (R_+ and R_-) of the positive and negative differences
(a) Arrange scores in columns as in table 8.12. In column 4 put the difference between each pair of scores. The difference is found by subtracting the sample 2 score from the sample 1 score. Delete any pair with zero difference from the calculations and reduce the number of matched pairs accordingly.

Table 8.12 Computational layout for Wilcoxon's matched pairs test (example in section 8.7)

1 Subject	2 Sample 1 (following light quiet sleep)	3 Sample 2 (following active sleep)	4 Difference	5 Ranks Positive	6 Negative
1	4.25	1.625	2.625	8	
2	2.50	1.75	0.75	3.5	
3	2.75	2.00	0.75	3.5	
4	2.375	2.875	−0.5		1.5
5	5.5	4.00	1.5	5	
6	17.00	15.00	2	7	
7	4.75	3.00	1.75	6	
8	3.5	3.00	0.5	1.5	
m', no of untied pairs 8				R_+ 34.5	R_- 1.5

(b) Rank the differences irrespective of sign (i.e. ignore the fact that a score is positive or negative when ranking). Put the ranks associated with a positive difference in column 5 and put the ranks associated with a negative difference in column 6.

(c) Find the sum of the ranks in column 5 (R_+) and the sum of the ranks in column 6 (R_-). In our case $R_+ = 34.5$ and $R_- = 1.5$.

(d) Check that the ranking has been carried out correctly. The total sum of ranks should equal $m'(m' + 1)/2$ where m' is the number of matched pairs after deletion of all tied pairs:

$$m'(m' + 1)/2 = R_+ + R_-$$
$$8(9)/(2) = 34.5 + 1.5$$
$$36 = 36$$

Step 3 - interpret the rank sums

The number and size of positive differences is much greater than the number and size of the negative differences. This means that the reactions following light quiet sleep are slower than those following active sleep.

Step 4 - evaluate our key statistics (R_- and H)

Exact method

(a) Evaluate R_- or R_+, whichever is the smaller, by consulting table J (in the appendix) for $m' = 8$ and $R_- = 1.5$. The appropriate row of table J is

m'		5%	2.5%	1%	0.1%
8	R_+ or $R_- \leqslant 3$	2	0	–

To be significant at a given level either R_- or R_+ must be *equal to or less* than the corresponding critical value. Our value of R_- (1.5) is therefore significant at better than the 2.5 per cent level ($p < 0.025$).

Approximate method
(a) Find H:

$$H = \frac{24[m'(m'+1)/4 - R_-]^2}{m'(m'+1)(2m'+1)}$$

$$= \frac{24[8(9)/(4) - 1.5]^2}{8(9)(17)} = 5.34 \qquad (8.5)$$

(b) Evaluate H by consulting tables of the chi-square distribution (table B in the appendix) for $H = 5.34$. In two-sample problems we have only one degree of freedom (df = 1). The appropriate row of table B is therefore

Degrees of freedom	5%	2.5%	1%	0.1%
1	$\chi^2 > 3.8$	5.0	6.6	10.8

To be significant our value of H must be equal to or greater than the critical χ^2 value for the appropriate level. Our value of H (5.34) is, therefore, significant at the 2.5 per cent level ($p < 0.025$). Notice that the approximate and exact values agree.

Step 5 - draw conclusions
Because our result is significant ($p < 0.025$), we can reject H_0. In simple terms, the data support a conclusion that people have faster reactions following spontaneous arousals when these arousals are preceded by active sleep than when they are preceded by quiet sleep.

8.8 Two-sample matched sets test (nonspecific)

The nonspecific version of this test is used when there is no prior basis for predicting the observed direction of the difference between the two treatments/conditions. See section 8.4 for further comments on this test.

Example

The data in the example in section 8.7 were averages of two scores. Using this same example, but replacing each average by its two component scores in table 8.13, we can illustrate the general matched sets procedure. See section 8.7 for details of the experiment.

Step 1 - formulate the problem
See section 8.7, step 1.

Step 2 - find the sample rank sums and means
(a) Put the scores within each cell into ascending order for easy ranking. This has already been done in table 8.13.
(b) Rank the scores within each matched set separately and note the ranks in a new table (table 8.14) - it would be confusing to mix raw scores and ranks for this design.
(c) Find the two sample rank sums (R_j) by totalling the ranks in the two columns. In our case, $R_1 = 46.5$ and $R_2 = 33.5$.
(d) Check that the ranking has been carried out correctly. The sum of ranks in each matched set should equal $b_i(b_i + 1)/2$, where b_i is the number of scores in the *i*th set. In our case $b_i = 4$ for each block.

Block	$b_i(b_i + 1)/2$	Expected rank sum	Actual rank sum
1	$4(5)/2 = 10$	7	7
2	10	7	7
3	10	7	7
4	10	7	7
5	10	7	7
6	10	7	7
7	10	7	7
8	10	7	7

When the cell frequencies (n) are equal throughout the data we expect

$$R_1 + R_2 = mn(2n + 1)$$
$$46.5 + 33.5 = (8)(2)(5)$$
$$80 = 80$$

(e) Note the cell frequencies $(n_{i1}$ and $n_{i2})$ in columns 4 and 5 of table 8.14. Then find the two sample sizes $(n_1$ and $n_2)$. In our case $n_1 = 16$ and $n_2 = 16$. Find the two sample rank means $(\bar{R}_j = R_j/n_j)$. In our example $\bar{R}_1 = 2.9$ and $\bar{R}_2 = 2.09$.

Table 8.13 Reaction times (seconds) to an alarm *following* a spontaneous arousal from two kinds of sleep for eight subjects (section 8.8, but see also section 8.7)

Subject	Light quiet sleep		Active sleep	
1	3.5	5.0	1.5	1.75
2	1.5	3.5	1.0	2.5
3	2.5	3.0	2.0	2.0
4	2.0	2.75	2.75	3.0
5	4.0	7.0	3.0	5.0
6	6.0	28.0	12.0	18.0
7	3.6	6.0	3.0	3.0
8	2.5	4.5	3.0	3.0

Table 8.14 Computational layout for example in section 8.8

1 Subject	2 Sample 1 Light quiet sleep (ranks)	3 Sample 2 Deep quiet sleep (ranks)	4 n_{i_1}	5 n_{i2}	6 $n_{i1}(b_i + 1)$	7 $n_{i1}n_{i2}(b_i + 1)$
1	3 4	1 2	2	2	10	20
2	2 4	1 3	2	2	10	20
3	3 4	1.5 1.5	2	2	10	20
4	1 2.5	2.5 4	2	2	10	20
5	2 4	1 3	2	2	10	20
6	1 4	2 3	2	2	10	20
7	3 4	1.5 1.5	2	2	10	20
8	1 4	2.5 2.5	2	2	10	20
rank sum R_1 46.5		R_2 33.5			Σ 80	Σ 160

sample
size n_1 16 n_2 16
rank sum $\bar{R}$ 2.9 $\bar{R}_2$ 2.09   NB $b_i = n_{i1} + n_{i2}$

Step 3 – interpret the rank means

Sample 1 has a higher rank mean than sample 2. This means that reactions following light quiet sleep are slower than those following active sleep.

Step 4 - compute and evaluate our key statistics (K and H)

(a) Find K, which in the two-sample case is simply

$$K = R_1 = 46.5$$

For each matched set, compute the values $n_i(b_i + 1)$ and $n_{i1}, n_{i2}(b_i + 1)$. Enter these values in columns 6 and 7 of table 8.14 respectively. In our example, these are the same for each matched set since the cell frequencies are constant. Find the totals of columns 6 and 7 and enter these at the foot of the column. In our case $\Sigma n_{i1}(b_i + 1) = 80$ and $\Sigma n_{i1} n_{i2}(b_i + 1) = 160$.
Now find H:

$$H = \frac{3[2K - \Sigma n_{i1}(b_{i1} + 1)]^2}{\Sigma n_{i1} n_{i2}(b_i + 1)}$$

$$= \frac{3[2(46.5) - 80]^2}{160} = 3.17$$

When the cell frequencies are all equal as they are in this example, we can use the following short-cut expression:

$$H = \frac{3[2K - mn(2n + 1)]^2}{mn^2(2n + 1)}$$

$$= \frac{3[2(46.5) - 8(2)(5)]^2}{8(4)(5)} = 3.17$$

where n is the cell frequency (in our case, 2). This method avoids the need to create columns 6 and 7.

(b) Evaluate H by consulting tables of the chi-square distribution (table B in the appendix). In two-sample problems we have only one degree of freedom (df = 1). The appropriate row of table B is therefore

Degrees of freedom	5%	2.5%	1%	0.1%
1	$\chi^2 \geqslant 3.8$	5.0	6.6	10.8

To be significant our value of H must be equal to or greater than the critical χ^2 value for the appropriate level. Our value of H (3.17) is, therefore, not significant ($p > 0.05$).

(c) Correct for ties:
Find the correction term T_i for each block separately and use it to correct the expression $\Sigma n_{i1} n_{i2}(b_i + 1)$ in column 7 of table 8.14:

Block	Tie lengths	T_i	$\Sigma n_{i1} n_{i2}(b_i + 1)$	
		$1 - \dfrac{\Sigma(t^3 - t)}{(b_i^3 - b_i)}$	Uncorrected	Corrected
1	–	1	20	20
2	–	1	20	20
3	2	$1 - \dfrac{(2^3 - 2)}{(4^3/4)} = 0.9$	20	18
4	2	$1 - \dfrac{(2^3 - 2)}{(4^3 - 4)} = 0.9$	20	18
5	–	1	20	20
6	–	1	20	20
7	2	$1 - \dfrac{(2^3 - 2)}{(4^3 - 4)} = 0.9$	20	18
8	2	$1 - \dfrac{(2^3 - 2)}{4^3 - 4} = 0.9$	20	18
			Σ	152

We may now recompute H:

$$H' = \frac{3[2K - \Sigma n_{i1}(b_i + 1)]^2}{\Sigma n_{i1} n_{i2}(b_i + 1) T_i}$$

$$= \frac{3[2(46.5) - 80]^2}{152} = 3.34$$

The new value of H is larger but still not large enough to be significant.

Step 5 - draw conclusions
Because our result is not significant ($p > 0.05$) we cannot reject H_0. In simple terms, our data do not support the conclusion that the particular sleep stage prior to waking affects speed of reaction following a spontaneous arousal.

Note This conclusion is in contradiction to the conclusion following the example in section 8.7 using Wilcoxon's matched pairs test. The short explanation of this difference is that different tests ask different questions, although it is often difficult to put into words just what these questions are. Some hint can be gained by looking in detail at the figures in tables 8.11 and 8.13. In table 8.11, all but one of the pairs of means shows a result in favour of faster reactions in the active sleep sample. However, in table 8.13 we can see that those means concealed the fact that some of the reactions in the active sleep sample were slower than reactions in the light quiet sleep group. The matched sets analysis in this section takes this fine structure into account and, in our example, this has markedly weakened the significance level. Notice that *in terms of ranks alone*, subject 4 goes against the hypothesis and subjects 1, 6 and 8 show an evenly balanced effect. Only four out of eight subjects show an effect in favour of faster reaction time following active sleep arousals.

8.9   Two-sample test (nonspecific) for multiple frequency tables

The nonspecific version of this procedure is used when there is no prior basis for predicting the direction of the difference between the two groups. See section 8.5 for further notes on this design.

Example

Green (1982) studied the probability concepts of school pupils in the age range 11–16 years. An incidental finding from this work was that girls appeared to lag behind boys in the development of probability concepts. The data for two of the

Table 8.15   Numbers of boys and girls in four categories of conceptual sophistication at two different ages (Green, 1982). See example in section 8.9

Probability concept level	Boys	Girls
Year 1 (ages 11–12 years)		
0 *(least sophisticated)*	48	58
1	163	192
2	99	67
3 *(most sophisticated)*	10	6
Year 2 (ages 12–13 years)		
0 *(least sophisticated)*	40	37
1	147	184
2	127	90
3 *(most sophisticated)*	20	20

five age ranges studied are given in table 8.15. Note that each child is rated in terms of one of four levels of probability concept attainment; level 3 is the most sophisticated. Because the children in each age range have a different average level of attainment, it is advisable to restrict the data to separate frequency tables (Green studied five age ranges but only two are shown here to keep the presentation manageable). There was no strong prior reason for expecting male superiority (in the same study, the girls were as good or better than the boys on tests involving 'combinatoric' ability). A nonspecific test is, accordingly, in order.

Step 1 - formulate the problem

(a) Specify the hypotheses:
Null hypothesis (H_0): there is no difference in the rate of development of probability concepts in boys and girls.
There is no specific alternative. The general alternative hypothesis is simply that H_0 is untrue.

(b) Specify the variables:
Independent variable: sex (male/female).
Dependent variable: achieved level of probability conceptualization (four-point scale 0–3).
Blocks: the data are subdivided into two age groups represented by separate frequency tables (years 1 and 2).

(c) Order the samples:
This is not necessary for this test.

Step 2 - find the sample rank sums and means

(a) Draw up the frequency tables as in table 8.16. Find the category totals (t_q) and the cumulative category totals (Σt_q).

(b) Find the shared ranks for each category. Do this separately for each table using the expression

$$r_q = \Sigma t_q - t_q/2 + 0.5$$

and enter these in table 8.16.

(c) Find the rank sum for each sample by multiplying the shared rank for each category by the number of scores in that category. For sample 1 this is

$$R_j = \Sigma(\text{frequency})(\text{shared rank})$$

$$R_1 = 48(53.5) + 163(284) + 99(544.5) + 10(635.5)$$
$$+ 40(39) + 147(243) + 127(517) + 20(645.5)$$
$$= 224\,970.5$$

Table 8.16 Computational layout for example in section 8.9

Probability concept level	Sex		Total t_q	Cumulative total Σt_q	Shared rank r_q	For tie correction $(t_q^3 - t)$
	Boys	Girls				
Year 1 (age 11–12 years)						
0	48	58	106	106	53.5	1 190 910
1	163	192	355	461	284	44 738 520
2	99	67	166	627	544.5	4 574 130
3	10	6	16	643	635.5	4 080
	320	323		b_i 643		Σ 50 507 640
	$n_{1,1}$	$n_{1,2}$				
Year 2 (age 12–13 years)						
0	40	37	77	77	39	456 456
1	147	184	331	408	243	36 264 360
2	127	90	217	625	517	10 218 096
3	20	20	40	665	645.5	63 960
	334	331		b_i 665		Σ 47 002 872
	$n_{1,1}$	$n_{1,2}$				
rank sum	R_1 244 970.5	R_2 203 520.5				
sample size	n_1 654	n_2 654				
rank mean	$\bar{R}_1$ 343.99	$\bar{R}_2$ 311.91				

$$R_1 = 58(53.5) + 192(284) + 67(544.5) + 6(635.5)$$
$$+ 37(39) + 184(243) + 90(517) + 20(645.5)$$
$$= 203\,520.5$$

(d) Check that all the ranking has been carried out correctly. The total sum of ranks $(R_1 + R_2)$ should equal the sum of ranks in each table:

$$R_1 + R_2 = \Sigma b_i(b_i + 1)/2$$
$$224\,970.5 + 203\,520.5 = 643(644)/2 + 665(666)/2$$
$$428\,491 = 428\,491$$

where b_i is the number of scores in the ith block (frequency table).

(e) Note the sample sizes ($n_1 = 654$, $n_2 = 654$) and find the sample rank means ($\bar{R}_j = R_j/n_j$). In our case $\bar{R}_1 = 344$ and $\bar{R}_2 = 312$.

Step 3 - interpret the rank means
The boys have a higher sample rank mean (344) than the girls (312). This means that the distribution of boys' scores is shifted towards the bottom of the table where the higher shared ranks are to be found. This, in turn, implies that boys have a tendency toward a more advanced development of the probability concept. Inspection of table 8.15 shows that this tendency is only slight, however. Our concern now is whether this slight tendency is statistically reliable.

Step 4 - compute and evaluate our key statistics (K and H)
(a) Find K, which is our case is simply

$$K = R_1 = 224\,970.5$$

Compute

$$\sum_i n_{i1}(b_i + 1) = 320(643 + 1) + 334(665 + 1)$$
$$= 428\,524$$
$$\Sigma n_{i1}n_{i2}(b_i + 1) = (320)(323)(643 + 1) + (334)(331)(665 + 1)$$
$$= 140\,192\,804$$
$$H = \frac{3[2K - \Sigma n_{i1}(b_i + 1)]^2}{\Sigma n_{i1}n_{i2}(b_i + 1)}$$
$$= \frac{(3)[2(224\,970.5) - 428\,524)]^2}{140\,192\,804}$$
$$= 9.82$$

(b) Evaluate H by consulting tables of the chi-square distribution (table B in the appendix). In two-sample problems we have only one degree of freedom ($df = 1$). The appropriate row of table B is therefore

Degrees of freedom	5%	2.5%	1%	0.1%
1	$\chi^2 > 3.8$	5.0	6.6	10.8

To be significant our value of H must be greater than or equal to the critical value of chi-square for the appropriate level. Our value of H (9.82) is therefore significant at better than the 1 per cent level ($p > 0.01$).

(c) Correct for ties:

Tie correction usually has a large effect on the value of H in frequency table analysis and is worth doing even if the compilations are tedious. We need a separate tie correction factor for each table. Firstly, compute $t^3 - t$ for each category in each table. In table 8.16, this is done in the extreme right hand column. Then compute $\Sigma(t^3 - t)$ for each table separately. We can now find the tie correction factors using the expression

$$T_i = 1 - \frac{\Sigma(t_{iq}^3 - t_{iq})}{b_i^3 - b_i}$$

where t_{iq} is the total in the qth category in the ith block and b_i is the total number of scores in the ith block. For year 1 in table 8.16 we have

$$T_1 = 1 - \frac{50\,507\,640}{(643^3 - 643)} = 1 - \frac{50\,507\,640}{265\,847\,064}$$

$$= 0.81$$

For year 2

$$T_2 = 1 - \frac{47\,002\,872}{(665^3 - 665)} = 1 - \frac{47\,002\,872}{294\,078\,960}$$

$$= 0.84$$

We must now recompute the denominator of expression (8.6):

$$\Sigma n_{i1} n_{i2}(b_i + 1) T_i = 320(323)(643 + 1)(0.81)$$

$$+ 334(331)(665 + 1)(0.84)$$

$$= 115\,765\,040.1$$

We can now compute the corrected value of H:

$$H' = \frac{3[2K - \Sigma n_{i1}(b_i + 1)]^2}{\Sigma n_{i1} n_{i2}(b_i + 1) T_i}$$

$$= \frac{3[2(224\,970.5) - 428\,524]^2}{115\,765\,040.1}$$

$$= 11.89$$

Our new value of H' (11.89) is considerably larger than the uncorrected value ($H = 9.82$), and our result is now significant at the 0.1 per cent level ($p < 0.001$).

Step 5 - draw conclusions
Because our result is statistically significant ($p < 0.001$) we may reject H_0 (see step 1). In simple terms, the data support the conclusions that boys have a slight but reliable tendency to acquire probability concepts earlier than girls.

Exercises

8.1 Reanalyse Bradley's data in section 8.4 using both a matched pairs test (section 8.2) and a Wilcoxon matched pairs, signed ranks test (section 8.3). Compare the results of both analyses with each other and with the result of the analysis given in section 8.2.

8.2 The example in section 8.3 reported an analysis by Taub et al. (1971) into the effect of extended sleep on performance at a vigilance task. In the same study they measured the motor coordination of their 12 subjects using a pinball game. The data in table 8.17 are the pinball scores (average of trials on two separate days) for each subject following control and extended sleep nights. Taub et al. expected *lower* pinball scores following extended sleep. Do the data justify their view that extra sleep impairs motor coordination?

8.3 In the example in section 8.2 we found using Dixon and Spitz's (1980) data that people would tolerate greater delays in a film soundtrack for 'speech' than for 'a hammer hitting a peg'. They explained this in terms of a learning effect: sound naturally trails behind light, and we have more opportunity to adapt to this effect for speech which is a very much more common experience. However, in the example in section 8.6 we also found that people tolerated greater *advancement* of the sound for speech than for a hammer hitting a peg. This could not be explained as a learning effect because we almost never experience situations where sound arrives before the visual component. It might be simpler to say that it is easier to spot asynchronism in the case of very specific events like a hammer hitting a peg than a continuous stream of words coming out of someone's mouth.

Table 8.17 Pinball scores for 12 subjects following control and extended sleep nights. Data from Taub et al. (1971)

Subject	Control night	Extended sleep
1	592	643.5
2	557.5	583.5
3	464	516.5
4	559.5	550
5	571	515.5
6	654.5	516
7	638	378
8	525	469
9	533	481
10	531.5	522.5
11	586	505.5
12	548	490.5

Where, then, does this leave the learning hypothesis? It is certainly not dead because Dixon and Spitz might reasonably point to the fact that subjects appear to show a much greater difference between speech and hammering for the delay than for the advancement condition. The data in table 8.18 show these difference scores. Is the effect statistically significant in the direction predicted by the researchers?

8.4 In the example in section 8.7, an experiment by Langford et al. (1973) was described in which volunteers were studied in a sleep laboratory by being given tests during the night after spontaneous awakenings. The aim of the investigation was to decide whether people were more mentally alert following awakenings from light sleep or from active (dreaming) sleep. One of the tests used by Langford was a small battery of eight simple items from an IQ test for able schoolchildren, and the measure of success was the time taken (in seconds) to complete them. If they were not all completed in five minutes the test was terminated and their time entered into the data table as greater than 300 seconds (>300). The data given in table 8.19 indicate that subjects perform more quickly after a spontaneous awakening from active sleep. Is this effect statistically significant? Prior to this study, the research literature contained conflicting indications as to the possible outcome.

8.5 Using the data in exercise 7.7, sort the psychiatric patients into four categories according to their depression ratings *before* treatment. Use the following categories:

(a) less than 40 points
(b) between 40 and 49 points

Table 8.18 'Speech'/'hammer hitting peg' differences in tolerance for soundtrack delay and advancement

Subject	Sound track	
	Delay	Advancement
1	238	76
2	99	181
3	3	−26
4	23	30
5	70	−10
6	56	0
7	39	9
8	74	72
9	66	2
10	−20	33
11	109	76
12	4	−3
13	70	54
14	95	23
15	200	196
16	46	66
17	159	24
18	93	92

Table 8.19 Time taken (seconds) to complete a simple eight-item intelligence test following spontaneous arousals from active and quiet sleep (Langford et al., 1973). 300 seconds was the maximum duration allowed before terminating the test

Subject	Active sleep		Quiet sleep	
1	192	213	155	230
2	150	157	230	240
3	112	207	150	>300
4	290	>300	235	>300
5	140	177	285	>300
6	180	290	215	>300
7	210	>300	200	210
8	262	286	>300	>300

(c) between 50 and 59 points

(d) more than 59 points.

Treat these four groups as separate blocks and compare the depression ratings *after* treatment for the real and simulated ECT samples.

9 Three or more independent samples

9.1 Introduction

There is no substantial difference between tests involving many samples and tests involving only two samples. The tests described in chapter 7 are merely simplified versions of the tests to be described in this chapter. Traditionally, they have been presented as quite different tests with different names, different justifications and even different notation. The aim of the presentation here is to emphasize their similarity. Strictly speaking, the tests in this chapter render those in chapter 7 quite redundant. It would be enough to learn the many-sample methods which could then be used to analyse two-sample methods. However, the two-sample techniques are so much faster that it would be a pity not to use them when appropriate. Chapter 14 shows how the two-sample methods are merely a simplification of the many-sample methods.

One aspect of the computational and conceptual economy of the two-sample methods of chapter 7 concerns the choice of the (trend/contrast) coefficients in tests using a specific alternative hypothesis. For two-sample tests, the convention was adopted that the first sample was the sample with the higher expected rank mean. This convention results in a considerable simplification which cannot be obtained in the many-sample case where a range of expected patterns of ordering of sample rank means is possible. Some method must be devised to allow the test user to express his prediction of the ordering of sample rank means. The use of coefficients for this purpose is discussed at length in section 3.2 and section 2.10 which should be consulted again in case of difficulty. In the two-sample case we adopt the coefficient pair $(1, 0)$ to express our hypothesis that sample 1 will have a higher rank mean than sample 2. In this case, we find that

$$L = \sum_j \lambda_j R_j = 1 \times R_1 + 0 \times R_2 = R_1$$

As a result, we were able to ignore coefficients and simply use R_1 as our estimate of L. For designs with more than two samples we need to be more explicit, and coefficients are always specified.

The computational procedures for coping with independent samples are also closely linked in a simple way with those described in chapter 10 for dealing with related samples. This becomes clear if we can regard the independent samples design as a single block of scores but the related samples design as a multiple block design. Once this is understood, it follows that the general procedures for analysing multiple blocks with unequal cell frequencies can happily cope with the single-block designs given in this section. Once again the benefit of a specific chapter for independent samples is largely one of computational convenience because multiple block techniques for unequal sample sizes are particularly cumbersome, especially when done by hand.

The procedures described below are a mixture of the familiar, the unfamiliar and the new. They represent an attempt to draw together the work of people such as Kruskal (1952), Marascuilo and McSweeney (1967) and Dunn (1964) and to fill in the gaps of a more rational structure of test procedures.

A   Tests against a specific alternative hypothesis

9.2   Exact test for trend in three unrelated samples

This exact test deals with the special case of simple trends across three groups. The trend coefficients always take the form ($\lambda_j = 1, 2, 3$) and no variation is permitted. The sample sizes must all be equal to or less than five because the tables do not extend beyond that point. The benefit of the tables is mainly one of computational convenience since the approximate test is normally quite satisfactory even for very small samples. The exact test is also a useful introduction for beginning students since it temporarily avoids the need to teach the theory behind the normal approximation.

The following example is a fictitious example taken from Siegel (1956, chapter 8) who used a nonspecific test to analyse the data. At that time no exact test existed to analyse a trend across his three groups, but it is likely that he would have used such a test if it had been available (the exact tables in the appendix have been computed by the author specifically to plug this gap). By using Siegel's example we can draw attention to the merits of using a trend test, when appropriate, for three unrelated samples (see step 5 in the example).

Example

A (fictitious) investigator measured the authoritarianism of teachers and school administrators using an appropriate questionnaire. The hypothesis to be tested was that administrators were more authoritarian. However, some of the teachers aspired eventually to become administrators and could be thought of as half-way between teaching and administrating. To allow for this, the teachers were classified as:

Table 9.1   Authoritarianism scores for three groups of educators in the example in section 9.2. Fictitious data taken from Siegel (1956, chapter 8)

Teachers	Teacher-administrators	Administrators
96	82	115
128	124	149
83	132	166
61	135	147
101	109	

(a) either teachers (i.e. planning to remain teachers)
(b) or teacher-administrators (i.e. planning to become administrators)
(c) or administrators.

The authoritarianism scores for these three groups are given in table 9.1. Do the data support the hypothesis that authoritarian attitudes are more often associated with administrators?

Step 1 – formulate the problem
(a) Specify the hypotheses:
   Null hypothesis (H_0); authoritarian attitudes are not associated with the teacher/administrator status of educators.
   Specific alternative hypothesis (H_1); authoritarian attitudes are more often associated with administrators. Administration-oriented teachers might be expected to lie halfway between teaching-oriented teachers and administrators. If this is the case, then we expect an ascending trend across the three groups and choose the following coefficients:

   λ_j 1   2   3

   Note that we might alternatively have expected administration-oriented teachers to be just as authoritarian as the administrators; in that case we could simply combine these two groups to form a single sample and then proceed as for a two-sample test.
(b) Specify the variables:
   Independent variable: educator status and leaning (teacher/teacher-administrator/administrator).
   Dependent variable: authoritarianism; measured by a questionnaire (the greater the score the greater the degree of authoritarianism).
(c) Order the samples:
   For the exact test the samples should be ordered so that the coefficient for sample 1 is 1, for sample 2 is 2 and for sample 3 is 3.

Table 9.2 Computational layout for example in section 9.2

Teachers		Teacher-administrators		Administrators			
Score	Rank	Score	Rank	Score	Rank		
61	1	82	2	115	7		
83	3	109	6	147	12		
96	4	124	8	149	13		
101	5	132	10	166	14		
128	9	135	11				
rank sum							
R_1	22	R_2	37	R_3	46		
sample size							
n_1	5	n_2	5	n_3	4	N	14
rank mean							
$\bar{R}_1$	4.4	$\bar{R}_2$	7.4	$\bar{R}_3$	11.5		
coefficients							
λ_1	1	λ_2	2	λ_3	3		
$\lambda_1 R_1$	22	$\lambda_2 R_2$	74	$\lambda_3 R_3$	138	$\Sigma \lambda_j R_j$	234

Step 2 – find the sample rank sums and means

(a) Arrange the scores in columns as in table 9.2. Put the scores in ascending order of size for easy ranking.

(b) Rank the scores *irrespective of sample*, remembering to give low scores low ranks. Write the ranks alongside the raw scores.

(c) Find the sample rank sums (R_j) by totalling the ranks in the columns. In our case $R_1 = 22$, $R_2 = 37$ and $R_3 = 46$.

(d) Check that the ranking has been carried out correctly. The total sum of ranks should equal $N(N + 1)/2$, where N is the total sample size.

$$N(N + 1)/2 = \sum_j R_j$$

$$(14)(15)/2 = 22 + 37 + 46$$

$$105 = 105$$

(e) Note the sample sizes ($n_1 = 5$, $n_2 = 5$ and $n_3 = 4$) and compute the sample rank means ($\bar{R}_j = R_j/n_j$). In our case $\bar{R}_1 = 4.4$ and $\bar{R}_2 = 7.4$ and $\bar{R}_3 = 11.5$.

Step 3 - interpret the rank means

When testing a specific alternative hypothesis, we must check that the rank means are ordered in accordance with our expectations. It is not meaningful to proceed with our analysis if they are not so ordered. The expected ordering is given by our set of coefficients ($\lambda_j = 1, 2, 3$) and we can see from table 9.2 that our sample rank means are ordered as we expect and we may proceed to step 4. Although the ordering is perfect in this case we are usually only looking for an approximate agreement.

Step 4 - compute and evaluate our key statistic (L)

(a) Find L:

$$L = \sum_j \lambda_j R_j = (1)(22) + (2)(37) + (3)(46) = 234$$

(b) Evaluate L by consulting table K in the appendix for $L = 234$ and $n_1 = 5$, $n_2 = 5$ and $n_3 = 4$. The critical row for our example is

N	n_1	n_2	n_3	5%	2.5%	1%	0.1%
14	5	5	4	$L \geqslant 224$	228	232	238

To be significant, L must be equal to or greater than the critical value at the appropriate level. In our case, L (234) is significant at better than the 1 per cent level ($p < 0.01$).

(b) Correct for ties:

No correction for ties is possible in the exact test. In our example there were no ties, so this is not a problem. When ties exist, the exact test becomes conservative. If ties are heavy, an approximate test may be a better guide. Fortunately, heavy ties are unlikely for total sample sizes (N) less than 15.

Step 5 - draw conclusions

Because our result is significant ($p < 0.01$), we may reject H_0 (see step 1) in favour of H_1. In simple terms, the data in this fictitious example support the conclusion that authoritarianism in schools is associated with administrators and aspiring administrators rather than teachers.

Siegel's (1956) analysis of the same data using a nonspecific test gives a result which is barely significant at the 5 per cent level. Notice that a specific test is always more powerful than a nonspecific test and should always be used, if at all appropriate.

9.3   Approximate test for trends and contrasts

The approximate test is slower than the exact test but is much more useful. It can deal with large samples, adjust for tied data and evaluate any simple hypo-

G

Table 9.3   Percentage of correct hits for three groups of volunteers in a sleep deprivation experiment

Control	Sleep deprived	Sleep deprived + incentives
87	65	85
91	58	60
85	47	61
80	38	38
80	47	81

thesis pattern. Unequal sample sizes are also acceptable. The name 'approximate' is a little misleading in that the approximation to the exact distribution is very good (at least near the important 1 per cent significance level) even for very small samples. It is much better, for example, than the chi-square approximation given in section 9.6.

Example

The data in table 9.3 are taken from an experiment by Horne and Pettit (1984) in which they investigated the effects of financial reward on the ability of volunteers to remain alert during long periods of loss of sleep. Three groups of volunteers were used: (i) controls who were not sleep deprived at all, (ii) volunteers who were sleep deprived but without financial incentive and (iii) volunteers who were sleep deprived but had some financial incentive to perform well on the tests of alertness. Alertness was measured using a 30 minute test in which participants listened to short monotone pulses delivered through headphones at a rate of one every two seconds. An average of one in four of these pulses was of slightly shorter duration than the others and participants had to press a button whenever they believed they had heard one of these short signals. One of the measures used was the percentage of these short signals which were correctly identified.

The data in table 9.3 represent the performance of all 15 volunteers at 10.00 a.m. on the second day of the experiment (i.e. after two nights without sleep). We expect the control group to produce the best performances. Do the results bear out our understanding of the situation?

Step 1 – formulate the problem
(a) Specify the hypotheses:
    Null hypotheses (H_0): sleep deprivation has no effect on alertness whether or not financial incentives are given.
    Specific alternative hypothesis (H_1): sleep deprivation will impair alertness but this impairment will be reduced if financial incentives are given.

(b) Specify the variables:
Independent variable: sleep deprivation, with and without financial incentives. The variable is represented by three samples: (i) control, (ii) sleep deprivation, (iii) sleep deprivation plus financial incentives.
Dependent variable: alterness. This is measured in terms of percentage correct hits.

(c) Choose coefficients:
Since we expect the control group (i) to be best and the simply sleep deprived group (ii) to be worst, we use the following coefficients:

Control	Sleep deprived (SD)	SD plus incentive
3	1	2

Step 2 - find the sample rank sums and means
(a) Rearrange the data table to form columns with scores in ascending order for easy ranking (see table 9.4).
(b) Rank the scores, irrespective of sample, remembering to give low ranks to low scores. Write the ranks alongside the original scores.
(c) Find the rank sums (R_j) by totalling the ranks in each column. In our case we have, $R_1 = 60.5, R_2 = 21.5$ and $R_3 = 38$.

Table 9.4 Computational layout for example in section 9.3

Control		Sleep deprived		Sleep deprived + incentives			
80	*9.5*	38	*1.5*	38	*1.5*		
80	*9.5*	47	*3.5*	60	*6*		
85	*12.5*	47	*3.5*	61	*7*		
87	*14*	58	*5*	81	*11*		
91	*15*	65	*8*	85	*12.5*	N	15
rank sum							
R_j	60.5		21.5		38	ΣR_j	120
sample size							
n_j	5		5		5		
rank mean							
$\bar{R}_j$	12.1		4.3		7.6		
λ_j	3		1		2		
$\lambda_j R_j$	181.5		21.5		76	$\Sigma \lambda_j R_j$	279
$n_j \lambda_j$	15		5		10	$\Sigma n_j \lambda_j$	30
$n_j \lambda_j^2$	45		5		20	$\Sigma n_j \lambda_j^2$	70

(d) Check that the ranking has been carried our correctly:

$$N(N + 1)/2 = \Sigma R_j$$
$$15(16)/2 = 60.5 + 21.5 + 38$$
$$120 = 120$$

(e) Note the sample sizes (n_j) and find the sample rank means $(\bar{R}_j)$. In our case $\bar{R}_1 = 12.1, \bar{R}_2 = 4.3$ and $\bar{R}_3 = 7.6$.

Step 3 - interpret the rank means
It is important to check that the rank means are ordered at least approximately as expected. In our case the control group has the best performance $(\bar{R}_1 = 12.1)$ and the sleep deprived without incentives have the worst performance $(\bar{R}_2 = 4.3)$.

Step 4 - compute and evaluate our key statistics (L and Z)
(a) Find L:

$$L = \Sigma \lambda_j R_j$$
$$= 3(60.5) + 1(21.5) + 2(38)$$
$$= 181.5 + 21.5 + 76 = 279$$

Find Z:

$$Z = \frac{L - E(L)}{\sqrt{\text{var}(L)}}$$

where

$$E(L) = (N + 1)(\Sigma n_j \lambda_j)/2 \qquad (9.1)$$

and

$$\text{var}(L) = (N + 1)(N \Sigma n_j \lambda_j^2 - (\Sigma n_j \lambda_j)^2)/12 \qquad (9.2)$$

First, we must compute

$$\Sigma n_j \lambda_j = 5(3) + 5(1) + 5(2) = 30$$
$$\Sigma n_j \lambda_j^2 = 5(9) + 5(1) + 5(4) = 70$$

We now have

$$Z = \frac{279 - 16(30)/2}{\sqrt{\{16[15(70) - 30^2]/12\}}}$$
$$= \frac{279 - 240}{\sqrt{200}} = 2.76$$

Table 9.5   Correction for ties computation, example in section 9.3

Score	Rank	Tie length t	$t^3 - t$
38	1.5	2	6
47	3.5	2	6
80	9.5	2	6
85	12.5	2	6
		Σ	24

$$T = 1 - \frac{\Sigma(t^3 - t)}{N^3 - N} = 1 - \frac{24}{15^3 - 15} = 1 - \frac{24}{3360} = 0.9929$$

$$Z' = Z/\sqrt{T} = 2.76/\sqrt{0.9929} = 2.76/0.9964 = 2.77$$

(b) Evaluate Z by consulting a table of the normal distribution (table A in the appendix). The critical values are reproduced here:

	5%	2.5%	1.0%	0.1%
$z \geqslant$	1.64	1.96	2.33	3.10

To be significant at any of the four levels, Z must be equal to or greater than the corresponding critical value. Our value of Z (2.76) is significant at better than the 1 per cent level ($p < 0.01$).

(c) Correct for ties:

The correction for ties will make our value of Z larger and, therefore, more significant. Although this is hardly necessary on this occasion, the exercise will be carried out by way of illustration. Table 9.5 gives a list of scores which were tied, their shared rank, the length of the tie (t) and ($t^3 - t$). We correct Z thus:

$$Z' = Z/\sqrt{T} \tag{9.3}$$

where

$$T = 1 - \frac{\Sigma(t^3 - t)}{N^3 - N}$$

The computations are given in table 9.5. Our new value of Z' is 2.77, which is hardly different from our original value of Z (2.76). In general, the correction for ties causes a useful increase in Z only when the ties are long relative to the total sample size.

Step 5 - draw conclusions

Because our result is significant ($p < 0.01$), we may reject H_0 (see step 1) in favour of H_1. In simple terms, the data support a conclusion that alertness is depressed by sleep deprivation despite financial incentives, but, in the absence of incentives, it depresses alertness even more.

9.4 $k \times Q$ frequency table test for trends and contrasts

This test can be used to analyse a single frequency table with any number of rows and columns. The only restriction is that the rows should constitute a set of rankable categories (of the dependent variable). In this respect the test differs from traditional chi-square (homogeneity) tests which do not require that the categories are ordered (see section 9.8). If the categories are indeed ordered, then the analysis of variance by ranks technique described in this section is the appropriate test. Note that when we have only two categories of the dependent variable, they are always considered to be rankable whatever they may be. For ordered categories, the rank sum procedures are superior to the chi-square homogeneity test given in section 9.8 because the order information is preserved in the former test and ignored in the latter. The rank sum test also lends itself to the evaluation of trends and contrasts, which is not always possible or meaningful using homogeneity tests.

Example

In the examples in sections 7.5 and 8.5 we analysed data from an investigation into headaches by Wilkins et al. (1984). They asked volunteers to say whether their recent headaches were localized on the left side, or right side, or both sides of the head. They were also shown a display of horizontal black and white stripes bisected by a single vertical line. This display generates both visual effects, such as shimmering and colours, and nonvisual effects such as feelings of ocular discomfort. The volunteers were asked to say whether they saw the visual effects on the left, right or both sides of the display. In sections 7.5 and 8.5 it was shown that lateralization of headachies was associated with lateralization of the visual effects. However, it was not shown that the *direction* of laterlization (i.e. left, right) was the same for the two variables. The researchers have kindly recast the data in a form suitable for reanalysis, and these are given in table 9.6. It might for present purposes be interesting to determine whether, when visual effects do occur exclusively on one side, this should be the side associated with unilateral headaches.

Table 9.6 Data for example in section 9.4

		Illusions in visual hemifields		
		Left	Bilateral	Right
	Left	3	3	1
Headaches	*Bilateral*	1	13	0
	Right	0	3	2

Step 1 - formulate the problem

(a) Specify the hypotheses:

Null hypothesis (H_0): lateralization of headaches is not associated with lateralization of visual effects on viewing horizontal striped displays.

Specific alternative hypothesis (H_1): the side on which headaches are reported is associated with the side on which the visual effects are reported.

(b) Specify the variables:

In this example it is difficult to decide which variable is dependent and which variable is independent, since both phenomena (headaches, visual effects) are presumed to be dependent upon a common neurological process. Arbitrarily, we have allocated 'lateralization of headaches' to the dependent variable status. We should not expect much difference from swapping the variables, but this is left to exercise 9.2.

Independent variable: side of visual effect (left, bilateral, right).

Dependent variable: side of heachaches (left, bilateral, right).

The categories of the dependent variable must be rankable. In this case we are assuming that bilateral lies intermediate between left and right. If this assumption were to be successfully challenged, then the analysis would be invalid.

(c) Choose the coefficients:

We choose the following coefficients to reflect our hypothesis:

$$\lambda_j \quad 1 \quad 2 \quad 3$$

This implies that sample 3 (right sided illusions) will have the highest sample rank mean. Remember that the highest ranks are associated with the lowest row of the table (in this case, right sided headaches). As a result our set of coefficients imply that right side visual effects will be associated with right sided headaches and vice versa for left sided phenomena.

Step 2 - find the same rank sums and means

(a) Draw up the frequency table as in table 9.7. Find the category totals (t_q) and the cumulative category totals (Σt_q).

Table 9.7   Computational layout for example in section 9.4

Headaches	Visual effects			t_q	Σt_q	r_q	$t^3 - t$
	Left	Bilateral	Right				
Left	3	3	1	7	7	4	336
Bilateral	1	13	0	14	21	14.5	2730
Right	0	3	2	5	26	24	120
							Σ 3186

	Left	Bilateral	Right		
sample rank sum					
R_j	26.5	272.5	52		
sample size					
n_j	4	19	3		
sample rank mean					
$\bar{R}_j$	6.625	14.34	17.3		
coefficients					
λ_j	1	2	3		
$\lambda_j R_j$	26.5	545	156	$L = \Sigma \lambda_j R_j$	727.5
$n_j \lambda_j$	4	38	9	$\Sigma n_j \lambda_j$	51
$n_j \lambda_j^2$	4	76	27	$\Sigma n_j \lambda_j^2$	107

(b) Find the shared ranks for each category using the expression

$$r_q = \Sigma t_q - t_q/2 + 0.5$$

In our example we have

$$r_1 = 7 - 7/2 + 0.5 = 4$$
$$r_2 = 21 - 14/2 + 0.5 = 14.5$$
$$r_3 = 26 - 5/2 + 0.5 = 24$$

To check these shared ranks find

$$N(N + 1)/2 = \Sigma(t_q r_q)$$
$$(26)(27)/2 = (7)(4) + (14)(14.5) + (5)(24)$$
$$351 = 351$$

where N is the total number of scores in the frequency table.

(c) Find the sample rank sums for each sample by multiplying the shared rank for each category by the number of scores in that category and adding:

$$R_1 = (3)(4) + (1)(14.5) + (0)(24) = 26.5$$
$$R_2 = (3)(4) + (13)(14.5) + (3)(24) = 272.5$$
$$R_3 = (1)(4) + (0)(14.5) + (2)(24) = 52$$

(d) Check that the ranking has been carried out correctly. The total sum of ranks should equal $N(N + 1)/2$:

$$N(N + 1)/2 = \Sigma R_j$$
$$(26)(27)/2 = 26.5 + 272.5 + 52$$
$$351 = 351$$

e) Note the samples sizes ($n_1 = 4$, $n_2 = 19$ and $n_3 = 3$) and compute the sample rank means ($\bar{R}_j = R_j/n_j$). In our example, $\bar{R}_1 = 6.625$, $\bar{R}_2 = 14.34$ and $\bar{R}_3 = 17.3$.

Step 3 – interpret the rank means
When testing a specific alternative hypothesis, we must check that the rank means are ordered at least approximately in accordance with our specific alternative hypothesis. If they are not so ordered, it is meaningless to continue with the analysis. In our example we predicted that the sample means would increase from left to right across the table, and this prediction does agree with the results. Inspection of table 9.6 does show that a majority of people with left sided visual effects have left sided headaches. Those with bilateral visual effects show no bias to left or right for headaches, whereas the three individuals with right sided visual effects are split 2:1 between right and left sided headaches in the predicted direction. Although the trend is in the right direction the effect does not seem to be very pronounced.

Step 4 – compute and evaluate our key statistics (L and Z)
(a) Find L:

$$L = \Sigma \lambda_j R_j$$
$$= (1)(26.5) + (2)(272.5) + (3)(52) = 727.5$$

Find Z:

$$Z = \frac{L - E(L)}{\sqrt{\text{var}(L)}}$$

where

$$E(L) = \tfrac{1}{2}(N + 1)\, \Sigma n_j \lambda_j \qquad\qquad (9.3)$$

and

$$\mathrm{var}\,(L) = \tfrac{1}{12}(N + 1)(N\Sigma n_j \lambda_j^2 - (\Sigma n_j \lambda_j)^2) \qquad\qquad (9.4)$$

Firstly, we must compute

$$\Sigma n_j \lambda_j = (4)(1) + (19)(2) + (3)(3) = 51$$
$$\Sigma n_j \lambda_j^2 = (4)(1) + (19)(4) + (3)(9) = 107$$

We now have

$$Z = \frac{727.5 - \tfrac{1}{2}(27)(51)}{\sqrt{\{\tfrac{1}{12}(27)[26(107) - (51)^2]\}}}$$

$$= \frac{727.5 - 688.5}{\sqrt{407.25}} = 1.93$$

(b) Evaluate Z by consulting a table of the normal distribution (table A in the appendix). The initial values are reproduced here:

5%	2.5%	1%	0.1%
$z \geqslant 1.64$	1.96	2.33	3.10

To be significant at any of the four levels, Z must be equal to or greater than the corresponding critical value. Our value of Z (1.93) is significant at better than the 5 per cent level ($p < 0.05$).

(c) Correction for ties:
The tie correction procedure will make our value of Z larger and more significant. The correction is usually worth while for frequency tables where ties are long relative to the total sample size. At the right of table 9.7 add the values for $(t^3 - t)$. Table D in the appendix will be found useful for finding values for $(t^3 - t)$. Find

$$\Sigma(t^3 - t) = 336 + 2730 + 120 = 3186$$

Find T:

$$T = 1 - \frac{\Sigma(t^3 - t)}{N^3 - N}$$

$$= 1 - \frac{3186}{26^3 - 26} = 1 - \frac{3186}{17\,550} = 0.818$$

We can now correct Z:

$$Z' = Z/\sqrt{T} = 1.93/\sqrt{0.818} = 2.13$$

Our new value of Z' is now significant at the 2.5 per cent level ($p < 0.025$).

Step 5 - draw conclusions
Because our result is significant we may reject H_0 (see step 1) in favour of H_1. In simple terms the data support a conclusion that the location reported for headaches (left side, both sides and right side) is systematically related to the location reported for visual effects on a horizontal stripes display. This conclusion will be qualified in example 9.7 when look yet again at the same data by adding the data for neurology patients.

B Tests against a nonspecific alternative hypothesis

9.5 Exact test for three unrelated samples (nonspecific)

This test deals with designs involving three very small samples – no sample should contain more than five scores. Unlike the corresponding exact test for trend (see section 9.2), the exact tables in table I in the appendix offer us more than just computational convenience. For very small samples the chi-square approximation is not very accurate, and exact tables are essential for handling significant or marginally significant results. Ideally the tables should be larger because the approximation continues to be less than accurate even for slightly larger samples, but the computation of an extension to table L has proved very onerous and, for the time being, we shall have to go without.

Example

Horne and Pettit (1984) report an experiment which investigated the effects of financial rewards on the ability of volunteers to remain alert during long periods of loss of sleep. Three groups of five volunteers were used: (i) sleep deprived with no financial incentive, (ii) sleep deprived with some financial incentive, and (iii) control volunteers who neither were sleep deprived nor had any financial incentive. Alterness was measured using a 30 minutes test in which participants listened to short monotone pulses delivered through headphones at a rate of one every two seconds. An average of one in four of these pulses was of slightly shorter duration than the others, and participants had to press a button whenever they believed they had heard one of these short signals. One of the measures used was the percentage of these short signals which were correctly identified.

The data given in table 9.8 represent the performance of all 15 volunteers before the beginning of the experiment, *i.e. before any sleep deprivation and before incentives were introduced.* If the experiment has been designed correctly, there should be no significant differences between the groups at the outset. Is this the case?

Table 9.8   Percentage of correct hits for three groups of subjects performing a vigilance task

Sleep deprived (no incentive)	Sleep deprived (incentive)	Control (no incentive)
75.0	65.6	84.2
90.6	68.5	81.9
77.7	79.9	76.4
57.6	63.4	72.5
68.3	77.3	72.0

Step 1 – formulate the problem
(a) Specify the hypothesis:
    Null hypothesis (H_0): the three groups do not differ in their success rate.
    There is no specific alternative hypothesis. The general alternative is that H_0 is untrue.
(b) Specify the variables:
    Independent variable: type of condition experienced (i) sleep deprivation, (ii) sleep deprivation with financial incentive, and (iii) control.
    Dependent variable: success in a vigilance task measured in terms of percentage of correct hits.

Step 2 – find the sample rank sums and means
(a) Arrange the scores in columns as in table 9.9. Put the scores in ascending order for easy ranking.
(b) Rank the scores, *irrespective of sample*, remembering to give low scores low ranks.
(c) Find the sample rank sums (R_j) by totalling the ranks in each column. In our case $R_1 = 39$, $R_2 = 32$ and $R_3 = 49$.

Table 9.9 Computational layout for example in section 9.5

Sleep deprived (no incentive)		Sleep deprived (incentive)		Control	
Score	Rank	Score	Rank	Score	Rank
57.6	1	63.4	2	72.0	6
68.3	4	65.6	3	72.5	7
75.0	8	68.5	5	76.4	9
77.7	11	77.3	10	81.9	13
90.6	15	79.9	12	84.2	14
rank sum	R_1 39		R_2 32		R_3 49
sample size	n_1 5		n_2 5		n_3 5
rank mean	$\bar{R}_1$ 7.8		$\bar{R}_2$ 6.4		$\bar{R}_3$ 9.8

(d) Check that the ranking has been carried out correctly. The total sum of ranks should equal $N(N + 1)/2$ where N is the total sample size ($n_1 + n_2 + \ldots n_k$).

$$N(N + 1)/2 = \Sigma R_j$$
$$15(16)/2 = 39 + 32 + 49$$
$$120 = 120$$

(e) Note the sample sizes ($n_1 = 5$, $n_2 = 5$ and $n_3 = 5$) and compute the sample rank means ($\bar{R}_j = R_j/n_j$). In our case, $\bar{R}_1 = 7.8$, $\bar{R}_2 = 6.4$ and $\bar{R}_3 = 9.8$.

Step 3 - interpret the rank means
The rank means for the three groups are not very different although the control group are performing slightly better than the other two groups.

Step 4 - compute and evaluate our key statistic (K)
(a) Find K:

$$K = \Sigma(R_j^2/n_j)$$
$$= \frac{39^2}{5} + \frac{32^2}{5} + \frac{49^2}{5}$$
$$= 989.2$$

(b) Evaluate K by consulting table L in the appendix for $K = 982.2$ and $n_1 = n_2 = n_3 = 5$. The critical row for our example is

n_1	n_2	n_3	5%	2.5%	1%	0.1%
5	5	5	$K \geqslant 1073.2$	1092.0	1119.6	1155.6

To be significant, K must be equal to or greater than the critical value at the appropriate level. In our case, K (989.2) is clearly not significant ($p > 0.05$).

(c) Correct for ties:

No correction for ties is possible in the exact test. In our case there were no ties, so this is not a problem. When ties exist, the exact test becomes conservative. If ties are heavy, an approximate test *may* be a better guide.

Step 5 - draw conclusions

Because our result is not significant ($p > 0.05$) we may not reject H_0 (see step 1). In simple terms there appears to be no reason for believing the groups to have different vigilance capacities before the experiment begins.

9.6 Approximate test (nonspecific)

This general-purpose test is often used as a first analysis on data from independent samples to see 'if there is anything there'. Its main function is to detect any source of nonrandomness, and many statisticians prefer to get a significant result using this test before attempting a specific test of a trend or contrast hypothesis. Although in principle this approach has much to commend it, it is unreasonably harsh. The nonspecific test is very much less powerful than the specific test, and we often find examples where a specific analysis yields a significant result when the nonspecific analysis does not. It would seem better to reserve the nonspecific test for problems either where the investigator has no specific hypothesis to test or where his specific hypothesis has failed but some other effects may yet be present.

A serious, and largely unacknowledged, problem with the nonspecific test is the limitation which attends the drawing of conclusions following a significant result. The only conclusion which can be drawn is that 'at least one of the samples has been drawn from a population which is significantly different from the others'. It is not possible to draw any specific conclusion about the reliability of the ordering of the individual samples (except when there are only two). Specific tests are necessary for this purpose. Nonspecific tests for three or more samples are therefore to be avoided, if possible on two counts: firstly that they are less sensitive than specific tests, and secondly that they do not lead to clear and useful conclusions.

Example

This example comes from an exploratory survey by Southall and Stone (1982) of many aspects of the vision of partially sighted children. The data in table 9.10 give glare thresholds for 24 partially sighted school children, 10 normally sighted children and 10 normally sighted adults. Glare is measured by adjusting the brightness of a central patch of light while the background luminance is held steady until glare discomfort is first reported. For these data the background luminance was fixed at 100 candles per square metre (cd/m^2). The glare source luminance at the discomfort threshold is given in 9.10. Is glare sensitivity a measure on which partially sighted children differ from other people?

Table 9.10   Brightness of glare source (cd/m^2) at the point where glare becomes uncomfortable for three groups (see example in section 9.6)

Partially sighted children	Fully sighted children	Fully sighted adults
1460	2028	1195
2367	1167	950
8500	3090	562
8500	321	593
370	728	1950
1220	2550	4100
7150	8500	1182
7217	2117	2258
2800	5450	1497
8500	2267	893
8000		
2583		
2850		
1100		
1657		
557		
119		
1570		
3200		
1310		
1460		
640		
8500		
2683		

Step 1 - formulate the problem
(a) Specify the hypothesis:
Null hypothesis (H_0): glare sensitivity does not vary across the three populations sampled.
There is no specific alternative hypothesis. The general alternative is that H_0 is untrue.
(b) Specify the variables:
Independent variable: age and visual status of subjects (note that this is an uncomfortable mix for a single variable – see step 5).
Dependent variable: glare discomfort threshold.

Step 2 - find the sample rank sums and means
(a) Arrange the scores in columns as in table 9.11. Put the scores in ascending order for easy ranking.
(b) Rank the scores, *irrespective of sample*, remembering to give low scores low ranks.
(c) Find the sample rank sums (R_j) by totalling the ranks in each column. In our case $R_1 = 598, R_2 = 234$ and $R_3 = 158$.
(d) Check that the ranking has been carried out correctly. The total sum of ranks should equal $N(N+1)/2$, where N is the total sample size ($n_1 + n_2 + \ldots n_k$).

$$N(N+1)/2 = \Sigma R_j$$

$$(44)(45)/2 = 598 + 234 + 158$$

$$990 = 990$$

(e) Note the sample sizes ($n_1 = 24$, $n_2 = 10$ and $n_3 = 10$) and compute the sample rank means ($\bar{R}_j = R_j/n_j$). In our case, $\bar{R}_1 = 24.92, \bar{R}_2 = 23.4$ and $\bar{R}_3 = 15.8$.

Step 3 -- interpret the rank means
The rank means for partially and normally sighted children are very similar ($\bar{R}_1 = 24.92$ and $\bar{R}_2 = 23.4$). However, glare thresholds do appear to be substantially greater than for the adults.

Step 4 - compute and evaluate our key statistic (H)
(a) Find H:

$$H = \frac{12}{N(N+1)} \Sigma \frac{R_j^2}{n_j} - 3(N+1)$$

$$= \frac{12}{(44)(45)} \left(\frac{598^2}{24} + \frac{234^2}{10} + \frac{158^2}{10} \right) - 3(45)$$

$$= 138.619 - 135 = 3.62$$

Table 9.11   Computational layout of example in section 9.6

Partially sighted children		Fully sighted children		Fully sighted adults	
Score	Rank	Score	Rank	Score	Rank
119	1	321	2	562	5
370	3	728	8	593	6
557	4	1167	12	893	9
640	7	2028	23	950	10
1100	11	2117	24	1182	13
1220	15	2267	26	1195	14
1310	16	2550	28	1497	19
1460	17.5	3090	33	1950	22
1460	17.5	5450	36	2258	25
1570	20	8500	42	4100	35
1657	21				
2367	27				
2583	29				
2683	30				
2800	31				
2850	32				
3200	34				
7150	37				
7217	38				
8000	39				
8500	42				
8500	42				
8500	42				
8500	42				
R_j	598		234		158
N_j	24		10		10
$\bar{R}_j$	24.92		23.4		15.8

(b) Evalate H by consulting tables of chi-square (table B in the appendix) for $H = 3.62$. The number of degrees of freedom is always one less than the number of samples:

$$\text{df} = k - 1 = 3 - 1 = 2$$

The appropriate row of table B is therefore

Degrees of freedom	5%	2.5%	1%	0.1%
2	$\chi^2 \geqslant 6.0$	7.4	9.2	13.8

To be significant, our values of H must be greater than or equal to the critical χ^2 value for the appropriate level. Our value of H (3.62) is clearly too small and is therefore not significant ($p > 0.05$).

(c) Correct for ties:

In this example there are only two ties, but the procedure for correcting will be illustrated none the less. Our two ties are at ranks 17.5 (length 2) and 42 (length 5). Now find T:

$$T = 1 - \frac{\Sigma(t^3 - t)}{N^3 - N}$$

$$= 1 - \frac{(2^3 - 2) + (5^3 - 5)}{44^3 - 44}$$

$$= 1 - \frac{126}{85\,140} = 0.9986$$

Use table D in the appendix to find values for $(t^3 - t)$ or $(N^3 - N)$. We can now find H', the corrected value of H:

$$H' = H/T = 3.62/0.9986$$

$$= 3.62$$

i.e. no change.

Step 5 – draw conclusions

Because our result is not significant, we cannot reject H_0. This means that we have not adduced any support from the data for the suggestion that partially sighted children, fully sighted children and fully sighted adults differ in terms of glare discomfort thresholds.

When analysed in this way the question confuses two issues, one of the effect of visual deficiency and the other of the effect of age. It might have been better to perform two separate tests, (a) comparing samples 1 and 2 and (b) comparing samples $(1 + 2)$ and 3. See exercise 9.6.

9.7 Approximate test (nonspecific) for a $k \times Q$ frequency table

Many large-scale investigations in the life and social sciences generate frequency tables which need analysing. Traditionally this has been achieved using a chi-square technique, which assumes that the categories of the dependent variable cannot be ordered in any natural way despite the fact that these categories often are obviously ordered. In the example to be discussed, various kinds of office workers were asked to rate their current visual discomfort on a five-point scale from 1 (no visual discomfort) to 5 (very bad visual discomfort). These five categories of the dependent variable are clearly ordered. A traditional chi-square (homogeneity) test (see section 9.8) would completely ignore the order information and constitute an insensitive and possibly meaningless statistical test. The point is discussed by Marascuilo and McSweeney (1967, p. 137) and various solutions proposed. Exercise 9.4 allows the user to compare chi-square homogeneity analyses with the analysis of variance by ranks technique. When we have only two categories of the dependent variable, the nonspecific test based on analysis of variance by ranks is the same as Pearson's chi-square homogeneity test. This equivalence is proved in section 14.3. The distinction between the two approaches is therefore only crucial when we have three or more categories of the dependent variable.

Example

The following analysis is taken from a much larger investigation by Istance and Howarth (1983) into the effect of visual display units (VDUs) on visual discomfort experienced by office workers. Among many other questions, a large group

Table 9.12 Frequency table summary of visual discomfort reports for 171 office workers belonging to four different occupational groups: (a) WP, word processor operators (b) DP, data preparation staff (c) TY, typists and (d) GC, general clerical staff. All reports are from Friday afternoon (example in section 9.7)

Discomfort scale	Staff groups			
	WP	DP	TY	GC
1	13	15	22	26
2	10	13	19	20
3	3	8	8	3
4	2	5	3	0
5	0	0	0	1

of office workers were asked to rate their current degree of discomfort in the eyes on a five-point scale from 1 (no visual discomfort) to 5 (very bad visual discomfort). They did this for five consecutive working days (a) on arrival at work and (b) just before leaving work. The data given in table 9.12 are the replies given on Friday afternoon only. There are four groups of workers: (a) word processor operators (WP) (b) data preparation staff (DP) (c) typists (TY) and (d) general clerical staff (GC). Both WP and DP groups worked routinely with VDUs whereas TY and GC groups did not. Are there are differences between the groups in the amount of reported eyestrain, on this occasion?

Step 1 - formulate the problem
(a) Specify the hypotheses:
    Null hypothesis (H_0): visual discomfort at the end of the working day does not differ between the four occupational groups studied.
    There is no specific alternative hypothesis. The general alternative is that H_0 is false.
(b) Specify the variables:
    Independent variable: occupational group (four different groups).
    Dependent variable: visual discomfort measured in terms of five ordered categories; higher scores stand for more discomfort.

Step 2 - find the same rank sums and means
(a) Draw up the frequency table as in table 9.13. Find the category (t_q) and the cumulative category totals (Σt_q).
(b) Find the shared ranks for each category using the expression

$$r = \Sigma t_q - t_q/2 + 0.5$$

In our example we have

$$r_1 = 76 - 76/2 + 0.5 = 38.5$$
$$r_2 = 138 - 62/2 + 0.5 = 107.5$$
$$r_3 = 160 - 22/2 + 0.5 = 149.5$$
$$r_4 = 170 - 10/2 + 0.5 = 165.5$$
$$r_5 = 171 - 1/2 + 0.5 = 171$$

(c) Find the sample rank sums for each sample by multiplying the shared rank for each category by the number of scores in that category and adding:

$$R_1 = 13(38.5) + 10(107.5) + 3(149.5) + 2(165.5) = 2355$$
$$R_2 = 15(38.5) + 13(107.5) + 8(149.5) + 5(165.5) = 3998.5$$
$$R_3 = 22(38.5) + 19(107.5) + 8(149.5) + 3(165.5) = 4582$$
$$R_4 = 26(38.5) + 20(107.5) + 3(149.5) + 1(171) = 3770.5$$

Table 9.13 Computational layout for example in section 9.7

Discomfort scale	Word processor WP	Data preparation DP	Typists TY	General clerical	t_q	Σt_q	r_q	$(t^3 - t)$
1	13	15	22	26	76	76	38.5	438 900
2	10	13	19	20	62	138	107.5	238 266
3	3	8	8	3	22	160	149.5	10 626
4	2	5	3	0	10	170	165.5	990
5	0	0	0	1	1	171	171	0
							Σ	688 782
R_j	2355	3998.5	4582	3770.5				
n_j	28	41	52	50				
$\bar{R}_j$	84.1	97.52	88.12	75.41				

(d) Check that the ranking has been carried out correctly:

$$N(N + 1)/2 = \Sigma R_j$$

$$(171)(172)/2 = 2355 + 3998.5 + 4582 + 3770.5$$

$$14\,706 = 14\,706$$

(e) Note the sample sizes ($n_1 = 28, n_2 = 41, n_3 = 52$ and $n_4 = 50$) and compute the sample rank means ($\bar{R}_j = R_j/n_j$). In our example $\bar{R}_1 = 84.1$, $\bar{R}_2 = 97.52$, $\bar{R}_3 = 88.12$, $\bar{R}_4 = 75.41$.

Step 3 - interpret the rank means
There is some variation among the rank means. The general clerical workers report least discomfort, the data preparation staff report the most, and word processor operators and typists are between the two.

Step 4 - compute and evaluate our key statistic H
(a) Find H:

$$H = \frac{12}{N(N + 1)} \Sigma \frac{R_j^2}{n_j} - 3(N + 1)$$

$$= \frac{12}{(171)(172)} \left(\frac{2355^2}{28} + \frac{3998.5^2}{41} + \frac{4582^2}{52} + \frac{3770.5^2}{50} \right) \quad (9.5)$$

$$- (3)(172)$$

$$= 4.65$$

(b) Evaluate H by consulting tables of the chi-square distribution (table B in the appendix) for $H = 4.65$. The appropriate value for the degrees of freedom parameter is always one less than the number of samples:

$$df = k - 1 = 4 - 1 = 3$$

The appropriate critical values for a four-sample problem are therefore

Degrees of freedom	5%	2.5%	1%	0.1%
3	$\chi^2 \geqslant 7.8$	9.3	11.3	16.3

To be significant H must be equal to or greater than the critical value of χ^2 for the appropriate level. Our value of H (4.65) is clearly too small and is therefore not significant ($p > 0.05$).

(c) Correct for ties:

A correction for ties usually has a large effect in frequency table examples where ties are often extensive. Find the values for $(t^3 - t)$ and enter them alongside the frequency table. Find T:

$$T = 1 - \frac{\Sigma(t^3 - t)}{N^3 - N} = 1 - \frac{688\,782}{171^3 - 171}$$

$$= 0.8622$$

Compute H':

$$H' = H/T = 4.65/0.86 = 5.39$$

Although much increased, our value of H still does *not* exceed the critical χ^2 value at the 5 per cent level.

Step 5 - draw conclusions

Because our result is not statistically significant ($p > 0.05$) we are unable to reject H_0. In simple terms, the data given in table 9.12 do not support the idea that there are any differences between our occupational groups in terms of visual discomfort at the end of the Friday afternoon on which the data were collected.

These data form only a small subset of the total data, and hasty conclusions should not be based on a single afternoon's study. We shall return to the total data set in section 10.7.

9.8 Chi-square homogeneity test for $k \times Q$ frequency table

Occasionally we find a $k \times Q$ frequency table where both k and Q are greater than two and where neither variable is composed of categories which can be naturally ordered. We must then leave analysis of variance by ranks to use a quite different procedure due to Karl Pearson. This is one of the oldest and most popular tests in the history of statistics; in common parlance it goes under the name of the 'chi-square test'. It is confusing name since there are many quite different statistical tests which use the chi-square distribution to assess the significance level of results. To avoid confusion, the expression 'chi-square homogeneity test' is used throughout this book.

The basic principle underlying the homogeneity test is simple. The values in the frequency table to be analysed are called the 'observed frequencies' and symbolized by O_{qj}. A second table is also created which indicates what the frequencies should be if H_0 is true. This is called the table of 'expected frequencies' and each element is characterized by E_{qj}. When H_0 is true, the corresponding observed and expected frequencies will be approximately equal and the following will be true:

$$\frac{O_{qj}}{E_{qj}} \simeq 1$$

and

$$\frac{O_{qj}^2}{E_{qj}} \simeq O_{qj}$$

where $\simeq$ means approximately equal. If we calculate this ratio for each pair of frequencies we expect the following to be true:

$$\sum_q \sum_j \frac{O_{qj}^2}{E_{qj}} \simeq \sum_q \sum_j O_{qj} = N$$

where N is the total of all the observed frequencies. When H_0 is false, this expectation is invalidated and we expect a large value for

$$H = \sum_q \sum_j \frac{O_{qj}^2}{E_{qj}} - N$$

The question arises as to how large H must be before we can reject H_0 with any confidence. Fortunately H is distributed as chi-square with $(k-1)(Q-1)$ degrees of freedom, where k is the number of columns and Q is the number of rows. This means that we can simply consult tables of the chi-square distribution to discover how often such a large value of H might occur if H_0 were, in fact, true.

The table of expected frequencies is computed using an assumption of homogeneity, i.e. that the scores in each column will be distributed across the rows in exactly the same proportions. Using the notation adopted in this book (primarily for rank sum analysis) we have

$$E_{qj} = \frac{t_q n_j}{N}$$

It is important to bear in mind that this homogeneity test is unaffected by swapping around the rows or the columns of the table because the ordering of the variables is ignored. This is the crucial difference between a homogeneity test and a rank sum analysis. If the variables could be ordered then this potential information is simply lost by the use of a homogeneity test and a less powerful analysis results. However, if the variables cannot be ordered in some natural way then it is simply a nonsense to use rank sum analysis, which assumes that ordering to be present. The chi-square homogeneity test is not a rank sum test, but it is included in this book to cope with a simple experimental design which rank sum analysis cannot handle. Note that when we have only two categories of the dependent variable both techniques yield the same result. Exercise 9.5 provides the reader with an oppportunity to check this rule.

Example

Cohen (1979) asked a group of students whether they would accept an offer of a (hypothetical) operation which would eliminate the need to sleep. The offer was made on the supposition that the operation would be free of charge, free of discomfort and free of risk, and that it would eliminate sleepiness and any of the negative consequences which normally follow upon not sleeping. Approximately 60 per cent of the students said that they would not accept the offer and their reasons were classified into three groups:

(1) negative reasons, e.g. need to escape from the burdens and miseries of life
(2) positive reasons, e.g. because sleeping and dreaming are pleasant
(3) operation invalid, i.e. rejection of the premise that the operation would not directly affect mental/physical health.

In table 9.14 Cohen has divided the student group into four categories: (a) male psychology students (2) female psychology students (3) male student outpatients (4) female student outpatients. Do they differ in terms of their reasons for rejecting the operation?

Step 1 - formulate the problem
(a) Specify the hypotheses:
Null hypothesis (H_0): the student groups sampled to not differ in terms of the type of reason given for rejecting the operation.
There is no specific alternative hypothesis. The general alternative is that H_0 is false.

Table 9.14 Data for example in section 9.8

| Respondents: | Reasons for rejecting operation | | |
	Negative	*Positive*	*Invalid*
Male psychology students	11	25	4
Female psychology students	17	16	9
Male outpatients	5	1	4
Female outpatients	10	6	0

(b) Specify the variables:
For this test it is not necessary to distinguish dependent and independent variables:

variable 1: student group (four groups used).
variable 2: reason for rejecting the operation (positive/negative/invalid).

Step 2 - find the expected frequencies
(a) Draw up the observed frequency table as in table 9.15, leaving adequate space for the expected frequencies and other computations.
(b) Find the row totals (t_q) and the column totals (n_j) and the grand total (N).
(c) Compute an expected frequency for each observed frequency using the row, column and grand totals:

$$E_{qj} = t_q n_j / N$$

Table 9.15   Computational layout for example in section 9.8 (*O* observed, *E* expected

		Negative $(q = 1)$	Positive $(q = 2)$	Invalid $(q = 3)$	Total t_q
Psychology students					
Male $(j = 1)$	O	11	25	4	40
	E	15.93	17.78	6.30	40.01
	O^2/E	(7.60)	(35.15)	(2.54)	(45.29)
Female $(j = 2)$	O	17	16	9	42
	E	16.72	18.67	6.61	42.00
	O^2/E	(17.28)	(13.71)	(12.25)	(43.24)
Outpatients (students)					
Male $(j = 3)$	O	5	1	4	10
	E	3.98	4.44	1.57	9.99
	O^2/E	(6.28)	(0.22)	(10.19)	(16.69)
Female $(j = 4)$	O	10	6	0	16
	E	6.37	7.11	2.52	16.00
	O^2/E	(15.70)	(5.06)	(0.00)	(20.76)
Totals n_j	ΣO	43	48	17	108
	ΣE	43	48	17	108
	$\Sigma O^2/E$	(46.86)	(54.14)	(24.98)	(125.98)

and enter these in table 9.15 beneath the observed frequency. For example,

$$E_{1,1} = t_1 n_1 / N = (43)(40)/108 = 15.93$$

(d) Check the accuracy of your calculations by finding the row and column totals of the expected frequencies. These totals should be the same as the row and column totals for the observed frequencies.

For example, the first column total for the expected frequencies is

$$15.93 + 16.72 + 3.98 + 6.37 = 43.00$$

Very small discrepancies due to rounding errors can be ignored.

(e) Compute O^2/E for each observed frequency. For example,

$$O^2_{1,1}/E_{1,1} = 11^2/15.93 = 7.60$$

and enter these values in the table.

Step 3 - interpret the frequency table
It is very difficult to interpret large frequency tables in a meaningful way, although it is tempting to look for large discrepancies between observed and expected frequencies. It can also be dangerous since it is tempting to interpret a statistically significant result as establishing the reliability of individual discrepancies. Such interpretations are quite invalid. A significant result simply means that the frequency table *as a whole* is an unlikely configuration given H_0 true. It is reasonable, however, to note the direction and size of the major discrepancies in order to develop a general feeling for the results.

Step 4 - compute and evaluate our key statistic H
(a) Find the sum of all the O^2/E values:

$$\sum_q \sum_j \frac{O^2_{qj}}{E_{qj}} = 7.60 + 17.28 + \ldots + 10.19 + 0.00 = 125.98$$

A useful way of checking this result is to find the row and column totals for O^2/E in table 9.15, and then the sum of both of these totals should agree.

Now find H:

$$H = \sum \sum O^2/E - N$$
$$= 125.98 - 108 = 17.98$$

(b) Evaluate H by consulting tables of the chi-square distribution (table B in the appendix) for $H = 17.98$. The appropriate value for the degrees of freedom parameter is

$$df = (Q - 1)(k - 1) = (3)(2) = 6$$

where Q is the number of rows and k is the number of columns.

The term 'degrees of freedom' is difficult to explain in introductory texts, but the topic is discussed in section 3.5. In our example we might say that when the row, column and grand totals are specified, then only $(Q - 1)(k - 1)$ scores in the body of the table are free to vary, since the remaining scores will be constrained by the need to sum appropriately in their respective rows and columns. We need to specify the degrees of freedom only because we need to choose the appropriate row of table B. In our example this row is

Degrees of freedom	5%	2.5%	1%	0.1%
6	$\chi^2 \geqslant 12.6$	14.5	16.8	22.5

To be significant, our value of H must be greater than or equal to the critical value at the appropriate level. Our value of H (17.98) is significant at better than the 1 per cent level ($p < 0.01$).

Step 5 - draw conclusions
Because our result is statistically significant ($p < 0.01$) we may reject H_0 and conclude that some differences do exist among the groups of respondents in terms of the reasons given for rejecting the sleep removing operation. No more specific conclusions may be drawn using the results of this analysis. Inferences concerning specific differences observed between the groups need to be tested using more specific tests. At this stage it is certainly unacceptable to point out interesting features in the table and imply that these are reliable features on the basis of the overall significance value obtained from the analysis.

Exercises

9.1 Any two sets of lambda coefficients which are linearly related to each other (see section 3.2) should produce the same result in terms of level of statistical significance. In the example in section 9.3 we used the following set of coefficients: $\lambda_j = 3, 1, 2$. Two other sets are possible candidates, namely (a) $\lambda_j = 1, -1, 0$ (b) $\lambda_j = 2, 0, 1$. Repeat the relevent part of the analysis of section 9.3 using these two sets of coefficients and show that the same result is obtained. Draw a graph using pairs of corresponding coefficients and show that they can be linked by a straight line.

9.2 In the example in section 9.4 either 'headache reports' or 'visual effects reports' could have been chosen as the independent variable. Repeat the analysis using 'headache reports' as the independent variable.

9.3 An investigation into hypnosis as a form of obedience behaviour (Hunt, 1979) was introduced in the example in section 7.2 as a two-group experiment. In fact she ran four different groups. In each condition a volunteer was hypnotized and given 12 different tests to measure the depth of the trance (Stanford hypnotic susceptibility scale). The final score was a count of the number of tests producing a positive response for that subject. Volunteers were alone when hypnotized in group A but were accompanied by a stooge in the other three groups. The stooge behaved in a manner which was scripted by the experimenter. In group B (obedient stooge) he pretended to produce positive responses to each test. In group C (defiant stooge) he cooperated up to test 6 then angrily left the room saying that the whole thing was ridiculous. In group D (defiant/acquiescent) he behaved just like the defiant stooge but agreed to remain quietly in the room after item 6 rather than leaving. Do the data given in table 9.16 support Hunt's theory that hypnotic susceptibility can be influenced by examples of obedience and defiance in a third party? Hunt used a nonspecific test for independent samples. In addition we might also explore the reasonable prediction which orders three of the groups as follows:

obedient stooge > alone > defiant stooge

Table 9.16 Hypnotic trance depth scores for 40 subjects (Hunt, 1979)

Alone	Obedient	Defiant stooge	Acquiescent stooge
12	12	6	12
8	12	6	8
9	9	7	8
8	11	8	8
10	10	6	12
7	11	6	9
12	12	6	11
12	8	12	8
8	8	9	12
5	12	6	12

For this purpose, delete group D from the analysis altogether and do a specific test for independent groups.

9.4 Marascuilo and McSweeney (1967, p. 136) analyse the data in table 9.17 using a chi-square homogeneity test, but agree that this was not the optimum procedure. Do *both* a chi-square homogeneity test *and* a specific (analysis of variance by ranks) test for trend on this data and compare the results. The data represent reports from students concerning their perception of the difficulty of a 30-item test. This difficulty was rated on a three-point scale (very easy, easy and very hard). The students had been given different amounts of practice trials prior to the test and it was expected that those who had received most practice would perceive least difficulty. (The numbers have been scaled down from the original example in order to emphasize the difference between the two tests, but the proportions have been approximately preserved.)

9.5 100 consecutive clients visiting a community counselling centre were classified by sex and by the main area of presenting problem (table 9.18). Do men differ from women with respect to the presenting problem? Note that rank sum analysis can only be used on this problem if 'presenting problem' is used as the independent variable. In this case, sex is the dependent vari-

Table 9.17 Fictitious data take from Marascuilo and McSweeney (1967, p. 136). See exercise 9.4

		Treatment condition			
		20 pretrials	15 pretrials	10 pretrials	0 pretrials
Perceived difficulty	Very easy	9	7	1	0
	Average	2	3	6	3
	Very hard	0	3	6	9

Table 9.18 Frequency of various kinds of problem presented to a community counselling centre (data supplied by Kathy Parkes, Department of Experimental Psychology, Oxford University)

		Marital	Personal relation- ships	Depression	Work problems	Life crises	Family	Other
Sex	Male	8	10	6	7	3	3	4
	Female	15	8	12	5	7	11	1

able and has only two categories, and both rank sum and homogeneity analysis should yield the same result. Attempt both methods of analysis and compare them for speed and ease of computation.

NB   Before starting your analysis, make an intelligent decision on what to do with the 'other' category. Will you keep it or delete it?

9.6 Reanalyse the data in the example in section 9.6 using a specific test to compare both groups of children with the group of adults. Use the coefficients $\lambda_j = +1, +1, -1$. To compare normal and partially sighted children with each other, remove adults from the analysis altogether, create new rankings and perform a test for two independent samples (see chapter 7).

10 Three or more related samples

10.1 Introduction

We say that a design has related samples when the total set of data points is subdivisible into m sets of scores which are matched in some way. The independent samples design discussed in chapter 9 is not subdivided in this way because the scores are not systematically linked to each other in any manner. The most obvious example of a related set of scores occurs with the repeated measures design, where each subject is studied under each of a variety of experimental conditions. In this case, all of the scores from a single subject constitute a matched set (or block) and we have as many matched sets as we have subjects. (If *all* of the scores come from one subject then the design is probably a single-block design; see chapter 9.) In another design we might use different subjects for each treatment but match them into subgroups in terms of some variable such as weight, intelligence or educational history. In this case, all of the subjects within a given weight (or intelligence) class constitute a matched set or block of scores. The statistical analysis of related samples concentrates on the relative values of the scores *within* the sets, i.e. the sets are kept separate during the calculations.

The advantages of the related samples analysis are immediately obvious. It is clearly advantageous to compare like with like when evaluating a treatment. For example, when studying the effects of alcohol on performance of a skilled task such as driving it would be advantageous to have subjects with all the same bodyweight. Variations of bodyweight would confuse the issue since heavier individuals are more tolerant of a fixed drug dosage. Although it is not always possible to have a large group of subjects with the same weight, it is quite feasible to subdivide a large group of variously heavy subjects into small sets of similarly heavy individuals. In this example weight is the matching variable and is not explicitly studied in the analysis.

When we match sets of data points in this way, we increase the efficiency of the analysis, so that fewer data points are needed to achieve the same degree of statistical significance. The gain in efficiency depends very much on the matching variable that is chosen. The more clearly the matching variable influences the effect under study, the greater the gain in efficiency. In our example, weight clearly influences the effect of a fixed dose of alcohol on performance of the skilled task. Social class is less influential and its use as a matching variable would not have improved the efficiency of the test by very much, if at all.

The disadvantage of matching is that it involves an extra measuring operation. Clearly it is not too onerous to measure, say, age or weight. However, the assessment of the intelligence of each subject could be slow and expensive. The cost in time and effort in making these measures may not justify the benefits of increased efficiency. It might be cheaper to run a few more subjects and use an independent samples design. In some circumstances, however, the treatments applied to the subjects may be very expensive or dangerous, such as new drugs or surgical operations. When this is the case, maximum efficiency must be sought even at great expense and many matching variables may be used simultaneously to increase the similarity of the subjects within the matched sets.

The related samples design can sometimes be used to rescue an independent samples design which has run into trouble with unequal pretest scores. This problem occurs when a number of subjects are assigned at random to the treatments (independent samples design) but the groups are later discovered to have different average scores on a pretest (i.e. a test administered before the treatment). These initial differences can complicate the evaluation of any differences which are present after the treatments are applied. One solution to this common problem is to restrict the analysis to difference or change scores which measure the extent of the effect of the treatment for each subject. However, scaling considerations may render change scores of dubious value. When this occurs we can use the pretest score as the matching variable and group the subjects into sets of individuals with similar pretest scores. The analysis can then be completed using the post-test scores.

Repeated measures

The most popular application of the related samples design comes in the form of repeated measures on a single subject. The design is clearly attractive for two reasons. Firstly, it is most economical to use a single subject many times rather than once. Secondly, each subject is his own best control, being more like himself than any other person, however carefully matched. There are dangers with this method, however, and it must be used with extreme caution. As subjects move from treatment to treatment they change in many ways. They may become more tired, more confident, more practised, more careless, more familiar with the hypothesis under test, more irritated etc. These effects can often produce greater effects on test scores than the treatments themselves. In such circumstances the gain in efficiency achieved by matching the scores (by using the same subject for each) is offset by a reduction in efficiency by introducing these extraneous sources of variation in the test scores. The data become contaminated with uncontrolled influences. This inefficiency leads to an increase in

H

the number of subjects needed to achieve the same level of statistical significance as a related samples design using different subjects in the same related sets. In this way, the economy of using the same subject repeatedly can be partially (or even completely) offset by the need to use more subjects.

The worst problem associated with repeated measures concerns the systematic effects of these new unwanted influences such as fatigue and practice. These systematic effects can be used as alternative explanations for the differences between the treatments so that, even if statistical significance is achieved, it may be unclear whether the differences between the samples are caused by the treatments or by such effects as practice and fatigue. The experienced researcher deals with this problem by counter-balancing (or randomizing) the order of the application of the treatments to each subject, and this can be effective when the number of subjects is reasonably large. The data are always open to challenge on this point, however, and the researcher must be sure _before collecting the data_ that he has worked out all of his counter-arguments to such challenges. Repeated measures designs require considerable vigilance.

Power/efficiency

The transformation from raw scores to ranks entails a loss of information which, in turn, results in a loss of efficiency compared with some ideal test. Loss of efficiency here means that more data points are required to achieve the same level of statistical significance when the null hypothesis is, in fact, false. When data points are cheap and conveniently acquired, efficiency is not a serious problem. Nor is it a problem when the data already to hand are clearly statistically significant or nonsignificant. When H_0 is false, a more efficient test applied to the data will simply make a small increase in statistical significance; this means that efficiency is only a problem when the data are _almost but not quite statistically significant_. Furthermore, there are many situations where hardly any information is lost by the use of ranks. This is the case, for example, where the original scores are ranks by another name (such as preference orderings) or belong to very short discrete scales (such as five-point scales).

The loss of information following rank transformation is greatest when the number of scores to be ranked is small. Unfortunately, the independent samples design increases this loss of information effect by subdividing the total data set into a number of smaller matched sets and the ranking takes place _within these small sets_. It is ironic that the matched set design might be chosen because it is an efficient design but this efficiency is then lost by the ranking procedure. The problem is most acute when there are only two samples and one score per sample per set (discussed in section 8.1). It becomes slowly less of a problem as the number of scores in the matched set increases. Until recently there have been few popular techniques which could process designs with multiple scores in each cell of the design or unequal cell frequencies. As a consequence, multiple scores were often combined (or averaged) to form a single score to fit the existing techniques, with a resulting loss of efficiency from having few scores to rank. The techniques given in this chapter do tolerate large and unequal cell frequencies (at some computational expense) and significantly increase the efficiency and extend the range of nonparametric techniques available for use by the unaided researcher.

A   Tests against a specific alternative hypothesis

10.2   Exact test for trends

This simple and often used test is due to Page (1963). It can only be used for designs

(a) with three or four samples
(b) with small numbers of blocks (matched sets)
(c) with only one score per sample per block and
(d) with the coefficients $\lambda_j = 1, 2, 3$ or $\lambda_j = 1, 2, 3, 4$.

However, it turns out to be frequently useful. It is especially to be recommended for its speed of application and ease of interpretation of the results.

Example

These data are taken from an undergraduate laboratory class where students were asked to rate, in order of *preference*, three methods of communicating a message:

(a) by letter
(b) by phone
(c) face to face.

Sixteen different messages were rated, one each by 16 groups of students. The data in table 10.1 are taken from the responses to the first five messages:

Table 10.1   Mean preference rating for three kinds of communication methods for five kinds of message

Message	Letter	Phone	Face to face
(a)	2.3	2.2	1.4
(b)	2.2	1.9	1.5
(c)	2.5	1.9	1.2
(d)	2.6	2.0	1.3
(e)	1.7	1.6	2.5

(a) reporting to someone, that you know, the death of their parent
(b) giving a classmate their examination results, in which they failed
(c) having it out with a friend who is spreading false rumours about you
(d) reprimanding a peer who has failed to fulfil a promise
(e) refusing an unreasonable request from someone who wants help.

The values in the table are mean preferences (1 = most preferred). We hypothesize that people will prefer more immediate (and hence controllable) communications for these delicate messages.

Step 1 - formulate the problem
(a) Specify the hypotheses:
Null hypothesis (H_0): there is no difference in preference for the three means of communication.
Specific alternative hypothesis (H_1): for these delicate messages people will most prefer face-to-face communication and least prefer the use of letters.

NB   For this exact test the specific alternative hypothesis must indicate a simple ordering of the three groups. The only coefficients we can use are $\lambda_j = 1, 2, 3$. *No other coefficients are permitted.*
(b) Specify the variables:
Independent variable: three types of communication varying in degree of personal contact.
Dependent variable: preference rank order. Each value in table 10.1 represents the mean preference across a group of subjects.
(c) Order the samples:
Rearrange the table so that the expected rank means are in ascending order from left to right. They will then correspond to the lambda coefficients which are always $\lambda_j = 1, 2, 3$. We expect more intimate forms of communication to be preferred for these difficult messages, and we rearrange the groups as in table 10.2.

Step 2 - find the sample rank sums and means
(a) Rank the scores *in each block separately*, remembering to assign low ranks to low scores.
(b) Find the sample rank sums (R_j) by totalling the ranks in each column. In our case $R_1 = 7, R_2 = 9, R_3 = 14$.
(c) Check that the ranking has been carried out correctly by finding ΣR_j, the sum of the rank sums:

$$mk(k + 1)/2 = \Sigma R_j$$

$$(5)(3)(4)/2 = 7 + 9 + 14$$

$$30 = 30$$

Table 10.2 Computational layout for example in section 10.2

Message	Face-to-face		Phone		Letter	
	Score	Rank	Score	Rank	Score	Rank
(a)	1.4	1	2.2	2	2.3	3
(b)	1.5	1	1.9	2	2.2	3
(c)	1.2	1	1.9	2	2.5	3
(d)	1.3	1	2.0	2	2.6	3
(e)	2.5	3	1.6	1	1.7	2
R_j		7		9		14
λ_j		1		2		3
$\lambda_j R_j$		7		18		42

(d) Note the sample sizes ($n = 5$) and compute the sample rank means ($\bar{R}_j = R_j/n_j$). In our case, $\bar{R}_1 = 1.4$, $\bar{R}_2 = 1.8$ and $\bar{R}_3 = 2.8$.

Step 3 - interpret the rank means
The rank means are in ascending order ($\bar{R}_1 < \bar{R}_2 < \bar{R}_3$), as predicted, with face-to-face communication the most preferred and the use of letters least preferred.

Step 4 - compute and evaluate our key statistic (L)
(a) Find L:

$$L = \Sigma \lambda_j R_j = (1)(7) + (2)(9) + (3)(14) = 67$$

(b) Evaluate L by consulting table M (in the appendix) for $L = 67$ and $m = 5$. The critical row for our problem is

k	m	5%	2.5%	1%	0.1%
3	5	$L \geqslant 67$	68	68	70

To be significant our value of L must be greater than or equal to the critical value for the appropriate level. Our value of L is, therefore, significant at the 5 per cent level ($p < 0.05$).

Step 5 - draw conclusions
Because our result is significant ($p < 0.05$) we may reject H_0 (see step 1) in favour of H_1. In simple terms, the data support the hypothesis that people prefer

immediate, such as face-to-face, communication when dealing with socially delicate messages.

10.3   Approximate test for trends and contrasts

This approximate test is restricted to designs with only one score per sample per block, but it will cope with any number of samples, any number of blocks and any set of coefficients. The approximation to the normal distribution is very good indeed even for quite small sample sizes. When these are very small, however, the exact test should be used (section 10.2) if possible. For more than one score per cell, when all cell frequencies are equal, see the supplementary note at the end of this section.

Example

Parsons (1982) meausured the vibration perception threshold of 12 young male subjects sitting on a hard seat which vibrated at 16 Hz. The vibration was switched on for brief periods varying from 0.0625 to 4 seconds. The vibration thresholds (lowest noticeable acceleration of the chair) was measured separately for each of seven different periods within this range. It was expected that the thresholds, given in table 10.3, would be lower for longer durations of the vibration.

Table 10.3   Perception threshold (m/s^2) for vibration at 16 Hz for seven durations of the stimulus

Subject	Stimulus duration (s)						
	0.0625	0.125	0.25	0.5	1	2	4
1	0.022	0.014	0.018	0.018	0.012	0.031	0.014
2	0.037	0.024	0.014	0.021	0.014	0.034	0.014
3	0.033	0.036	0.043	0.028	0.017	0.014	0.018
4	0.048	0.031	0.038	0.036	0.033	0.021	0.025
5	0.026	0.031	0.028	0.019	0.021	0.018	0.023
6	0.035	0.046	0.020	0.017	0.016	0.015	0.018
7	0.050	0.038	0.030	0.032	0.029	0.024	0.018
8	0.025	0.027	0.021	0.018	0.022	0.020	0.015
9	0.023	0.040	0.031	0.027	0.024	0.025	0.015
10	0.035	0.035	0.030	0.030	0.031	0.018	0.022
11	0.061	0.041	0.054	0.033	0.018	0.032	0.018
12	0.031	0.019	0.022	0.017	0.019	0.012	0.014

Step 1 - formulate the problem
(a) Specify the hypotheses:
Null hypothesis (H_0): the vibration perception threshold is unrelated to the duration of the vibration stimulus.
Specific alternative hupothesis (H_1): the vibration perception threshold will be lower for longer durations.
(b) Specify the variables:
Independent variable: stimulus duration (0.0625 to 4 s).
Dependent variable: vibration perception threshold measured in terms of the acceleration of the chair (m/s^2) at the point where the subject could just notice the vibration.
(c) Choose the coefficients:
Because we are dealing with a simple trend across seven samples, we can use the numbers 1-7. We expect the longer-duration samples to have lower thresholds, so the coefficients are arranged from right to left: $\lambda_j = 7, 6, 5, 4, 3, 2, 1$. We can use the linearly related (and, therefore, equivalent) coefficients $\lambda_j = 3, 2, 1, 0, -1, -2, -3$ with some reduction in computational load by using Marascuilo and McSweeney's (1967) formula. Both methods will be illustrated in the following.

Step 2 - find the sample rank sums and means
(a) Arrange the scores as in table 10.4. All of the raw scores have been increased by a factor of 100 to remove the decimal point and simplify the data layout. This change does not, of course, affect their ranks.
(b) Rank the scores *in each block separately*, remembering to assign low ranks to low scores.
(c) Find the sample rank sums (R_j) by totalling the ranks in each column.
(d) Check that the ranking has been carried out correctly by finding ΣR_j, the sum of the rank sums:

$$mk(k + 1)/2 = \Sigma R_j$$
$$(12)(7)(8)/2 = 71.5 + 66.5 + 60 + 45 + 36 + 31 + 26$$
$$336 = 336$$

(e) Note the sample sizes ($n = 12$) and compute the sample rank means ($\bar{R}_j = R_j/n$), which are given in table 10.4.

Step 3 - interpret the rank means
The rank means are arranged in almost perfect descending order as predicted by the specific alternative hypothesis. This means that the vibration perceptual threshold gets lower as the duration of the stimulus is lengthened.

Table 10.4 Computational layout for example in section 10.3. Note that all thresholds have been increased by a factor of 100 to simplify the layout

Subject	Stimulus duration (s)													
	0.625		0.125		0.25		0.5		1		2		4	
	Score	Rank	Score	Rank	Score	Rank	Score	Rank	Score	Rank	Score	Rank	Score	Rank
1	22	6	14	2.5	18	4.5	18	4.5	12	1	31	7	14	2.5
2	37	7	24	5	14	2	21	4	14	2	34	6	14	2
3	33	5	36	6	43	7	28	4	17	2	14	1	18	3
4	48	7	31	3	38	6	36	5	33	4	21	1	25	2
5	26	5	31	7	28	6	19	2	21	3	18	1	23	4
6	35	6	46	7	20	5	17	3	16	2	15	1	18	4
7	50	7	38	6	30	4	32	5	29	3	24	2	18	1
8	25	6	27	7	21	4	18	2	22	5	20	3	15	1
9	23	2	40	7	31	6	27	5	24	3	25	4	15	1
10	35	6.5	35	6.5	30	3.5	30	3.5	31	5	18	1	22	2
11	61	7	41	5	54	6	33	4	18	1.5	32	3	18	1.5
12	31	7	19	4.5	22	6	17	3	19	4.5	12	1	14	2
R_j		71.5		66.5		60		45		36		31		26	ΣR_j 336
$\bar{R}_j$		5.96		5.54		5.00		3.75		3.00		2.58		2.17	
λ_j		7		6		5		4		3		2		1	
$\lambda_j R_j$		500.5		399		300		180		108		62		26	$\Sigma \lambda_j R_j$ 1575.5

Step 4 - compute and evaluate our key statistics (L and Z)
(a) Find L:

$$L = \Sigma\lambda_j R_j$$
$$= (7)(71.5) + (6)(66.5) + (5)(60) + (4)(45) + (3)(36)$$
$$+ (2)(31) + (1)(26)$$
$$= 1575.5$$

Find Z:

$$Z = \frac{L - E(L)}{\sqrt{\text{var}(L)}}$$

where

$$E(L) = \tfrac{1}{2}m(k + 1)\,\Sigma\lambda_j$$
$$= (\tfrac{1}{2})(12)(8)(7 + 6 + 5 + 4 + 3 + 2 + 1) = 1344$$
$$\text{var}(L) = \tfrac{1}{12}m(k + 1)[k\Sigma\lambda_j^2 - (\Sigma\lambda_j)^2]$$
$$= \tfrac{1}{12}(12)(8)[(7)(7^2 + 6^2 + 5^2 + 4^2 + 3^2 + 2^2 + 1^2) - (28)^2]$$
$$= 1568$$

So

$$Z = (1575.5 - 1344)/\sqrt{1568}$$
$$= 5.85$$

Notes
(1) When the coefficients are the numbers 1 to k (*and only then*), Page (1963) gives us a quicker calculation procedure:

$$Z = [L - \tfrac{1}{4}mk(k + 1)^2]/\sqrt{[\tfrac{1}{144}mk^2(k^2 - 1)(k + 1)]}$$
$$= [1575.5 - \tfrac{1}{4}(12)(7)(8)^2]/\sqrt{[\tfrac{1}{144}(12)(7)^2(48)(8)]}$$
$$= 5.85$$

(2) When the coefficients are so chosen that they sum to zero (*and only then*), we can use Marascuilo and McSweeney's (1967) formula:

$$Z = L/\sqrt{[\tfrac{1}{12}km(k + 1)(\Sigma\lambda_j^2)]}$$

We can illustrate this using the following coefficients:

R_j	71.5	66.5	60	45	36	31	26	
λ_j	3	2	1	0	−1	−2	−3	
$\lambda_j R_j$	214.5	133.0	60	0	−36	−62	−78	
λ_j^2	9	4	1	0	1	4	9	$\Sigma \lambda_j^2 = 28$

$$L = \Sigma \lambda_j R_j = 231.5$$

so that

$$Z = 231.5/\sqrt{[(\tfrac{1}{12})(12)(7)(8)(28)]} = 5.85$$

(b) Evaluate Z by consulting tables of the normal distribution (table A in the appendix). The relevant values are

5%	2.5%	1%	0.1%
$z \geqslant 1.64$	1.96	2.33	3.10

To be significant at any of the four levels, Z must be equal to or greater than the corresponding critical value. Our value of Z (5.85) is greater than the highest critical value and is therefore significant at the 0.1 per cent level ($p < 0.01$).

(c) Correct for ties:

This correction is tedious for the many-block design, and often influences the result hardly at all. It is given here merely for completeness. A separate correction for ties T_i needs to be computed for each block. This is given in table 10.5. Note that, when there are no ties in a given block, $T_i = 1$ and, if all the scores are tied, $T_i = 0$. In our example $\Sigma T_i = 11.892$.

We now find the corrected values of Z:

$$Z' = Z \sqrt{\frac{m}{\Sigma T_i}} = 5.85\sqrt{(12/11.892)}$$

$$= (5.85)(1.004) = 5.87$$

which is a very slight increase on the uncorrected value and remains highly statistically significant.

Step 5 - draw conclusions

Because our result is statistically significant ($p < 0.01$), we may reject H_0 in favour of H_1. In simple terms, the data provide support for the conclusion that the vibration threshold gets lower as the stimulus duration gets longer (over the range of durations studied).

Table 10.5   Computation of T (correction for ties) in example in section 10.3

Correction for ties:

$$T_i = 1 - \frac{\Sigma(t^3 - t)}{k^3 - k}$$

Block 1: ties at 2.5 ($t = 2$) and 4.5 ($t = 2$):

$$T_1 = 1 - \frac{(2^3 - 2) + (2^3 - 2)}{7^3 - 7} = 1 - \frac{12}{336} = 0.964$$

Block 2: tie at 14 ($t = 3$):

$$T_2 = 1 - \frac{(3^3 - 3)}{7^3 - 7} = 1 - \frac{24}{336} = 0.929$$

Blocks 3 to 9: no ties: $T_i = 1$

Block 10: ties at 3.5 ($t = 2$) and 6.5 ($t = 2$):

$$T_{10} = 1 - \frac{(2^3 - 2) + (2^3 - 2)}{7^3 - 7} = 1 - \frac{12}{336} = 0.964$$

Block 11: tie at 1.5 ($t = 2$):

$$T_{11} = 1 - \frac{(2^3 - 2)}{7^3 - 7} = 1 - \frac{6}{336} = 0.982$$

Block 12: tie at 4.5 ($t = 2$):

$$T_{12} = 1 - \frac{(2^3 - 2)}{7^3 - 7} = 1 - \frac{6}{336} = 0.982$$

$$T = \Sigma T_i = 0.964 + 0.929 + 1 + 1 + 1 + 1 + 1 + 1 + 1 + 0.964$$
$$+ 0.982 + 0.982$$
$$= 11.821$$

Supplementary note. When the number of scores in each cell are all equal but in excess of 1, the following expressions may be used to compute $E(L)$ and $\mathrm{var}(L)$, ignoring ties:

$$E(L) = \tfrac{1}{2}mn(nk + 1) \Sigma \lambda_j$$

$$\mathrm{var}(L) = \tfrac{1}{12}mn^2(nk + 1)[k\Sigma\lambda_j^2 - (\Sigma\lambda_j)^2]$$

where n is the number of scores per cell.

If the coefficients are adjusted to sum to zero we have:

$$E(L) = 0$$

$$\mathrm{var}(L) = \tfrac{1}{12}kmn^2(nk + 1)\,\Sigma\lambda_j^2$$

In all other respects the computations are the same as before.

10.4   Dichotomous data (trends and contrasts)

Data very often come in the form of yes/no dichotomies, as either answers to questions or the presence or absence of some property. Cochran (1950) devised a method for dealing with such dichotomous data in a related samples design, but this was suitable for testing nonspecific hypotheses only (see section 10.10). Marascuilo and McSweeney (1967) later devised a variant of the test which permits the testing of a specific alternative hypothesis, and their formulation will be used in this section. Although the rank sums are not explicitly calculated, this is a rank sum technique and does not differ in any essential way from the example given in section 10.3. The only difference arises from a computation short cut which takes us directly to Z without passing through the ranking stage. The same kind of short cut can be seen in section 7.5 which dealt with 2×2 frequency tables. For those interested, in section 14.2 it is shown how the computational procedures are derived from general analysis of variance by ranks formulae.

To keep the computations even simpler, this procedure requires that the coefficients used should sum to zero. This will not present any special problems since any set can be transformed to a new set with the zero sum property by simply subtracting from each coefficient the average value of all the coefficients. This does not in any way change the sense of the alternative hypothesis being tested.

Example

The data in table 10.6 are taken from an investigation by Jane Fulton (1983) into the visibility of motorcyclists using various kinds of visibility aids:

(1) control
(2) headlight switched on
(3) one running light (RL) (sidelight) switched on
(4) two running lights switched on
(5) fluorescent jacket.

Table 10.6 Data and computational layout for example in section 10.4

Subject	Control	Headlight	Running lights		Fluorescent jacket	Row totals S_i.
			1	2		
1	0	1	0	0	1	2
2	1	1	1	1	1	5
3	1	1	1	1	1	5
4	1	1	1	1	1	5
5	0	1	1	0	1	3
6	0	1	1	1	1	4
7	1	1	0	1	1	4
8	1	0	1	1	0	3
9	0	1	1	1	0	3
10	0	1	0	0	0	1
11	0	1	0	0	1	2
12	0	1	1	1	1	4
13	1	1	1	1	1	5
14	0	1	1	0	1	3
15	1	1	1	1	1	5
16	0	1	1	1	0	3
17	0	1	0	0	0	1
18	1	1	0	1	1	4
19	0	1	1	1	1	4
20	0	1	1	1	1	4
21	1	0	0	1	1	3
22	0	0	0	0	0	0
23	1	1	1	1	0	4
24	1	1	1	1	1	5
25	1	1	0	1	1	4
26	1	1	1	1	1	5
27	0	1	1	1	0	3
28	0	1	0	0	1	2
29	0	0	1	0	0	1
30	0	1	1	1	1	4
31	1	1	1	0	1	4
32	1	1	1	1	1	5
33	1	1	1	1	0	4
34	1	1	1	1	0	4
35	1	1	1	0	0	3
36	0	0	0	0	0	0
37	1	0	1	1	1	4
38	1	1	1	1	1	5

Table 10.6   continued

Subject	Control	Headlight	Running lights		Fluorescent jacket	Row totals $S_i.$
			1	2		
39	0	0	1	1	1	3
40	0	1	1	0	1	3
41	0	0	1	1	1	3
42	0	1	1	1	1	4
43	0	1	1	1	1	4
44	0	1	1	1	1	4
45	0	1	1	1	0	3
46	0	1	0	1	0	2
47	1	1	0	0	0	2
48	0	1	1	1	0	3
49	1	0	0	1	1	3
50	0	1	0	1	1	3
51	0	0	0	0	1	1
52	0	1	1	0	0	2
53	1	1	0	1	1	4
54	1	1	1	1	1	5
55	0	1	0	0	0	1
56	0	0	0	0	0	0
57	0	0	0	0	1	1
58	1	1	1	1	1	5
59	0	0	1	1	1	3
60	1	1	0	1	1	4
61	0	0	0	0	1	1
62	0	1	1	1	1	4
63	0	1	1	1	1	4
64	1	1	0	1	0	3
65	1	1	1	1	0	4
$S_{\cdot j}$	28	51	42	45	43	

Observers were sat in a parked car during daylight hours facing oncoming traffic but the view was obscured by a screen in front of the observer. When a motorcyclist was, along with other traffic, in the field of view, a hole in the screen was opened briefly and the observer was then asked to report what traffic he had seen. A score of one was given if the motorbike was reported, otherwise a zero was given. If the brightness of the visibility aid is used to predict the frequency of detections of the cyclist we might order the groups (in terms of increasing visibility) as follows: (1) control (5) fluorescent jacket (3) one running light (4) two running lights and (2) headlight. The data are for a single site only. Data for another site are given in section 10.10. Do the observed frequencies of detection agree with our expectations?

Step 1 - formulate the problem

(a) Specify the hypotheses:
Null hypothesis (H_0): the detection of motorcyclists is unaffected by any of the visibility aids.
Specific alternative hypothesis (H_1): the greater the brightness of the visibility aid, the more frequent the detections of the motorcyclist.

(b) Specify the variables:
Independent variable: brightness of the visibility aid.
Dependent variable: detection of the motorcyclist. This is a dichotomous variable (yes = 1, no = 0).

(c) Choose the coefficients:
Using the predicted order of visibility given in the preamble, we should use the following set of coefficients:

Sample	1	2	3	4	5
	control	headlight	$1 \times$ RL	$2 \times$ RL	fluorescent jacket
λ_j	1	5	3	4	2

To simplify the computations, we must adjust the coefficients to sum to zero. We can do this by subtracting 3 (the average value of the coefficients) from each value, leaving us with this set:

λ_j	-1	$+2$	0	$+1$	-1

Step 2 - find the column and row frequencies

In this design, the information contained in each score is low and large amounts of data are often required for a fair test of any hypothesis. As a consequence table 10.6 is relatively long, but the size of the table does not greatly add to the computational labour since the main effort merely involves totalling the 1s in each column. Some new notation is required to explain the computations:

S_i.   number of 1s in the ith row ($i = 1, \ldots, m$).
$S_{.j}$ the number of 1s in the jth column ($j = 1, \ldots, k$).

Before finding S_i. and $S_{.j}$ we can delete all tied lines, that is lines with all 1s or all 0s. These lines contribute nothing to the value of Z to be calculated and the same result is obtained if we keep them or delete them. To avoid the need to present table 10.6 twice, these lines have been left in, but removing them would reduce the table by 14 lines. Exercise 10.1 offers an opportunity to repeat the calculations with these lines removed.

(a) Find $S_{.j}$, $\Sigma S_{.j}$ and $\Sigma S_{.j}^2$:
The column totals ($S_{.j}$) are 28, 51, 42, 45 and 43.
Therefore

$$\Sigma S_{.j} = 28 + 51 + 42 + 45 + 43 = 209$$
$$\Sigma S_{.j}^2 = 28^2 + 51^2 + 42^2 + 45^2 + 43^2 = 9023$$

(b) Find $S_{i.}$, $\Sigma S_{i.}$, and $\Sigma S_{i.}^2$:
There are 65 rows, which might make this operation tedious. However, there are only six possible values for S_i, which means that we can simplify our work by counting the numbers of examples of each score:

$S_{i.}$	Frequency	$S_{i.}^2$
0	3	0
1	7	1
2	6	4
3	17	9
4	21	16
5	11	25

so that

$$\Sigma S_{i.} = (0)(3) + (1)(7) + (2)(6) + (3)(17) + (4)(21) + (5)(11)$$
$$= 209$$

which is the same as $\Sigma S_{.j}$. This is a useful check on the counting process, an operation which can easily go wrong with inattention.
Now find

$$\Sigma S_{i.}^2 = (0)(3) + (1)(7) + (4)(6) + (9)(17) + (16)(21) + (25)(11)$$
$$= 795$$

Step 3 – interpret the column totals
Large column totals indicate a high detection rate. In ascending order of frequency of detection we have; control (28), one running light (42), fluorescent jacket (43), two running lights (45) and headlight (51). These totals are very much in line with the prediction given above. The only exception is the fluorescent jacket sample total which should have been less than that for the single running light condition.

Step 4 - compute and evaluate our key statistics (L and Z)

(a) Find Z:

First find

$$\Sigma\lambda S_{\cdot j} = (-2)(28) + (2)(51) + (0)(42) + (1)(45) + (-1)(43)$$
$$= 48$$

and

$$\Sigma\lambda_j^2 = (-2)^2 + (2)^2 + (0)^2 + (1)^2 + (-1)^2 = 10$$

$$Z = \sqrt{\left\{\frac{k(k-1)(\Sigma\lambda_j S_{\cdot j})^2}{(k\Sigma S_{i\cdot} - \Sigma S_i^2)(\Sigma\lambda_j^2)}\right\}}$$

$$= \sqrt{\left\{\frac{5(4)(48)^2}{[5(209) - 795](10)}\right\}}$$

$$= 4.29$$

(b) Evaluate Z by consulting tables of the normal distribution (table A in the appendix). The critical values are

5%	2.5%	1%	0.1%
$z \geqslant 1.64$	1.96	2.33	3.10

To be significant at any of the four levels, Z must be greater than or equal to the appropriate critical value. Our value of Z (4.29) is, therefore, significant at better than the 1 per cent level ($p < 0.01$).

(c) Correction for ties:

No correction is required since the above computations automatically correct for ties.

Step 5 - draw conclusions

Because our result is statistically significant ($p < 0.01$), we may reject H_0 in favour of H_1. In simple terms, the data support the hypothesis that visibility aids for motorcyclists will be effective in proportion to the brightness of the aid.

10.5 Balanced incomplete blocks (trends and contrasts)

Related samples often require that individuals be tested under every one of a number of conditions or that a single person rank a large number of different items. Too many tests can lead to subject fatigue and too many items can be impossible to rank sensibly. To avoid these problems it is sometimes necessary to expose subjects to only a limited number of the possibilities. If every combination of conditions (or items) is experienced by exactly one subject, then we have a design with balanced but incomplete blocks. When the blocks are balanced, we have exactly the same number of entries in each column and this greatly simplifies the computations. If the blocks are not balanced, the method given in section 10.6 must be used and this will be found to be computationally much more onerous.

The computational formulae used in this section are not derived explicitly in chapter 14 but are based directly on Durbin's (1951) formulation as given in Marascuilo and McSweeney (1967). The formulae have been adapted to agree with the standard practice used in this book. To use these formulae we must agree with the restriction that the coefficients used must sum to zero. This is usually easy to arrange and results in a considerable saving in computational effort.

Example

The data given in table 10.7 are taken from an investigation by Geyer *et al.* (1983) into the relative comfort of six different industrial respirators. Three of the respirators were variants of the face mask principle, where air is sucked through a filter close to the wearer's mouth. The other three respirators were more elaborate arrangements involving a motor, hung on the user's back, which pumped air through a filter into a hood which completely covered the user's head. Subjects had to rate three respirators for comfort while performing heavy industrial work. The order of testing the three respirators was chosen at random, but the actual respirators chosen were selected according to a balanced design whereby each subject was exposed to one of the 20 possible combinations of three from six respirators. (In fact, each subject was tested using all six respirators but only data for the first three used are given here. The investigation also used more subjects than are reported here.) The comfort rating scale varied from 1 (not at all uncomfortable) to 5 (very uncomfortable). The investigators were mainly interested to test the hypothesis that pump driven respirators were less uncomfortable than conventional mask respirators.

Step 1 – formulate the problem
(a) Specify the hypotheses:
Null hypothesis (H_0): the respirators do not differ in terms of discomfort.
Specific alternative hypothesis (H_1): pump ventilated respirators (4, 5 and 6) are less uncomfortable than conventional respirators (1, 2 and 3).
(b) Specify the variables:
Independent variable: respirator type; six different respirators were used.

Table 10.7 Discomfort ratings for six different respirators tested using an incomplete blocks design

Subject	Respirator	Conventional			Pump ventilated		
		1	2	3	4	5	6
1		4	3	2			
2		4	3		4		
3		4	2			2	
4		1	4				1
5		4		4	3		
6		4		4		4	
7		1		4			4
8		4			5	3	
9		5			3		4
10		4				3	4
11			4	5	3		
12			2	4		5	
13			4	5			1
14			4		2	4	
15			4		2		1
16			4			3	2
17				4	4	4	
18				2	4		4
19				4		1	2
20					5	4	4

Dependent variable: reported discomfort; five-point scale was used (1 = not at all uncomfortable, 5 = very uncomfortable).
(c) Choose the coefficients:
  Since respirators 4, 5 and 6 are predicted to have lower discomfort scores than respirators 1, 2 and 3, we may use the coefficients $\lambda_j = 1, 1, 1, -1, -1, -1$.

  NB For this design we require that $\Sigma\lambda_j = 0$.

Step 2 – find the sample rank sums and means
(a) Before attempting an incomplete blocks analysis check that the following basic conditions are met:
    (i) All rows have the same number of scores. Let b be the number of scores in each row. In our case, $b = 3$.

(ii) Each row must contain scores in a different combination of columns and there must be

$$\frac{k!}{b!(k-b)!}$$

rows. In our case, there must be

$$\frac{6!}{3!(6-3)!}\ \frac{6\times5\times4\times3\times2\times1}{3\times2\times1\,(3\times2\times1)}=20\ \text{rows}$$

(iii) Each column must contain exactly the same number of scores. In our case, each column contains 10 scores.

When these points have been checked, draw up a computational layout as in table 10.8.

(b) Rank the scores in each block (row) separately, remembering to assign low ranks to low scores.

(c) Find the sample rank sums (R_j) by totalling the ranks in each column. In our case $R_1 = 22, R_2 = 21.5, R_3 = 22, R_4 = 19.5, R_5 = 17.5$ and $R_6 = 17.5$.

(d) Check that the ranking has been carried out correctly by finding ΣR_j, the sum of the rank sums:

$$mb(b+1)/2 = \Sigma R_j$$

$$20(3)(4)/2 = 22 + 21.5 + 22 + 19.5 + 17.5 + 17.5$$

$$120 = 120$$

(e) Note the sample sizes $(n = 10)$ and compute the sample rank means $(\bar{R}_j = R_j/n)$, which in our case are 2.2, 2.15, 2.2, 1.95, 1.75 and 1.75 respectively.

Step 3 - interpret the rank means
As predicted, the rank means for respirators 4, 5 and 6 are less than those for respirators 1, 2 and 3, which implies that they are more comfortable.

Step 4 - compute and evaluate our key statistics (L and Z)
(a) Find L:

$$L = \Sigma \lambda_j R_j$$

$$= (+1)(22) + (+1)\,(21.5) + (+1)(22) + (-1)19.5$$

$$+(-1)17.5 + (-1)17.5$$

$$= 11$$

Find Z:

$$Z = \frac{L - E(L)}{\sqrt{\text{var}(L)}}$$

Table 10.8 Computational layout for example in section 10.5

Subject	Respirator					
	1	2	3	4	5	6
1	3	2	1			
2	1.5	1.5		3		
3	3	1.5			1.5	
4	1.5	3				1.5
5	2.5		2.5	1		
6	2		2		2	
7	1		2.5			2.5
8	2			3	1	
9	3			1		2
10	2.5				1	2.5
11		2	3	1		
12		1	2		3	
13		2	3			1
14		2.5		1	2.5	
15		3		2		1
16		3			2	1
17			2	2	2	
18			1	2.5		2.5
19			3		1	2
20				3	1.5	1.5
R_j	22	21.5	22	19.5	17.5	17.5
n_j	10	10	10	10	10	10
$\bar{R}_j$	2.2	2.15	2.2	1.95	1.75	1.75
λ_j	+1	+1	+1	−1	−1	−1

where

$$E(L) = 0 \text{ when } \Sigma\lambda_j = 0$$

$$\text{var}(L) = \frac{nk(b^2 - 1)}{12(k - 1)} \Sigma\lambda_j^2$$

$$= \frac{10(6)(3^2 - 1)}{12(5)} [(+1)^2 + (+1)^2 + (+1)^2 + (-1)^2 + (-1)^2 + (-1)^2]$$

$$= 48$$

NB $n = 10$, the number of scores in each sample (column); $b = 3$, the number of scores in each block (row).

Whence

$$Z = 11/\sqrt{48} = 1.59$$

(b) Evaluate Z by consulting tables of the normal distribution (table A in the appendix). The relevant values are

5%	2.5%	1%	0.1%
$z > 1.64$	1.96	2.33	3.10

To be significant at any of the four levels, Z must be equal to or greater than the corresponding critical value. Our value of Z (1.59) is almost but not quite significant at the 5 per cent level ($p > 0.05$).

(c) Correction for ties:

No correction is available using this method. This is unfortunate in our particular case because the correction procedure would almost certainly result in a Z value inside the 5 per cent significance region. A tie correction procedure can be applied if we use the more laborious techniques of section 10.6. This is left to exercise 10.6.

Step 5 – draw conclusions
Because our result is not statistically significant ($p > 0.05$), we cannot reject H_0. However, using more sophisticated procedures involving tie correction factors, we might reasonably expect to have a significant result in favour of a difference between conventional and pump ventilated respirators. In the larger and complete data set of Geyer et al. (1983) the difference between the two was quite clear.

10.6   Unequal cell frequencies (trends and contrasts)

A repeated measures (or matched sets) design with unequal cell frequencies is a traditional bogey among researchers because it is so difficult to analyse using available tests. As a consequence, every effort is made to keep the cell frequencies equal. Often, however, this is not possible. Data do not always arrive in neat boxes and, even in the best contrived experiments, accidents will happen and data are lost. This section gives the procedures for handling unbalanced designs in the form of raw scores. The procedure copes well with occasionally empty cells but the researcher must not stretch this facility too far. Certainly, care must be taken to avoid the nonoverlapping sets problem, where the blocks can be subdivided into two or more subsets such that the samples represented in one set contain no scores at all in the other set and vice versa.

For this complicated design the computations become a little more involved, and care must be taken to understand the notation used. Nevertheless, at this stage the operation is purely mechanical; no new principles are introduced.

For this example, I have allowed myself the luxury of a fictional experiment. Unbalanced blocks designs are difficult to find when one is needed, and the small sample sizes allow a simple exposition.

Example

On three successive nights in the George and Dragon public house between 10.00 p.m. and 11.00 p.m., the author invited as many people as were willing to play a simple darts game. Each player had to throw darts for as long as necessary to hit a single bull's-eye. The number of darts needed was counted. Each player was then asked about his drinking for the previous three hours. This is expressed in table 10.9 in terms of the approximate number of pints of beer containing an equivalent amount of alcohol. Does alcohol affect performance? Do the data support a simple monotonic trend hypothesis, i.e. the more beer the worse the performance?

Step 1 – formulate the problem
(a) Specify the hypotheses:
     Null hypothesis (H_0): dart throwing skill is unaffected by alcohol consumption up to six pints of beer.

Table 10.9   Darts skill assessment for 17 players on three nights at different levels of alcohol consumption – example in section 10.6

Night	Samples (equivalent number of pints of beer)			
	None	*Two*	*Four*	*Six*
1	10	15	8	40
	8	23		
2	–	20	36	34
		19		28
3	16	5	21	–
		24	43	
			47	

Specific alternative hypothesis (H_1): dart throwing skill deteriorates as alcohol consumption increases.

(b) Specify the variables:

Independent variable: alcohol consumption, four levels: none, two pints, four pints and six pints.

Dependent variable: darts skill measured in terms of the number of darts thrown before a bull's-eye is hit.

(c) Choose the coefficients:

We expect performance to be worse (i.e. more throws, higher rank means) for the groups with the higher consumption of alcohol. We can therefore test for a simple trend ($\lambda_j = 1, 2, 3, 4$).

Step 2 – find the sample rank sums and means

(a) Draw up a new data table (see table 10.10) for the ranks.

(b) Rank the scores separately for each block, remembering to assign low ranks to low scores.

(c) Find the sample rank sums (R_j) by totalling the ranks in each column. In our case $R_1 = 6.5, R_2 = 17, R_3 = 20.5, R_4 = 13$.

Table 10.10   Computational layout for example in section 10.6

Block	Samples				Block frequencies b_i
	None	*Two pints*	*Four pints*	*Six pints*	
1	3	4	1.5	6	6
	1.5	5			
2	–	2	5	4	5
		1		3	
3	2	1	3	–	6
		4	5		
			6		
R_j	6.5	17	20.5	13	
n_j	3	6	5	3	
$\bar{R}_j$	2.17	2.83	4.1	4.3	
λ_j	1	2	3	4	
$\lambda_j R_j$	6.5	34	61.5	52	$L = \Sigma \lambda_j R_j$ 154

(d) Check that the ranking has been carried out correctly:

$$\Sigma b_i(b_i + 1)/2 = \Sigma R_j$$

$$(6)(7)/2 + (5)(6)/2 + (6)(7)/2 = 6.5 + 17 + 20.5 + 13$$

$$57 = 57$$

(e) Note the sample sizes $(n_1 = 3,\ n_2 = 6,\ n_3 = 5,\ n_4 = 3)$ and compute the sample rank means $(\bar{R}_j = R_j/n_j)$. In our case these are $\bar{R}_1 = 2.17, \bar{R}_2 = 2.83, \bar{R}_3 = 4.1, \bar{R}_4 = 4.3$.

Step 3 – interpret the rank means
The rank means are neatly arranged in ascending order from left to right, indicating less skilful dart throwing with greater consumption of alcohol. This is in agreement with our specific alternative hypothesis.

Step 4 – compute and evaluate our key statistics (L and Z)
(a) Find L:

$$L = \Sigma \lambda_j R_j = (1)(6.5) + (2)(17) + (3)(20.5) + (4)(13)$$

$$= 154$$

Find Z, where

$$Z = \frac{L - E(L)}{\sqrt{\operatorname{var}(L)}}$$

and

$$E(L) = \frac{1}{2} \sum_i^m (b_i + 1) \sum_j^k n_{ij}\lambda_j$$

$$\operatorname{var}(L) = \frac{1}{12} \sum_i^m \left\{ (b_i + 1) \left[b_i \sum_i^k n_{ij}\lambda_j^2 - \left(\sum_{j=1}^k n_{ij}\lambda_j \right)^2 \right] T_i \right\}$$

The complications in these formulae are caused by the irregular pattern of cell frequencies. As a result, we need to compute expressions separately for each block in the table. Some systematization is necessary. It helps to draw up a cell frequency (n_{ij}) matrix (table 10.11) which also shows the coefficients (λ_j) and the block frequencies.

Table 10.11   Computational layout for handling unequal cell frequency problems

Block		λ_j				b_i	$b_i + 1$	$\sum\limits_j n_{ij}\lambda_j$	$\sum\limits_j n_{ij}\lambda_j^2$
		1	2	3	4				
1	n_{1j}	2	2	1	1	6	7	13	35
2	n_{2j}	0	2	1	2	5	6	15	49
3	n_{3j}	1	2	3	0	6	7	14	36

(i) Compute

$$\sum_j^k n_{ij}\lambda_j$$

for each block; e,g, for block 1 ($i = 1$),

$$\sum_j^k n_{ij}\lambda_j = (2)(1) + (2)(2) + (1)(3) + (1)(4) = 13$$

(ii) Compute

$$\sum_j^k n_{ij}\lambda_j^2$$

for each block; e.g. for block 1 ($i = 1$),

$$\sum_j^k n_{ij}\lambda_j^2 = (2)(1) + (2)(4) + (1)(9) + (1)(16) = 35$$

and enter these results in table 10.11.

(iii) We can now find

$$E(L) = \frac{1}{2} \sum_{i=1}^m (b_i + 1) \sum_{j=1}^k n_{ij}\lambda_j$$

$$= \tfrac{1}{2}[(7)(13) + (6)(15) + (7)(14)]$$

$$= 139.5$$

(iv) Before calculating var(L) we need the tie correction factors. If ties are absent or to be ignored, set all $T_i = 1$. In our example ties are present only for the first block (one tie of length 2):

$$T_1 = 1 - \frac{\Sigma(t^3 - t)}{b_i^3 - b_i} = 1 - \frac{2^3 - 2}{6^3 - 6} = 1 - \frac{6}{210} = 0.9714$$

and $T_2 = 1, T_3 = 1$.

(v) We can now find

$$\text{var}(L) = \frac{1}{12} \sum_i^m \left\{ (b_i + 1) \left[b_i \sum_j^k n_{ij}\lambda_j^2 - \left(\sum_j^k n_{ij}\lambda_j \right)^2 \right] T_i \right\}$$

$$= \tfrac{1}{12}\{(7)[(6)(35) - (13)^2](0.9714)$$

$$+ (6)[(5)(49) - (15)^2](1)$$

$$+ (7)[(6)(36) - (14)^2](1)\}$$

$$= \tfrac{1}{12}(278.8 + 120 + 140)$$

$$= 44.9$$

(vi) Now

$$Z = \frac{L - E(L)}{\sqrt{\text{var}(L)}} = \frac{154 - 139.5}{\sqrt{44.9}} = 2.16$$

NB If ties had been ignored (i.e. all $T_i = 1$), Z would have been 2.14 - only slightly less.

(b) Evaluate Z by consulting tables of the normal distribution (table A in the appendix). The relevant values are

5%	2.5%	1%	0.1%
$z > 1.64$	1.96	2.33	3.10

To be significant at any of the four levels, Z must be equal to or greater than the corresponding critical value. Our value of Z (2.16) is therefore significant at better than the 2.5 per cent level ($p < 0.025$).

Step 5 - draw conclusions

Because our result is statistically significant ($p < 0.025$), we may reject H_0 (see step 1) in favour of H_1. In simple terms, the data support the conclusion that dart throwing skill is impaired by consumption of alcoholic liquor. Within the range tested (0–6 pints of beer) the trend appears to be monotonic. The reader is reminded that this study is wholly fictitious.

10.7   Repeated $k \times Q$ frequency tables

One of the most useful contributions of rank sum analysis is the ability to handle multiple frequency tables for the purpose of evaluating trend and contrast hypotheses in a manner which is a simple logical extension of all of the other tests in this book. No new principles are involved. The key to understanding this test is simply that each table represents a single block. Once this is grasped, then it can be seen that there is no substantial difference between this test and that given in the previous section.

It is, undoubtedly, a tedious test to perform by hand since each frequency table requires a fair amount of work in itself. Computer analysis is clearly to be preferred. However, a systematic approach will speed up the operation and reduce errors; manual processing certainly gives the user a 'feel' for the data he is handling!

The following is particularly challenging because there are four tables. It was chosen because it could also be used to illustrate a solution to a common problem which arises when the samples had been shown to be different from each other even before the treatments are applied. The solution is to divide the subjects into sets or blocks which are comparable from the outset and then to analyse these sets separately in a multiblock design.

Example

Istance and Howarth (1983; see also section 9.7) asked four groups of office workers to rate their current degree of eye discomfort on a five-point scale from 1 (no visual discomfort) to 5 (very bad visual discomfort), both in the morning on arrival at work and in the evening before departure.

The four groups were

(a) word processor operators (WP)
(b) data preparation staff (DP)
(c) typists (TY)
(d) general clerical staff (GC).

The purpose of the study was to look for possible harmful effects of the use of visual display units (VDUs). WP staff spend most time looking at a VDU and the DP staff spend some time doing it, but neither TY nor GC staff do. We can use these facts to predict most deterioration in eye discomfort for WP, followed by DP, and with both TY and GC showing least.

To test this Istance and Howarth rated each person by comparing morning and evening scores as either 'worse', 'same' or 'better'. Unfortunately, the data also showed that the groups were already different on the morning scores. These differences meant that there would be different degrees of scope for improvement or deterioration in the different groups. One solution to this problem is to match the subjects into five sets in terms of their morning score and treat each set separately as a block. In fact, no one rated themselves as 5 on the eye discomfort scale on the morning in question, so that we have only four blocks to deal with. Because we are dealing with so many subjects it is necessary to cast each block as a frequency table to speed the ranking process.

The data in table 10.12 refer only to the data for Thursday morning, which are not necessarily representative of the other four days of the investigation.

Table 10.12 Changes in eye discomfort level for office workers beginning the day at different levels of eye discomfort (1 = low discomfort, 4 = high discomfort). Each table contains results for a group of workers who started the day at the same level of discomfort

Initial eye discomfort level		Word processor	Data preparation	Typist	General clerical
1	Worse	0	0	0	0
	Same	15	12	21	16
	Better	3	7	16	10
2	Worse	2	1	4	3
	Same	5	7	3	8
	Better	4	3	6	8
3	Worse	1	2	1	2
	Same	2	3	2	3
	Better	2	2	1	0
4	Worse	0	1	1	1
	Same	0	2	0	0
	Better	0	0	0	0

Step 1 - formulate the problem
(a) Specify the hypotheses:
   Null hypothesis (H_0): the four groups of office workers do not differ in their tendency to change their ratings of eye discomfort during the working day.
   Specific alternative hypothesis (H_1): deterioration in eye discomfort levels will be greater for those groups with exposure to VDUs during their work.
(b) Specify the variables:
   Independent variable: exposure to VDUs. Three levels of exposure are represented across four occupational groups: TY and GC have no exposure, DP have some exposure and WP a great deal.
   Dependent variable: eye discomfort change, worse, same or better.
   Blocking variable: initial eye discomfort in the morning, four levels.
(c) Choose the coefficients:
   We expect the groups exposed to VDUs to show greater deterioration in eye discomfort, i.e. a tendency to occupy cells toward the top of each table. Since the top of a frequency table always gets the lower ranks, we can use the following coefficients:

	WP	DP	TY	GC
λ_j	1	2	3	3

Note that TY and GC have the same coefficient because we have no reason to differentiate between them.

   However, an alternative set of coefficients can be found which is linearly related to these but which promises to simplify the arithmetic enormously. This set is $\lambda_j = -1, 0, 1, 1$. Although, for reasons of clarity, we have avoided negative coefficients so far in the book, they will be used here for reasons of computational simplicity.

Step 2 - find the sample rank sums and means
(a) Draw up the frequency tables as in table 10.13. Note the category totals (t_q) and the cumulative category totals (Σt_q).
(b) Find the shared ranks for each category using the expression

$$r_q = \Sigma t_q - (t_q - 1)/2$$

(c) Find the rank sum (R_j) for each sample by multiplying the shared rank for each category by the number of scores in that category and adding. It is wise to do this separately for each block (table).

$$R_{ij} = \sum_q f_{ijq} r_{iq}$$

Table 10.13 Computational layout for example in section 10.7

Block		WP	DP	TY	GC	t_q	Σt_q	r_q
		0	0	0	0	0	0	—
1		15	12	21	16	64	64	32.5
		3	7	16	10	36	100	82.5
	R_{1j}	735	967.5	2002.5	1345			
	n_{1j}	18	19	37	26			
		2	1	4	3	10	10	5.5
2		5	7	3	8	23	33	22.0
		4	3	6	8	21	54	44.0
	R_{2j}	297	291.5	352	544.5			
	n_{2j}	11	11	13	19			
		1	2	1	2	6	6	3.5
3		2	3	2	3	10	16	11.5
		2	2	1	0	5	21	19.0
	R_{3j}	64.5	79.5	45.5	41.5			
	n_{3j}	5	7	4	5			
		0	1	1	1	3	3	2.0
4		0	2	0	0	2	5	4.5
		0	0	0	0	0	5	—
	R_{4j}	0	11	2	2			
	n_{4j}	0	3	1	1			
	R_j	1096.5	1349.5	2402	1933			
	n_j	34	40	55	51			
	$\bar{R}_j$	32.25	33.74	43.67	37.9			
	λ_j	−1	0	1	1			
	$\lambda_j R_j$	−1096.5	0	2402	1933	$L = \Sigma\lambda_j R_j$	3238.5	

where f_{ijq} is the frequency in the ith block, the jth sample and the qth category. r_{iq} is the shared rank for the qth category in the ith block.

The sample rank sums are 1096.5, 1349.5, 2402 and 1933 respectively.

(d) Check that the ranking has been carried out correctly:

$$\frac{1}{2}\sum_i^m b_i(b_i + 1) = \Sigma R_j$$

$$\tfrac{1}{2}[(100)(101) + (54)(55) + (21)(22) + (5)(6)]$$

$$= 1096.5 + 1349.5 + 2402 + 1933$$

$$6781 = 6781$$

where b_i is the total number of scores in the ith table.

(e) Note the sample sizes ($n_1 = 34, n_2 = 40, n_3 = 55, n_4 = 51$) and compute the sample rank means ($\bar{R}_j = R_j/n_j$). In our case, $\bar{R}_1 = 32.25, \bar{R}_2 = 33.74, \bar{R}_3 = 43.67, \bar{R}_4 = 37.9$.

Step 3 – interpret the rank means

The lowest sample rank mean belongs to the word processors (32.25), which is as predicted since low ranks imply a tendency to occupy categories toward the top of the table (i.e. same or worse). The typists (43.67) and the general clerical workers (37.9) have the highest sample rank mean, which is again as predicted by the specific alternative hypothesis.

Step 4 – compute and evaluate our key statistics (L and Z)
(a) Find L:

$$L = \Sigma\lambda_j R_j = (-1)(1096.5) + (0)(1349.5) + (1)(2402) + (1)(1933)$$

$$= 3238.5$$

Find Z using the following expressions:

$$Z = [L - E(L)]/\sqrt{\mathrm{var}(L)}$$

$$E(L) = \frac{1}{2}\sum_i^m \left[(b_i + 1)\sum_j^k n_{ij}\lambda_j\right]$$

$$\mathrm{var}(L) = \frac{1}{12}\sum_i^m \left\{(b_i + 1)\left[b_i \sum_j^k n_{ij}\lambda_j^2 - \left(\sum_j^k n_{ij}\lambda_j\right)^2\right]T_i\right\}$$

Table 10.14 illustrates the computation of $\Sigma n_{ij}\lambda_j$, $\Sigma n_{ij}\lambda_j^2$ and T_i. Using these values we may now compute:

$$E(L) = \tfrac{1}{2}[(101)(45) + (55)(21) + (22)(4) + (6)(2)]$$

$$= 2900$$

$$\mathrm{var}(L) = \tfrac{1}{12}\{(101)[(100)(81) - (45)^2](0.69)$$

$$+ (55)[(54)(43) - (21)^2](0.86)$$

$$+ (22)[(21)(14) - (4)^2](0.86)$$

$$+ (6)[(5)(2) - (2)^2](0.75)\}$$

$$= \tfrac{1}{12}(423\,366.75 + 88\,971.3 + 5259.76 + 27)$$

Table 10.14   Subsidiary steps in the computation of $E(L)$ and var(L) in example in section 10.7

λ_j	-1	0	1	1	b_i	$\Sigma n_{ij}\lambda_j$	$\Sigma n_{ij}\lambda_j^2$
λ_j^2	1	0	1	1			
n_{ij}	18	19	37	26	100	45	81
	11	11	13	19	54	21	43
	5	7	4	5	21	4	14
	0	3	1	1	5	2	2

Ties correction factors

$$T_1 = 1 - \frac{(64^3 - 64) + (36^3 - 36)}{100^3 - 100} = 1 - \frac{262\,080 + 46\,620}{999\,900} = 0.69$$

$$T_2 = 1 - \frac{(10^3 - 10) + (23^3 - 23) + (21^3 - 21)}{54^3 - 54} = 1 - \frac{990 + 12\,144 + 9240}{157\,410}$$

$$= 0.86$$

$$T_3 = 1 - \frac{(6^3 - 6) + (10^3 - 10) + (5^3 - 5)}{21^3 - 21} = 1 - \frac{210 + 990 + 120}{9240} = 0.86$$

$$T_4 = 1 - \frac{(3^3 - 3) + (2^3 - 2)}{(5^3 - 5)} = 1 - \frac{24 + 6}{120} = 0.75$$

$$= 517\,624.81/12 = 43\,135.40$$

$$Z = [L - E(L)]/\sqrt{\text{var}(L)}$$

$$= (3238.5 - 2900)/\sqrt{43\,135.4}$$

$$= 1.63$$

(b) Evaluate Z by consulting tables of the normal distribution (table A in the appendix). The relevant values are

	5%	2.5%	1%	0.1%
$z \geqslant$	1.64	1.96	2.33	3.10

To be significant at any of the four levels, Z must be equal to or greater than the corresponding critical value. Our value of Z (1.63) is almost but not quite significant at the 5 per cent level ($p > 0.05$).

Step 5 - draw conclusions

Because our result is not statistically significant ($p > 0.05$), we may not reject H_0 (see step 1) in favour of H_1. In simple terms the data do not support the conclusion that exposure to VDUs during working hours causes a deterioration in eye discomfort from morning to evening.

B   Tests against a nonspecific alternative hypothesis

10.8   Exact test (nonspecific) for three related samples

When we have

(a) three or four samples
(b) nine or fewer blocks in a related samples design with
(c) one score only per cell

we can use exact tables in a test due to Friedman (1937). Although the approximate test given in section 10.9 will often lead to the same decision, the approximation between the exact distribution of H and the chi-square distribution is much less close than for the specific test described in section 10.3. Even if the approximation were very good, an exact test is always welcome because it reduces the amount of computation necessary. The only disadvantage of the exact test is that tied data are not catered for and the test becomes increasingly conservative as the number of ties gets larger. However, in the very small designs catered for here this is rarely a problem.

Example

Joseph et al. (1984) studied the food intake increasing effect of the chemical tryptophan in six patients by comparing their total food intake during two days on tryptophan with two other two-day periods when the patients were given (a) casilan and (b) valine. The data given in table 10.15 show the total food intake on the second day for all three treatments. Is there any indication that food intake varies with the treatments?

Step 1 - formulate the problem

(a) Specify the hypotheses:
Null hypothesis (H_0): there are no differences in food intake following administration of casilan, tryptophan and valine.

Table 10.15   Total energy intake (kilocalories) for six subjects on the second day of administration of three different drugs. See example in section 10.8

Subject	Casilan (control)	Tryptophan	Valine (second control)
1	1658	1419	1401
2	1379	1465	1758
3	2147	1470	2078
4	2674	2487	2508
5	765	325	590
6	1076	1018	1152

There is no specific alternative hypothesis. The general alternative hypothesis is simply that H_0 is untrue. (See exercise 10.2 for the evaluation of a specific alternative hypothesis for this data.)

(b) Specify the variables:
Independent variable: drug treatment.
Dependent variable: energy intake (calories).

Step 2 – find the sample rank sums and means

(a) Create a computational layout as in table 10.16.
(b) Rank the scores separately for each block, remembering to give low scores to low ranks.

Table 10.16   Computational layout for example in section 10.8

Subject	Casilan (control)		Tryptophan		Valine		
	Score	Rank	Score	Rank	Score	Rank	
1	1658	*3*	1419	*2*	1401	*1*	
2	1379	*1*	1465	*2*	1758	*3*	
3	2147	*3*	1470	*1*	2078	*2*	
4	2674	*3*	2487	*1*	2508	*2*	
5	765	*3*	325	*1*	590	*2*	
6	1076	*2*	1018	*1*	1152	*3*	
rank sum R_j	15		8		13		ΣR_j^2 458
n_j	6		6		6		
rank mean $\bar{R}_j$	2.5		1.33		2.17		

(c) Find the sample rank sums (R_j) by totalling the ranks in each column. In our case $R_1 = 15, R_2 = 8$ and $R_3 = 13$.

(d) Check that the ranking has been carried out correctly:

$$mk(k + 1)/2 = \Sigma R_j$$

$$6(3)(4)/2 = 15 + 8 + 13$$

$$36 = 36$$

(e) Note the sample sizes $(n = 6)$ and compute the sample rank means $(\bar{R}_j = R_j/n)$. In our case $\bar{R}_1 = 2.5, \bar{R}_2 = 1.33, \bar{R}_3 = 2.17$.

Step 3 – interpret the rank means
The mean rank for tryptophan was less than either of the two control groups. It appears that tryptophan may have been exerting a hunger suppressive effect.

Step 4 – compute and evaluate our key statistic (K)
(a) Find K:

$$K = \Sigma R_j^2 = 15^2 + 8^2 + 13^2 = 458$$

(b) Evaluate K by consulting exact tables of K for $m = 6$ (table N in the appendix). The relevant row for our purposes is

m	5%	2.5%	1%	0.1%
6	$K \geqslant 474$	482	486	504

To be significant at any of the four levels, K must be equal to or greater than the appropriate critical value. Our value of K (458) is less than all of them and is, therefore, not significant $(p > 0.05)$.

Step 5 – draw conclusions
Because our result is not significant $(p > 0.05)$ we are not justified in rejecting H_0. In simple terms, results of this test do not support the idea that the drugs affected food intake to different degrees.

It might be argued that Joseph et al. (1984) expected a reduced food intake relative to the two controls and that a specific test of this hypothesis was required. See exercise 10.2.

10.9 Approximate test (nonspecific) for three or more related samples

The test given in this section is one of the oldest rank sum tests, and was first presented by Friedman (1937) – without the correction for ties. The test is simple, useful and, as a result, popular with researchers. Its usefulness has probably inspired the development of many other tests in this area, so that now it occupies only a very small corner in the universe of nonparametric testing. It is restricted to testing nonspecific hypotheses in designs with *only a single score per cell* - although a supplementary note at the end of this section generalizes Friedman's formulation to cover designs with more than one score per cell as long as the cell frequencies are equal.

Example

Parkes (1982) gave the Middlesex Hospital questionnaire (MHQ) to student nurses with a view to measuring changes in depression levels as they moved from ward to ward. The data in table 10.17 give depression ratings for 14 nurses at the midpoint and the end of two 13-week ward allocations, as well as a depression rating found before either of the allocations. Do the different duties affect the nurses' depression ratings at all?

Step 1 - formulate the problem
(a) Specify the hypotheses:
 Null hypothesis (H_0): nurses' depression ratings are not affected by the particular ward allocations.

Table 10.17 MHQ depression ratings of student nurses before, during and following two ward allocations

Nurse	Before ward allocation	Male medical		Female surgical	
		Mid	End	Mid	End
1	2	4	2	1	0
2	3	3	4	2	3
3	7	4	3	2	6
4	1	2	4	5	7
5	4	4	2	4	1
6	3	4	2	4	2
7	2	3	1	3	0
8	1	1	1	0	0
9	2	5	8	5	3
10	4	6	5	4	3
11	0	1	2	0	1
12	2	2	4	3	5
13	0	3	2	1	1
14	0	0	0	0	0

There is no specific alternative hypothesis. The nonspecific general hypothesis is that H_0 is untrue.

(b) Specify the variables:

Independent variable: ward allocation; male medical and female surgical wards.

Dependent variable: MHQ depression rating (high scores imply greater depression).

Step 2 – find the sample rank sums and means

(a) Draw up a computational layout as in table 10.18. To avoid a complicated layout, omit the raw scores and simply replace them with ranks.

(b) Rank the scores separately for each block, remembering to give low ranks to low scores.

(c) Find the rank sums (R_j) for each sample by totalling the ranks in each column. In our case $R_1 = 37, R_2 = 51.5, R_3 = 48, R_4 = 38.5, R_5 = 35$.

Table 10.18 Computational layout for example in section 10.9. All scores are ranks

Nurse	Before ward allocation	Male medical		Female surgical	
		Mid	End	Mid	End
1	3.5	5	3.5	2	1
2	3	3	5	1	3
3	5	3	2	1	4
4	1	2	3	4	5
5	4	4	2	4	1
6	3	4.5	1.5	4.5	1.5
7	3	4.5	2	4.5	1
8	4	4	4	1.5	1.5
9	1	3.5	5	3.5	2
10	2.5	5	4	2.5	1
11	1.5	3.5	5	1.5	3.5
12	1.5	1.5	4	3	5
13	1	5	4	2.5	2.5
14	3	3	3	3	3
R_j	37	51.5	48	38.5	35
n	14	14	14	14	14
$\bar{R}_j$	2.64	3.68	3.43	2.75	2.5

(d) Check that the ranking has been carried out correctly:

$$mk(k + 1)/2 = \Sigma R_j$$

$$(14)(5)(6)/2 = 37 + 51.5 + 48 + 38.5 + 35$$

$$210 = 210$$

(e) Note the sample sizes ($n = 14$) and compute the sample rank means ($\bar{R}_j = R_j/n$). In our case $R_1 = 2.64, R_2 = 3.68, R_3 = 3.43, R_4 = 2.75, R_5 = 2.5$.

Step 3 - interpret the rank means
Depression ratings are highest for the period on the male medical ward (samples 2 and 3) and lowest for the period on the female surgical ward (samples 4 and 5) and before the ward allocations (sample 1).

Step 4 - compute and evaluate our key statistic (H)
(a) Find H:

$$H = \frac{12}{mk(k + 1)} \Sigma R_j^2 - 3m(k + 1)$$

$$= \frac{12}{(14)(5)(6)} (37^2 + 51.5^2 + 48^2 + 38.5^2 + 35^2) - (3)(14)(6)$$

$$= 6.07$$

(b) Evaluate H by consulting tables of chi-square (table B in the appendix) for $(k - 1)$ degrees of freedom. In our case

$$df = k - 1 = 5 - 1 = 4$$

The relevant row of table B is

Degrees of freedom	5%	2.5%	1%	0.1%
4	$\chi^2 \geqslant 9.5$	11.1	13.3	18.5

To be significant at any of the four levels, H must be greater than or equal to the critical chi-square value. Our value of H (6.07) is, therefore, not significant ($p > 0.05$).

(c) Correct for ties:
In multiple block designs, tie correction procedures can be tedious. However, if the correction is approached in a systematic fashion as in table 10.19, it is not too daunting. Compute T_i for each block separately:

$$T_i = 1 - \frac{\Sigma(t^3 - t)}{k^3 - k}$$

Table 10.19   Computation of the correction factor ΣT_i for example in section 10.9

1 Block	2 Ties length (rank)	3 $\Sigma(t^3-t)$	4 k^3-k	5 T_i $=1-\Sigma(t^3-t)/(k^3-k)$
1	2(3.5)	6	120	0.95
2	3(3)	24	120	0.80
3	–			1.00
4	–			1.00
5	3(3)	24	120	0.80
6	2(1.5) 2(4.5)	6 + 6	120	0.90
7	2(4.5)	6	120	0.95
8	2(1.5) 3(4)	6 + 24	120	0.75
9	2(3.5)	6	120	0.95
10	2(2.5)	6	120	0.95
11	2(1.5) 2(3.5)	6 + 6	120	0.90
12	2(1.5)	6	120	0.95
13	2(2.5)	6	120	0.95
14	5(3)	120	120	0.00

$$\Sigma T_i \quad 11.85$$

These values are given in column 5 of table 10.19. Then find T:

$$T = \Sigma T_i = 11.85$$

We can now correct H for ties:

$$H' = H\frac{m}{T} = 6.07\ \frac{14}{11.85} = 7.17$$

Our value of H' is larger but remains nonsignificant.

If the correction for ties is not to be attempted, then all completely tied rows (e.g. nurse 14) should be eliminated from the data before any calculations are attempted. Exercise 10.5 looks at the effect of omitting nurse 14. The correction for ties procedure automatically deals with completely tied blocks, which in some applications can be very common.

Step 5 - draw conclusions
Because our result is not significant ($p > 0.05$), we are not justified in rejecting H_0 (see step 1). In simple terms, the results of our analysis do not support the

idea that depression in nurses is affected by their current ward allocation. This conclusion does not rule out the possibility that a significant effect might emerge from an analysis of the much larger sample of depression ratings accumulated by Parkes. Moreover, a specific test of the reasonable alternative hypothesis, that the medical wards are more depressing, might well prove significant even on the limited sample of scores in this example. Nonspecific tests are very much less powerful than a well aimed specific test on the same data.

Supplementary note. If the number of scores per cell are all equal but in excess of 1, use the following expression for H (ignoring ties):

$$H = \frac{12}{nmk(nk + 1)} \sum_{k}^{k} R_j^2 - 3(nk + 1)\, m$$

where n is the number of scores per cell.

10.10   Dichotomous data (nonspecific test)

This test is used when the following conditions are met:

    (1) The data are exclusively 1s and 0s.
    (2) The data are arranged in blocks.
    (3) There is exactly one score per cell.
    (4) There is no specific alternative hypothesis.

It was originally presented by Cochran (1950). When there are only two samples this test is equivalent to the matched pairs test given in section 8.6.

Completely tied blocks (that is, all 0s or all 1s) may be deleted or retained at the researcher's discretion. They do not affect the final result. In the example in section 10.4 they were retained, but in this example they will be omitted.

Example

In the example in section 10.4, data were analysed from an investigation into the visibility of motorcycles in traffic. In that example, the data from one site accorded well with the expected likelihoods of reporting the presence of the motorcycle. Data from another site (see table 10.20) yielded results which do not agree closely with our expectations. Although they do not support our hypothesis, it is still worth asking the question whether there is any evidence that the various visibility aids did affect the likelihood that the motorbike will be seen by the observers. Sixty-seven observers were used altogether, but 11 of them failed to spot the motorcycle on any of the five occasions and five of them saw the motorcycle on each occasion. These 16 completely tied blocks have been omitted from table 10.20.

Table 10.20   Data and computational layout for example in section 10.10

Observer	Control	Headlight	Running light		Fluorescent jacket	$S_i.$
			1	2		
1	1	0	1	0	0	2
2	1	1	0	0	1	3
3	0	1	0	1	0	2
4	1	0	0	0	0	1
5	1	0	0	0	0	1
6	0	0	1	0	0	1
7	1	1	1	0	0	3
8	1	0	0	0	0	1
9	0	1	1	1	0	3
10	1	1	1	0	1	4
11	1	0	0	0	0	1
12	0	0	0	1	0	1
13	0	0	1	1	0	2
14	0	1	0	0	0	1
15	0	1	0	0	0	1
16	0	1	0	1	0	2
17	0	0	1	1	0	2
18	1	0	1	0	1	3
19	1	1	0	1	1	4
20	1	1	0	1	0	3
21	0	0	1	0	0	1
22	1	1	0	1	1	4
23	0	1	0	0	1	2
24	0	1	1	1	1	4
25	1	0	1	1	1	4
26	0	1	1	1	1	4
27	0	1	1	1	1	4
28	1	1	1	0	1	4
29	0	1	0	1	0	2
30	0	1	0	0	1	2
31	1	1	0	0	0	2
32	0	0	0	0	1	1
33	1	1	0	1	1	4
34	0	1	0	1	0	2
35	0	1	0	0	0	1
36	1	0	0	0	1	2
37	0	0	0	0	1	1
38	1	0	1	0	1	3

Table 10.20 continued

Observer	Control	Headlight	Running light		Fluorescent jacket	S_i.
			1	2		
39	0	0	0	1	1	2
40	1	1	1	0	1	4
41	1	1	0	1	0	3
42	1	0	1	1	0	3
43	0	1	1	0	0	2
44	0	0	0	0	1	1
45	1	0	0	1	0	2
46	0	0	0	1	0	1
47	0	0	0	1	0	1
48	1	1	0	1	1	4
49	1	1	0	0	1	3
50	1	1	0	1	0	3
51	0	0	1	1	1	3
$S_{.j}$	24	29	20	26	23	

Step 1 – formulate the problem
(a) Specify the hypotheses:
Null hypothesis (H_0): the four visibility aids do not influence the number of times the motorcycle will be reported.
There is no specific alternative hypothesis. The general alternative hypothesis is that H_0 is untrue.
(b) Specify the variables:
Independent variables: visibility aid. There are four visibility aids: (a) headlight (b) fluorescent jacket (c) one running light and (d) two running lights. There is also a control condition.
Dependent variable: whether or not the motorcycle is spotted by the observer (1 = yes, 0 = no).

Step 2 – find the row and column frequencies
We use the notation given in section 10.4:

S_i.   the number of 1s in the *i*th row.
$S_{.j}$ the number of 1s in the *j*th column.

(a) Find $S_{.j}$, $\Sigma S_{.j}$ and $\Sigma S_{.j}^2$:
The column totals ($S_{.j}$) are 24, 29, 20, 26 and 23.

Therefore

$$\Sigma S_{.j} = 24 + 29 + 20 + 26 + 23 = 122$$

$$\Sigma S_{.j}^2 = 24^2 + 29^2 + 20^2 + 26^2 + 23^2 = 3022$$

(b) Find $S_{i.}$, $\Sigma S_{i.}$ and $\Sigma S_{i.}^2$:

To simplify the enumeration of $S_{i.}$, we note that there are only four possible values for $S_{i.}$. We therefore count how many there are of each:

$S_{i.}$	*Frequency*	$S_{i.}^2$
1	14	1
2	14	4
3	12	9
4	11	16

Thus

$$\Sigma S_{i.} = (1)(14) + (2)(14) + (3)(12) + (4)(11) = 122$$

which is the same as $\Sigma S_{.j}$. This is a useful check on the counting process.

$$\Sigma S_i^2 = (1)(14) + (4)(14) + (9)(12) + (16)(11) = 354$$

Step 3 - interpret the column totals

The column totals are not very widely spread, having a range of only 20–29. The headlight sample is the most visible, as we might expect, but the least visible is the single running light condition. It is surprising that the single running light should be less visible than the control condition. This alone suggests that there is possibly something wrong with these data.

Step 4 - compute and evaluate our key statistic (H)

(a) Find H:

$$H = \frac{(k-1)[k\Sigma S_{.j}^2 - (\Sigma S_{.j})^2]}{k\Sigma S_{i.} - \Sigma S_{i.}^2}$$

$$= \frac{(4)[(5)(3022) - (122)^2]}{(5)(122) - (354)} = 3.53$$

H is automatically corrected for ties by this computational procedure.

(b) Evaluate H by consulting tables of the chi-square distribution (table B in the appendix) with $(k - 1)$ degrees of freedom. In our case

$$\text{df} = k - 1 = 5 - 1 = 4$$

The relevant row of table B is therefore

Degrees of freedom	5%	2.5%	1%	0.1%
4	$\chi^2 \geqslant 9.5$	11.1	13.3	18.5

To be significant, H must be greater than or equal to χ^2 at the appropriate level. Our value of H (3.63) is clearly not significant at any of the levels ($p > 0.05$).

Step 5 - draw conclusions
Because our result is not significant ($p > 0.05$), we are not justified in rejecting H_0. In simple terms, our analysis has not detected any reliable tendency for the visibility aids to affect the likelihood that observers will report seeing the motorcycles in traffic conditions at this site.

This is a surprising result since, at another site, perfectly reasonable results were obtained (see the example in section 10.4). The results in this section are subject to question since the control condition with no visibility aids was more conspicuous than the condition with a single running light. If the control condition is ignored, then the other four conditions are ordered in the same way as in section 10.4.

10.11   Balanced incomplete blocks (nonspecific)

This test is the nonspecific analogue of the balanced incomplete blocks (trends and contrasts) test described in section 10.5, which should be consulted for its general notes. The example in section 10.5 will be used again, and tables 10.7 and 10.8 should be studied in conjunction with the following example. The formulae given are taken from Durbin (1951) and do not permit for a correction for ties. If a correction for ties is required then the methods of section 10.12 may be required, and this would be best approached using the computer program given in chapter 15.

Example

This is taken from the example in section 10.5, which should be consulted. The data are given in table 10.7 and represent the discomfort ratings of six different respirators. For this example we merely wish to ask whether there is any difference between the respirators in terms of discomfort, and a nonspecific test is appropriate.

Step 1 - formulate the problem
(a)   Specify the hypotheses:
   Null hypothesis (H_0): there are no differences between the respirators in terms of expected discomfort levels.

There is no specific alternative hypothesis. The nonspecific general hypothesis is that H_0 is untrue.

(b) Specify the variables:
Independent variable: respirator type; six different respirators were used.
Dependent variable: reported discomfort; a five-point scale was used (1 = not at all uncomfortable, 5 = very uncomfortable).

Step 2 - find the sample rank sums and means
This had already been done in section 10.5, which should be consulted, and the results are to be found in table 10.8.

Step 3 - interpret the rank means
Respirators 4, 5 and 6 (pump ventilated) have lower discomfort means than respirators 1, 2 and 3 (conventional ventilation).

Step 4 - compute and evaluate our key statistic (H)
(a) Find H:
When $n = 10$, the number of scores in each sample (column), and $b = 3$, the number of scores in each block (row),

$$H = \frac{12(k-1)}{n(b^2-1)k} \Sigma R_j^2 - \frac{3n(k-1)(b+1)}{b-1}$$

$$= \frac{12(5)}{10(8)(6)} (22^2 + 21.5^2 + 22^2 + 19.5^2 + 17.5^2 + 17.5^2)$$

$$- \frac{3(10)(5)(4)}{3}$$

$$= 2.87$$

(b) Evaluate H by consulting tables of chi-square (table B in the appendix) for $(k-1)$ degrees of freedom. In our case

$$\text{df} = k - 1 = 6 - 1 = 5$$

The relevant row of table B is

Degrees of freedom	5%	2.5%	1%	0.1%
5	$\chi^2 \geqslant 11.1$	12.8	15.1	20.5

To be significant at any of the four levels, H must be greater than or equal to the critical chi-square value. Our value of H (2.87) is therefore not significant ($p > 0.05$).

(c) Correct for ties:

No correction is available using this procedure. The methods referred to in section 10.12 are required if a correction is necessary.

Step 5 - draw conclusions

Because our result is not significant ($p > 0.05$), we are not justified in rejecting H_0 (see step 1). In simple terms, the results of our analysis do not support the idea that the respirators differ in terms of perceived discomfort.

10.12   Nonspecific tests for unequal cell frequencies and repeated frequency tables

Computational procedures will not be shown here for these tests. They are extremely cumbersome because they involve generating and inverting variance-covariance matrices. The procedure is, however, outlined in section 14.3. Computer programs for doing the work are given in chapter 15. If specific alternative hypotheses are available for testing, proceed immediately to sections 10.6 and 10.7.

If access to an appropriate program is not available and a researcher simply must perform a nonspecific test on data with a multiple block design with unequal cell frequencies, then a reasonable estimate can be obtained using the following method. Proceed as for a test of a specific alternative hypothesis (sections 10.6 or 10.7) but use the sample rank means as substitutes for coefficients. Take the resulting Z value, square it and treat it as H. Evaluate H by consulting chi-square tables with $k - 1$ degrees of freedom. The result is not guaranteed to be accurate but it can often give a fair idea. When this technique is applied to the darts skill data in section 10.6, we obtain $H = 4.51$. When the full procedure (involving the inversion of the variance-covariance matrix) is used we obtain $H = 4.70$.

In the relatively unlikely event that the unequal cell frequencies follow the same proportional pattern in each block, there is a formulation which will avoid the inversion of the variance-covariance matrix:

$$H = \frac{12N \sum_{j}^{k} \frac{1}{n_j} \left\{ R_j^2 - R_j \sum_{j}^{m} n_{ij}(b_i + 1) + \frac{1}{4}\left[\sum_{i}^{m} n_{ij}(b_i + 1) \right]^2 \right\}}{\sum_{j}^{m} b_i^2 (b_i + 1) T_i}$$

This formula can be applied whenever the following condition is true:

$$n_{ij} = n_j b_i / N \qquad \text{for all } n_{ij}$$

That is, the cell frequency is always the product of its sample and block frequency divided by the total frequency. Table 10.21 gives an example of a design which meets this condition. The analysis using this formula is left to the reader as an exercise.

The expression for H just given is rarely used in practice but is the grandfather of all the simpler formulae for computing H in the nonspecific tests given in chapters 7–10. The matter is discussed further in chapter 14.

Exercises

10.1 (a) Repeat the computations for the example in section 10.4, but this time omit all rows which are completely tied (all 0s or all 1s). The result should remain unchanged. With a little thought, the exercise can be completed quickly. It should not be necessary to draw up table 10.6 again.
(b) Using the same data, evaluate the hypothesis that headlights are a better visibility aid than the others. For quickest computation use the coefficients $\lambda_j = 0, 1, 0, 0, 0$.

10.2 (a) A nonspecific exact test was used in the example in section 10.8, where an appetite suppressing drug was compared with two other control

Table 10.21   An example of a design with unequal cell frequencies which meets the requirement of $n_{ij} = n_j b_i / N$

Block	Sample		
	1	2	3
1	27	17	19
	33		
	$(n_{1,1} = 2)$	$(n_{1,2} = 1)$	$(n_{1,3} = 1)$
2	18	12	21
	24	16	13
	67		
	32		
	$(n_{2,1} = 4)$	$(n_{2,2} = 2)$	$(n_{2,3} = 2)$
3	33	22	18
	42		
	$(n_{3,1} = 2)$	$(n_{3,2} = 1)$	$(n_{3,3} = 1)$

conditions. However, a specific test might have been more appropriate since Joseph et al. (1984) could have predicted that food intake would be lower for tryptophan than for either casilan or valine. If we assume that the two control conditions do not influence food intake, we might use the coefficients $\lambda_j = 0, -1, 0$. Use an approximate test despite the small sample sizes, because the exact test given in section 10.2 assumes a monotonic trend hypothesis which is not the case here. Comment on the difference between your result and the result in section 10.8.

(b) When two coefficients are the same, as they are in (a), this is equivalent to combining their two samples into a single sample. Repeat the example in section 10.8 as a two-sample specific test with valine and casilan samples combined into a single sample. It should give the same result.

10.3 Gazely and Stone (1981), as part of a much larger survey of visual defects in partially sighted schoolchildren, asked 32 such children to read a passage of text at five levels of luminance. The data in table 10.22 are the performance scores (based on speed of reading) in a high contrast condition (very black letters against a very white background). Does the luminance affect performance? A high performance score reflects a faster/more accurate reading speed.

10.4 Fleming, Unwin and Meehan (1982) asked a large group of patients admitted to hospital with an alcohol problem to indicate whether or not they had experienced any of 35 common symptoms of alcoholism. They were then asked to rank the symptoms in the order in which they first appeared. The data in table 10.23 refer only to the seven most commonly acknowledged symptoms:

(1) giving up interests because drinking interferes
(2) spending more time drinking
(3) needing more than companions
(4) trembling after drinking the day before
(5) morning drinking
(6) decreased tolerance
(7) amnesia.

In a previous study, other researchers had found that the symptoms occurred, on average, in the following sequence:

3, 2, 7, 1, 4, 5, 6

Do the results of Fleming et al. (1982) confirm the results of this previous study in terms of the sequence of symptom onset from 'needing more than companions' to 'decreased tolerance'?

Missing data in the table signify that the patient did not acknowledge the symptom. Each data point is the rank order of occurrence among all

Table 10.22   Performance scores for partially sighted children reading high contrast text under fine luminance conditions

Subject	Luminance (cd/m^2)				
	13	32	80	200	500
1	57.6	72.7	87.9	90.9	100
2	69.2	81.7	100	88.8	84.6
3	71.1	71.1	86.8	89.5	100
4	74.6	84.5	90.1	94.4	100
5	92.4	99.0	99.9	99.9	100
6	76.0	100	76.0	60.0	60
7	29.3	44.8	62.1	79.3	100
8	81.6	100	99.5	96.0	96.5
9	76.0	83.6	93.8	93.8	100
10	68.4	100	81.6	74.3	44.9
11	58.2	79.1	92.5	100	97.0
12	87.4	97.8	99.1	98.2	100
13	76.7	86.9	95.8	100	98.8
14	99.0	98.4	99.3	99.3	100
15	82.7	94.7	94.3	91.3	90.1
16	55.5	61.0	78.5	99.1	100
17	74.2	83.3	96.9	100	100
18	88.4	92.3	100	83.8	74.6
19	71.1	72.1	81.1	97.1	96.7
20	56.1	70.3	88.9	92.8	100
21	64.5	92.1	100	86.8	69.7
22	59.1	76.6	88.7	93.1	100
23	77.0	91.1	99.8	99.8	100
24	80.7	100	98.7	90.7	84.6
25	50.0	60.5	83.7	97.6	100
26	44.5	67.8	99.3	91.7	83.6
27	86.4	92.7	98.9	94.8	100
28	88.6	94.0	96.0	100	83.3
29	73.1	85.4	100	85.4	65.8
30	55.6	81.2	88.1	100	98.3
31	69.5	93.1	100	98.9	97.7
32	77.2	86.0	100	95.4	68.2

Table 10.23   Rank ordering of seven common symptoms of alcoholism for 13
patients. The data are taken from a larger data set involving 35 symptoms and 46
patients (see exercise 10.4)

Patient	Symptom						
	1	2	3	4	5	6	7
1	14	4	5	16	–	–	10
2	–	6	3	10	12	–	–
3	–	–	–	1	3	–	2
4	12	2	10	–	4	13	14
5	5	7	1	19	22	26	14
6	5	7	15	–	–	–	1
7	5	6	9	13	12	1	11
8	4	–	1	–	13	14	15
9	19	2	1	5	3	11	17
10	8	10	5	3	6	–	–
11	–	–	2	3	–	4	–
12	23	9	1	19	17	10	22
13	3	7	20	15	13	1	5

symptoms experienced including those not given in the table. Low values
indicate a relatively early occurrence of the symptom. Only the first 13
patients are represented in table 10.23. Fleming et al. used 46 patients
altogether.

10.5 (a) Repeat the example in section 10.9, omitting all tied rows (in this case
row 14 only), and compare the result with section 10.9.

(b) Using the data in section 10.9, test the hypothesis that the medical
wards have a more depressing effect on nurses than the surgical wards.

10.6 Apply the methods of section 10.6 to the data given in the example in
section 10.5. Compare the results with those given in section 10.5. What is
the effect of the tie correction procedure?

10.7 Geyer (1983) asked 20 subjects to work in a dusty environment using
respirators which were being evaluated. Each subject was given the opportunity to use three out of six respirators, and was subsequently invited to
rate each respirator for ease of breathing on a five-point scale (1 = breathing with great difficulty, 2 = breathing as easily as without). The design
was balanced so that each possible trio of respirators was evaluated by a
single subject. The data are given in table 10.24. Attempt a nonspecific
analysis of the data.

Table 10.24   Ease of breathing ratings for six respirators. Data from exercise 10.7

Subject	Respirators					
	1	2	3	4	5	6
1	3	3	1			
2	3	3		2		
3	3	2			1	
4	3	1				1
5	3		2	1		
6	2		3		1	
7	3		2			1
8	2			1	1	
9	3			1		1
10	4				1	1
11		2	2	1		
12		2	2		1	
13		3	3			1
14		4		1	2	
15		3		1		1
16		2			1	1
17			3	1	1	
18			1	1		2
19			3		1	2
20				3	2	2

Three of the respirators (4, 5, 6) were ventilated by a battery driven pump. The experimenter anticipated that these respirators would prove better for breathing. Evaluate his expectation using a specific test.

Note   In the original study, more subjects were used and each subject evaluated all six respirators. In table 10.24 the data for the last three respirators tried have been deleted from each subject.

11 Correlation and concordance

11.1 Introduction

Correlation between variables is often treated quite separately from the evaluation of differences between populations, especially in introductory texts. This is unfortunate because, at a computational level, they can both be seen as reverse sides of the same coin, and the methods for evaluating one can readily be adapted for investigating the other. This produces a reduction in the complexity of the business of applying statistics to research data, and from this will follow a more intelligent usage of statistical procedures. The close conceptual ties between the assessment of correlation and the evaluation of differences between populations is particularly easy to demonstrate using rank techniques.

Look again at section 10.8. Here, Joseph et al. (1984) measured the food intake of volunteers following the consumption of three different drugs. Note that each volunteer was studied following all three treatments. The analysis of the data focused on the differences *between* the three treatments. However, after only brief reflection, it should be clear that obvious differences between the samples can only occur if most of the volunteers *agree* in the manner of their response to the three drugs. If each subject showed a different response pattern to the three treatments, then no clear average difference between the drugs would emerge. *Thus, agreement between the subjects is reflected in clear differences between the treatments.* Lack of agreement between the subjects is reflected in little or no differentiation between the treatments. In this way, correlation between the blocks and differences between the samples amount to the same thing.

In terms of our computational procedures we can say that a large value of H is indicative of both (a) large differences between samples and (b) good agreement between blocks. Traditionally, however, correlation (or more generally concordance) is expressed in terms of a coefficient which has a maximum value of 1.0, where 1.0 represents perfect agreement and 0 represents no agreement. This can be done easily enough using the following trick:

$$W = \frac{H}{H_{max}}$$

where W is the measure of correlation: $0 < W < 1$.

H_{max} is the *highest possible* value of H given the number of blocks and samples.

H is the usual statistic for measuring differences between samples. H_{max} differs from one design to another but, normally, we restrict our attention to designs *with only one score per cell*. When this is the case we have

$$H_{max} = m(k - 1)$$

so that

$$W = \frac{H}{m(k - 1)}$$

This is the expression used in section 11.4.

The most popular application of correlation analysis occurs in designs with only two blocks ($m = 2$), so that

$$W = \frac{H}{2(k - 1)}$$

However, when we have only two blocks we can talk about positive and negative correlation. Negative correlation occurs when high scores in one block are associated with corresponding low scores in another block. Spearman (1906), therefore, devised a correlation coefficient ρ which could be either positive or negative. In our notation,

$$\rho = 2W - 1$$

$$= \frac{H}{k - 1} - 1$$

where ρ is the Spearman rank correlation coefficient: $1 > \rho > -1$. Once again a coefficient value of zero implies an absence of any correlation. See section 14.4 for a more detailed discussion.

Correlation is computed between two variables. For example we compute the corelation between (a) motion sickness rating and (b) extraversion score in section 11.2. To do this we put the scores on the first variable in block 1 and the scores on the second variable in block 2. At first it feels strange to put two different variables in the same calculation in this way. There is no need to worry; the two sets of scores are ranked quite independently and never become mixed up. In fact this operation makes it quite clear, for the first time, that our statistical procedures are quite insensitive to the nature of the material we put into them. The procedures merely compute the agreement between blocks of ranks, or differences between samples of ranks, whichever you prefer. What we infer from such a correlation is our business and has little to do with the procedure itself.

In conclusion, the rank sum analysis procedures can be used either for correlating the scores in the blocks or for evaluating differences between the samples. The distinction is merely one of interpretation. Similarly the distinction between concordance and correlation is simply to be understood in terms of the number of blocks in the design. If you have more than two blocks, then you are dealing with concordance. Section 11.3 looks at a novel but natural extension of the rank sum technique to cope with frequency tables. It has much in common with the evaluation of trends across frequency tables, but has the advantage of permitting an interpretation in terms of correlation between variables.

11.2 Rank sum correlation (Spearman's ρ)

Simple correlation is based upon a collection of pairs of scores. Typically each pair of scores comes from a single subject and represents measurements made on two variables. The aim of statistical analysis is to decide whether or not there is a reliable degree of association between the two variables. In the following example each subject presents two scores: one is a measure of his liability to suffer from motion sickness when travelling, the other is a personality measure (Eysenck's extroversion assessment). Just prior to the investigation it had been suggested that people high on extroversion would be less prone to motion sickness. In terms of ranks, this means that subjects with a high rank on extroversion should have a low rank on the motion sickness measure. We are therefore looking for an inverse association between the two variables.

To construe this problem in terms of rank sum analysis, we must treat each subject as a sample. All of the scores from the first variable belong to the first block and the scores from the second variable belong to the second block. Each rank sum, therefore, belongs to a different subject. These rank sums are then used to compute ρ, which measures the degree and direction of the correlation.

It is important to be clear whether we are attempting a specific or a nonspecific test. *If a prediction can be made concerning the sign of ρ (either positive or negative), then we are dealing with a specific test.* This prediction must follow from past experience or some reasonable theory which is currently being investigated. *If no prediction is made or if the prediction we make fails, the test becomes nonspecific.* Nonspecific tests require a stronger degree of association between two variables to achieve the same degree of significance. The example below begins as a specific test but the prediction fails, and it is necessary to proceed as for a nonspecific test. After all, the correlation may be statistically significant but in the opposite direction.

An exact test for this design exists but the approximate method illustrated in the following is so good and so easy to compute as to render the exact tables superfluous.

Example

Wilding and Meddis (1972) tested a suggestion by Reason (1968) that extroverts would be less susceptible to motion sickness than introverts. In a small pilot study they administered Reason's motion sickness questionnaire (MSQ) and Eysenck's personality inventory (EPI) to a class of 22 young psychology undergraduates. Their scores on the MSQ and their extroversion (E) score from the EPI are given in table 11.1. Reason would predict a negative association

Table 11.1   Motion sickness and extroversion ratings for 22 subjects (example in section 11.2)

Subject	Motion sickness	Extroversion
1	0.75	14
2	0	12
3	1	17
4	0.77	15
5	4.11	9
6	0.44	14
7	3.75	20
8	3.44	10
9	1.63	15
10	2.33	13
11	3.43	16
12	2.00	5
13	2.50	13
14	3	17
15	20.2	22
16	0.87	15
17	0.55	7
18	1.33	6
19	9.75	12
20	3.33	17
21	5.33	13
22	5.00	12

between extroversion and the tendency to report motion sickness. Do the data support this view?

Step 1 - formulate the problem
(a) Specify the hypotheses:
   Null hypothesis (H_0): there is no association between E and MS (i.e. $\rho = 0$).
   Specific alternative hypothesis (H_1): there is an inverse association between E and MS (i.e. $\rho < 0$).
(b) Specify the variables:
   Independent variable: motion sickness (MS). High scores reflect an enhanced tendency to be sick or nauseated when travelling.
   Dependent variable: extroversion (E) as measured by the EPI.

   NB   When evaluating correlation, these two variables are treated as two blocks.

Step 2 - find the rank sums (R_j) and correlation coefficient (ρ)
(a) Draw up the computational layout as in table 11.2. Remember that k is the number of *subjects* for this design, and $m = 2$ (is the number of variables being correlated.
(b) Rank the scores in each block (i.e. for each variable) separately, assigning low ranks to low scores in the usual way.
(c) Find the rank sums, R_j, by adding each pair of ranks.
(d) Find the sum of ranks, ΣR_j, and the sum of squares of the rank sums, $(\Sigma R_j)^2$. In our case $\Sigma R_j = 506$ and $(\Sigma R_j)^2 = 13\,623.5$.
(e) Check that the ranking has been carried out correctly:

$$\Sigma R_j = k(k + 1)$$

$$506 = 22(23) = 506$$

(f) Compute the correction terms for ties, T_1 and T_2.
   There are no ties in block 1 (motion sickness), so $T_1 = 1$.
   There are a number of ties in block 2 (extroversion):

Rank	t (tie length)	$t^3 - t$
7	3	24
10	3	24
12.5	2	6
15	3	24
19	3	24
		$\Sigma(t^3 - t) = 102$

$$T_2 = 1 - \frac{\Sigma(t^3 - t)}{k^3 - k} = 1 - \frac{102}{22^3 - 22} = 0.9904$$

Table 11.2   Computational layout for example in section 11.2

Subject	MS (ranks)	E (ranks)	R_j
1	4	12.5	16.5
2	1	7	8
3	7	19	26
4	5	15	20
5	18	4	22
6	2	12.5	14.5
7	17	21	38
8	16	5	21
9	9	15	24
10	11	10	21
11	15	17	32
12	10	1	11
13	12	10	22
14	13	19	32
15	22	22	44
16	6	15	21
17	3	3	6
18	8	2	10
19	21	7	28
20	14	19	33
21	20	10	30
22	19	7	26

$$\Sigma R_j = 506$$
$$\Sigma R_j^2 = 13\,632$$

We can now find:

$$\Sigma T_i = T_1 + T_2 = 1 + 0.9904 = 1.9904$$

(g) Compute ρ:

$$\rho = \left[\frac{6\Sigma R_j^2}{k^3 - k} - \frac{6(k+1)}{k-1} \right] \frac{2}{\Sigma T_i} - 1$$

$$= \left[\frac{6(13\,623.5)}{22^3 - 22} - \frac{6(23)}{22} \right] \frac{2}{1.9904} - 1$$

$$= 0.13$$

If there are no ties, omit $2/\Sigma T_i$ from the above expression, leaving

$$\rho = \frac{6\Sigma R_j^2}{k^3 - k} - \frac{6(k+1)}{k-1} - 1$$

Note that the omission of the tie correction factor when ties are present leads to a smaller result (i.e. less significant) for positive correlation coefficients but a *larger* result (i.e. more significant) for negative correlation coefficients. The correction factor should, therefore, only be omitted in casual computations and should not be omitted as a general rule.

Step 3 – interpret the correlation coefficient
Our specific alternative hypothesis predicts a negative coefficient of correlation, which is contradicted by the outcome of our calculations ($p = +0.13$). We must therefore abandon our specific hypothesis, which has been contradicted.

We may continue, however, with a nonspecific test. It is possible that the observed correlation is still statistically significant even though in the opposite direction to our original prediction. The nonspecific alternative hypothesis is that there is some association between the two variables (either positive or negative).

Step 4 – compute and evaluate our key statistics (Z or χ^2)
(a) For a *specific* test (where the sign of ρ was correctly predicted), compute

$$Z = \sqrt{[\rho^2(k-1)]}$$

We can then evaluate Z by consulting tables of the normal distribution (table A in the appendix). The relevant critical values are reproduced here:

	5%	2.5%	1%	0.1%
$z \geqslant$	1.64	1.96	2.33	3.10

To be statistically significant at a given level, Z must be greater than or equal to the appropriate critical value.
(b) For a *nonspecific* test, compute

$$H = \rho^2(k-1)$$

We can then evaluate H by consulting tables of the chi-square distribution (table B in the appendix) for one degree of freedom (df = 1). The relevant values are reproduced here.

Degrees of freedom		5%	2.5%	1%	0.1%
1	$\chi^2 \geqslant$	3.8	5.0	6.6	10.8

To be statistically significant at a given level, H must be greater than or equal to the appropriate critical value of χ^2.

In our case, we need to compute H for a nonspecific test:

$$H = (0.13)^2(22 - 1) = 0.35$$

Our value of H is clearly not statistically significant.

Step 5 - draw conclusions
The data clearly do not support Reason's prediction of an inverse correlation between proneness to motion sickness and extroversion.

Exercise 11.2 gives data for a repeat investigation using a much larger sample of subjects. Do they present the same picture as the pilot study data analysed above?

Historical note

Spearman's computational procedure used the *difference* between each pair of ranks (d_i) rather than the rank sum. He calculates the correlation coefficient thus:

$$\rho = 1 - \frac{6\Sigma d^2}{k^3 - k}$$

which is slightly less computationally onerous than the method given here. However, when ties are present, the correction procedure is quite complicated and requires a new computational formula. Since ties are usually present and should not be ignored, Spearman's method becomes less attractive. Both methods give the same result when the correction for ties is applied.

11.3   Correlation using frequency tables

When there are a large number of pairs of scores to be correlated, the ranking procedure can become very slow and prone to error. It may be expedient to recast the data into frequency table form and base the computations on the table alone. To do this we need to convert the scales of both variables to a small number of categories. This involves a loss of information, which is traded against an increase in computational speed and accuracy. When there is a large number of data points, this loss of information is not usually serious. However, if the result is marginally nonsignificant, the researcher always has the option of repeating the calculations using the original data and the procedure given in the previous section. Often, the data arrive already in the form of ordered categories, in which case there is no loss of information involved in using the procedures given here.

The method for frequency tables is different only in terms of arithmetic from the method in section 11.2; there are no new principles. All of the comments at the head of section 11.2, therefore, apply to this section too. This method also has similarities with techniques for evaluating simple trends across frequency tables (see section 9.4 for computational formulae) for those who wish to experiment with comparisons. However, beginners should note an important computational difference. In correlation analysis for frequency tables k refers to the number of elements in the frequency table, since each element refers to a pair of scores. Similarly, the variables represented by the rows and columns of the frequency table are *blocking variables*. In correlation analysis we are looking for agreement between two blocking variables rather than differences between samples (i.e. pairs of scores). At a deep level these become opposite sides of the same coin (see section 11.1) but they must not be confused at a computational level.

Example

One hundred and ten trainee nurses were asked to complete a general health questionnaire (GHQ) prior to their first ward allocation (Parkes, 1982). A high GHQ score indicates a poor psychoneurotic condition. During the following six months of ward experience, a record was kept of the number of days of short-term sickness absence. It was expected that GHQ score would be positively correlated with days of sickness absence. The GHQ scores were divided into three categories (high, medium and low) and the days of absence scale was divided into four categories (0, 1-2, 3-6 and 7+ days). This permitted the production of the frequency table given in table 11.3. Do the data support the hypothesis?

Table 11.3 Frequency table presentation of data for 110 nurses on a measure of general health and sickness absence (see example in section 11.3)

| | | *Total number of days short-term sickness absence over first six months of ward experience* | | | |
		0	1-2	3-6	7+
	High ≥12	0	5	9	6
GHQ scores	*Medium* 5-11	4	13	15	8
	Low 0-4	11	19	9	11

Step 1 - formulate the problem
(a) Specify the hypotheses:
     Null hypothesis (H_0): there is no association between GHQ score and days
     of sickness absence ($\rho = 0$).
     Specific alternative hypothesis (H_1): GHQ is positively associated with days
     of sickness absence ($\rho > 0$).
(b) Specify the variables:
     Independent variable: GHQ score (three categories).
     Dependent variable: days of sickness absence (four categories). For the
     purposes of correlation computation these are treated as *blocking* variables.

Step 2 - find the rank sums (R_j) and the correlation coefficient (ρ)
(a) Draw up a new frequency table, leaving room for two entries in each cell
     (see table 11.4). If necessary rearrange the rows or the columns so that low
     scores will get low ranks. Remember that the top row gets the lowest shared
     rank among the rows and the first column gets the lowest shared rank
     among the columns. In our case, we must make the low GHQ category the
     top row and the low rate of absence category the first column.
         Enter the row totals (t) and the cumulative row totals (Σt). Do this also
     for the columns.

Table 11.4   Computational layout for example in section 11.3. The values in
parentheses are the rank sums (R_j) for each subject featured in that cell

GHQ	Days of sickness absence						
	0	1-2	3-6	7+	t	Σt	r
Low	11 (33.5)	19 (59.5)	9 (94.5)	11 (123.5)	50	50	25.5
Medium	4 (78.5)	13 (104.5)	15 (139.5)	8 (168.5)	40	90	70.5
High	0 (108.5)	5 (134.5)	9 (169.5)	6 (198.5)	20	110	100.5
column total t	15	37	33	25	k	110	
cumulative total Σt	15	52	85	110			
shared rank r	8	34	69	98			

(b) Find the shared ranks for *both* rows and columns using the standard procedure

$$r = \Sigma t - t/2 + 0.5$$

and enter these in the table.

(c) Find the rank sum (R_j) for each cell by adding the shared rank for the row to the shared rank for the column. Enter this value into the cell (in parentheses). For example the R_j for the top left hand cell is

$$R_j = 25.5 + 8 = 33.5$$

The other R_j values are shown in table 11.4. It is important to appreciate the meaning of the values in each cell. In the top left hand cell the values 11 and (33.5) mean that 11 nurses have a rank sum of 33.5. Once this is understood we can proceed as for section 11.3.

(d) Find ΣR_j by multiplying each rank sum by its corresponding frequency and adding:

$$\Sigma R_j = 11(33.5) + 19(59.5) + 9(94.5) + 11(123.5) + 4(78.5)$$
$$+ 13(104.5) + 15(139.5) + 8(168.5) + 0(108.5) + 5(134.5)$$
$$+ 9(169.5) + 6(198.5)$$
$$= 12\,210$$

Check that the ranking has been carried out correctly:

$$\Sigma R_j = k(k + 1)$$
$$12\,210 = 110(111) = 12\,210$$

Remember that k is the number of rank sums (nurses).

(e) Find ΣR_j^2 by multiplying the *square* of each rank sum by its corresponding frequency and adding:

$$\Sigma R_j^2 = 11(33.5)^2 + 19(59.5)^2 + 9(94.5)^2 + 11(123.5)^2 + 4(78.5)^2$$
$$+ 13(104.5)^2 + 15(139.5)^2 + 8(168.5)^2 + 0(108.5)^2$$
$$+ 5(134.5)^2 + 9(169.5)^2 + 6(198.5)^2$$
$$= 1\,598\,847.5$$

(f) Compute the correction for ties $(\Sigma T_i = T_1 + T_2)$, where

$$T_i = 1 - \frac{\Sigma(t^3 - t)}{k^3 - k}$$

This must be done twice, once for the rows (T_1) and once for the columns (T_2):

$$T_1 = 1 - \frac{\Sigma(t^3 - t)}{k^3 - k}$$

$$= 1 - \frac{(50^3 - 50) + (40^3 - 40) + (20^3 - 20)}{110^3 - 110}$$

$$= 0.852$$

$$T_2 = 1 - \frac{(15^3 - 15) + (37^3 - 37) + (33^3 - 33) + (25^3 - 25)}{110^3 - 110}$$

$$= 0.948$$

See table D (in the appendix) for help with values of $(x^3 - x)$.

$$\Sigma T_i = T_1 + T_2 = 0.852 + 0.948 = 1.802$$

(g) Compute ρ:

$$\rho = \left[\frac{6\Sigma R_j^2}{k^3 - k} - \frac{6(k+1)}{k-1} \right] \frac{2}{\Sigma T_i} - 1$$

$$= \left[\frac{6(1\,598\,847.5)}{110^3 - 110} - \frac{6(111)}{109} \right] \frac{2}{1.802} - 1$$

$$= 0.22$$

Step 3 - interpret the correlation coefficient
The correlation coefficient is positive, indicating that high GHQ ranks tend to be associated with high absence ranks. A positive correlation was predicted. We may, therefore, proceed with a specific test of statistical significance.

Step 4 - compute and evaluate our key statistic (Z or χ^2)
(a) For a *specific test* (where the sign of ρ was correctly predicted) compute:

$$Z = \sqrt{[\rho^2(k-1)]}$$

where k is the sum total of all frequencies in the table.
    We can then evaluate Z by consulting tables of the normal distribution (table A in the appendix). The relevant critical values are reproduced here:

5%	2.5%	1%	0.1%
$z \geqslant 1.64$	1.96	2.33	3.10

To be statistically significant at a given level, Z must be greater than or equal to the appropriate critical value.

(b) For a *nonspecific test*, compute

$$\chi^2 = \rho^2(k-1)$$

where k is the sum total of all the frequencies in the table. We can then evaluate χ^2 by consulting tables of the chi-square distribution (table B in the appendix) for one degree of freedom (df $= 1$). The relevant values are reproduced here:

Degrees of freedom	5%	2.5%	1%	0.1%
1	$\chi^2 \geqslant 3.8$	5.0	6.6	10.8

To be statistically significant at a given level χ^2 must be greater than or equal to the appropriate critical value.

We need a *specific test* for our example. Compute

$$Z = \sqrt{[\rho^2(k-1)]}$$
$$= \sqrt{[0.22^2(110-1)]}$$
$$= 2.30$$

Our result is, therefore significant at the 2.5 per cent level ($p < 0.025$) and almost significant at the 1 per cent level.

Step 5 - draw conclusions

Because our result is significant ($p < 0.025$) we may reject H_0 (see step 1) and conclude that a positive association probably exists between GHQ score and days of sickness absence among trainee nurses on their first six months of ward experience. A correlation coefficient of $\rho = 0.22$ was obtained, but a higher value (and a more significant value) is likely to have been achieved if the full procedure of section 11.2 had been used.

11.4 Concordance

Rank sum correlation between two variables is applied to a simple design involving a large number of subjects each ranked twice, one on each variable. The subjects are treated as samples and the rankings for the two variables are located in two separate blocks. The special feature of correlation analysis is that our attention is focused on the agreement between the two sets of ranks rather than the differences between the samples. The same concept can be expanded to the many-variable case where we are interested in the agreement between any number of sets of rankings. This is called the evaluation of *concordance*.

The method is due to Kendall, who devised the statistic W which reflects the degree of concordance among the m variables. W varies between 0 and 1; a low value of W indicates little concordance, a high value indicates more concordance, and a value of 1 specifies perfect agreement between all the sets of ranks. W can be interpreted as a reflection of the average of the rank correlation coefficients between all possible pairs of the variables being studied. In fact, we can compute this average from W:

$$\bar{\rho} = (mW - 1)/(m - 1)$$

where m is the number of variables and $\bar{\rho}$ is the average value of ρ for all possible pairs of the m variables.

The design in question is in no way different from that analysed by the nonspecific test (given in section 10.9) for three or more related samples. This design has more than one block and only one score per cell. However, the interest there was directed toward the differences between the samples rather than the agreement between the blocks. Of course, the two interests are complementary since a clear pattern of differences between the samples can only become apparent if the blocks agree in the pattern of rankings. In section 10.9. the measure of differences between the ranks was our statistic H. As we might expect, H and W are closely related:

$$H = m(k - 1) W$$

or

$$W = \frac{H}{m(k - 1)}$$

We clearly do not need a new procedure for computing W. All we need to do is find H, using the methods of section 10.9, and then convert to W. The example given concerned depression ratings of 14 nurses at five points before and during their allocation to a male medical and a female surgical ward. In chapter 10 we wanted to know if the wards had a different effect on the nurses' tendency to be depressed. We might wish to turn the question on its side and ask to what extent the nurses' moods fluctuated in agreement as they moved from ward to ward. W would appear to be a suitable measure:

$$W = \frac{H}{m(k - 1)} = \frac{7.15}{14(5 - 1)} = 0.13$$

where m is the number of blocks (nurses) and k is the number of samples (times at which measurements were made). The value of W is low, indicating little agreement.

W is, of course, just a measure. To evaluate its reliability we need a test of significance. The obvious test is to use H and consult tables of the chi-square distribution (table B in the appendix). This we have already done in the example in section 10.9, and found our result not to be statistically significant ($p > 0.09$).

Kendall's W adds little to the analysis already performed in section 10.9. It supplies us with a statistic which reflects the extent of the agreement between blocks on a scale from 0 to 1, but this statistic is not very illuminating in that it does not refer beyond itself to any useful properties (as the parametric multiple correlation coefficient does). The key statistic is H, which can be evaluated in a useful manner. Perhaps the true value of Kendall's statistic W is to make clear the complementary nature of correlation analysis (reviewed in this chapter) and

Table 11.5   Neuroticism and motion sickness measures for 22 students (see exercise 11.1)

Subject	Neuroticism	Motion sickness
1	7	0.75
2	7	0.00
3	8	1.00
4	10	0.77
5	10	4.11
6	11	0.44
7	11	3.75
8	11	3.44
9	12	1.63
10	13	2.33
11	14	3.43
12	15	2.00
13	16	2.50
14	17	3.00
15	17	20.25
16	17	0.87
17	18	0.55
18	18	1.33
19	19	9.75
20	17	3.33
21	20	5.33
22	21	5.00

dispersion analysis (reviewed in the section on nonspecific tests in chapter 10). They are the same thing looked at differently.

Exercises

11.1   In the example in section 11.2 the results of a motion sickness question-naire were correlated with an extroversion measure from the Eysenck personality inventory. The same inventory gives a measure of neuroticism whose results are given in table 11.5. Is neuroticism correlated with motion sickness?

11.2   Section 11.2 and exercise 11.1 refer to a pilot study based on only 22 students. The main study involved 60 mature students studying psychology in the summer vacation (table 11.6). Do their results bear out the conclusions of the pilot study? The published paper (Wilding and Meddis, 1972) analysed the data using the long method of section 11.2, but the frequency table method of section 11.3 may be used for a conservative approximation.

Table 11.6   Motion sickness, extroversion and neuroticism scores for 60 students (exercise 11.2)

Subject	Motion sickness	Extroversion	Neuroticism
1	0	14	14
2	0	15	16
3	0	15	7
4	0	17	4
5	0	20	3
6	0	4	10
7	0	13	4
8	3.4	6	12
9	3.4	15	11
10	4	13	4
11	4	9	8
12	4.5	15	6
13	4.5	14	8
14	6.8	5	18
15	7.9	18	2
16	8	20	16
17	8	8	11
18	8	6	7
19	9	17	8
20	9.9	11	6
21	11.3	10	11

Table 11.6   continued

Subject	Motion sickness	Extroversion	Neuroticism
22	12	15	5
23	13.5	17	13
14	13.5	4	14
25	15	15	14
26	15	12	13
27	15.7	11	1
28	16	16	15
29	18	11	13
30	18	18	3
31	18	11	10
32	19.1	4	15
33	20.3	11	14
34	21	11	11
35	21.6	21	12
36	22	9	13
37	22	15	3
38	22.5	15	8
39	23	14	6
40	26	4	16
41	27	17	18
42	27	10	20
43	27	6	7
44	32.6	10	15
45	34	11	14
46	38.3	13	9
47	39.8	17	9
48	41	8	17
49	41	13	20
50	45	15	5
51	46	17	10
52	51	15	18
53	56	10	14
54	58.5	22	6
55	60	8	10
56	60.8	13	16
57	63	14	15
58	64.3	5	18
59	86	17	17
60	93	11	17

12 Multiple and *post hoc* hypothesis testing

12.1 Introduction

Chapters 7-10 presented a range of procedures for testing the null hypothesis against a single specific hypothesis. On many occasions, a researcher will want to test *a number of specific hypotheses*. These include comparisons between each treatment and a control, or comparisons between each possible pair of treatments, or simply a short list of trends or contrasts across the treatments. If these tests are planned *before* the data are collected – or can be construed as following as a direct consequence of the researcher's theoretical drift – they are called *a priori* or planned multiple tests of specific hypotheses. On a different set of occasions, a researcher may wish to test one or more hypotheses which occur to him *after* the data have been collected and scrutinized. These are called *post hoc* analyses. This chapter will consider the adjustment needed to the procedures of chapters 7-10 when such analyses are attempted. In both cases, *a priori* and *post hoc* hypothesis testing follow the methods presented in chapters 7-10 exactly as given up to step 4, when slightly different techniques are used to 'compute and evaluate our key statistics'.

Multiple hypotheses

The need for a change in step 4 arises from a consideration of the meaning of the term 'significance levels'. A result which is significant at the 5 per cent significance level is a result which could have occurred on 5 per cent of occasions even if H_0 were true. To reject H_0 is to risk confusing one of those occasions with a situation where H_0 is false. This risk accompanies every attempt to test the same null hypothesis. Thus, if we have multiple attempts to reject H_0, the risk of falsely rejecting H_0 *on at least one occasion* increases with each attempt. For example, if we adopt the 5 per cent significance level as our criterion for rejecting H_0 on each attempt, then after two attempts we will have accumulated a 10 per cent risk, after three attempts we will have accumulated a 15 per cent risk etc., if the specific alternative hypotheses are statistically independent. At this point our concept of 'significance level' becomes fragmented. On the one hand, we may say that we are taking a 5 per cent risk of falsely rejecting H_0 as far as each test is concerned, whereas on the other hand we seem to be taking a 15 per cent risk of falsely rejecting H_0 on an overall basis. In statistical jargon, the former is called the 'comparison-wise' error rate and the latter is called the 'experiment-wise' error rate.

In chapters 7–10 we have been exclusively concerned with comparison-wise error rates. From now on we shall look to experiment-wise error rates. Of course, if only one specific alternative hypothesis is tested in an experiment then the experiment-wise and the comparison-wise error rate are the same thing. The experiment-wise error rate is a simple function of the number of specific alternative hypotheses being tested. If these hypotheses are statistically independent we can say that

$$P' = p_1 + p_2 \ldots + p_i \ldots \tag{12.1}$$

where P' is the experiment-wise error rate and p_i is the probability of rejecting H_0 falsely for the ith comparison. (Two hypotheses are statistically independent if the outcome of a statistical analysis of one is totally unrelated to the outcome of a statistical analysis of the other.) In general, the hypotheses are not completely statistically independent and we use the following expression:

$$P' \not> p_1 + p_2 \ldots p_i = \ldots \tag{12.2}$$

which states that P' is not greater than the sum of the individual error rates. This expression is known generally as the *Bonferroni inequality*. To calculate P' exactly we would need to know the exact degree of the statistical inter-dependence between each pair of p_i. Usually this is not known and we content ourselves with the estimate, which is often a slightly larger than (but never less than) the true value of P'. P' is therefore a *conservative* estimate of the true value of the experiment-wise error rate.

The Bonferroni inequality is not, however, in a form which is immediately useful to the researcher. Firstly, he is not usually in possession of the exact values of p_i (although they could, in principle, be calculated). Secondly, H_0 is not usually rejected on the basis of the collective results of all the tests – rejection typically follows a single 'significant' hypothesis test. The solution is to fix P' at a particular value, say 0.05 (5 per cent experiment-wise significance level),

$$0.05 = p_1 + p_2 \ldots + p_i \ldots \tag{12.3}$$

and to equalize the p_i so that

$$0.05 = Qp$$

or

$$p = 0.05/Q \qquad (12.4)$$

Our new value p is the error rate – which, if sustained over Q hypotheses, would lead to a 5 per cent experiment-wise error rate. For example, a test of one specific hypothesis among three needs to achieve a probability level of $0.05/3 = 0.0167$ before being statistically significant at the 5 per cent (experiment-wise) level.

One result of this is that we need new tables for Z which take into account the number of hypotheses being tested. From the above we can see that the 5 per cent significance level corresponds to an error rate of 1.67 per cent per comparison. We therefore need a value of Z which corresponds to a probability of 0.0167. Tables of the normal distribution show that this value is 2.13. This is the value of Z which needs to be exceeded on any single test of a specific hypothesis (when three are being tested simultaneously) if the result is to be deemed significant at the 5 per cent (experiment-wise) significance level. By using this simple logic it has been possible to construct the complete version of table C in the appendix, which gives critical values of Z for our main significance levels (5, 2.5, 1.0 and 0.1 per cent) for a range of values of Q, the number of specific hypotheses being simultaneously considered against the same null hypothesis.

To use table C we need merely specify the number of alternative hypotheses being tested. This alone determines the appropriate row of the table. For example, if we want to compare every possible *pair of treatments* when there are k treatments altogether, then we are testing $k(k-1)/2$ hypotheses because that is the number of possible pairs. Alternatively, if one of the k treatments is a control and we wish to *compare each of the other treatments with the control*, then we are testing $k-1$ hypotheses (see section 12.2 for an example). Often the experimenter simply approaches the data with two or three hypotheses in mind. In that case, it is enough that these are explicitly listed in advance of the analysis so that they can be counted.

Multiple nonspecific tests

So far we have assumed that we are dealing with specific hypotheses which correctly imply the direction of the difference between two groups or the direction of a trend. This is not always the case, especially in pairwise testing when any difference in either direction might be of interest. It also happens when looking at a series of trends (e.g. linear, quadratic, cubic etc.) where the direction of the trend might not have been predicted. In this case each test is made up of two hypotheses (one for each direction) and must count as two when totalling the number of hypotheses under test. Thus, when testing all possible pairs among k treatments, we might be dealing with $k(k-1)$

hypotheses even though there are only $k(k-1)/2$ pairs. Only the researcher knows whether he is interested in only one direction for each pair or both directions. Sometimes, he will have a definite prediction for some pairs but not for others. In that case, count one for each hypothesis associated with a predicted direction and count two for each hypothesis with no directional prediction.

Post hoc *analysis*

When a hypothesis is chosen after the data have been scrutinized, the researcher is in the position of someone who decided in advance to consider every possible hypothesis and report only the most significant. This is because he allowed the data to suggest his hypothesis, probably on the grounds that this hypothesis was most likely to yield a significant result. For example, he might choose the two treatments whose sample rank means are furthest apart and test them, or he might rearrange the samples so that their rank means formed an ascending trend and then test for a trend. If we knew how many possible hypotheses there were, we could simply recommend that the researcher use that number to access table C. That is a complex estimation and, fortunately, an easier method is available.

Scheffé (1953) has shown that the significance level associated with a given result can be found using the following transformation:

$$H = Z^2$$

and then using the fact that H is distributed approximately as chi-square with $(k-1)$ degrees of freedom. This elegant solution allows the researcher to try any alternative hypothesis he likes without any restriction. He can even take the sample rank means and use them as coefficients when computing the statistic L (which is guaranteed to produce the largest possible value of Z for that data set). If he does that, he is quite likely to obtain exactly the same value of H as he would have done if he had simply performed a nonspecific test! There is justice in such a result since, in both cases, the researcher approached the data initially without any specific alternative hypothesis. *The result also implies that there is little point in searching for a* post hoc *hypothesis that will yield a significant result if the original nonspecific test did not yield a significant result.* See section 12.4 for an example.

Comparing pairs of samples

There are two situations in which researchers wish to select pairs of samples for comparison. These are (a) comparisons of each treatment with a control and (b) comparing all possible pairs of treatments. Both methods require a choice of the best procedure. Do we (a) apply a set of coefficients which are all zero except for the pair of samples to be compared (procedure I) or do we (b) take each pair quite separately from the others and submit them to a two sample test (procedure II)? Procedure II is considerably more effortful from a computational point of view, but is purer in the sense that the comparison of two samples is not confounded by the presence of the other samples. These two procedures have been compared formally and informally by Dunn (1964), who concludes that 'either procedure seems good'.

In fact, it is easy to create situations which will yield different results for a given pair of samples using the two methods. The problem has already been discussed in section 2.10, where it was pointed out that the coefficients

$$\lambda_j \qquad -1 \quad 0 \quad 0 \quad 0 \quad +1$$

imply the hypothesis

$$M_1 < M_2 = M_3 = M_4 < M_5$$

where M_k is the median of the kth sample.

We can see this more easily if we create an equivalent set of coefficients by adding 2 to each of the original coefficients:

$$\lambda_j' \qquad 1 \quad 2 \quad 2 \quad 2 \quad 3$$

This illustrates the common error of thinking that a zero coefficient is an instruction to ignore a particular sample. Our example is clearly not a simple comparison between samples 1 and 5 but a trend across samples 1 to 5, where samples 2, 3 and 4 are hypothesized to be between samples 1 and 5. Clearly, if we leave samples 1 and 5 unchanged but manipulate the other samples, we shall influence the result, making the comparison between 1 and 5 more or less significant. This is not what we would expect from a simple contrast between the two samples.

In practice, it is probably best to attempt all contrasts using procedure I just to gain an overall impression of the possibilities. This can be followed up by using procedure II to check any particular comparison which was either significant or almost significant during the first phase. In many situations the result will be very similar, but occasionally the analysis using procedure I will prove misleading. See section 12.3 for an example.

Sensitivity of post hoc *and multiple test procedures*

Critical values of Z are increased in table C (appendix) as the number of hypotheses to be tested is increased. This means that any individual trend or contrast is less likely to lead to the rejection of H_0 for a given set of data. Of course, the overall likelihood of rejecting H_0 remains the same but our attention is normally focused on the fate of individual alternative hypotheses. Researchers should, therefore, be quite clear that specific alternative hypotheses are less likely to lead to a significant result if other specific alternative hypotheses are simultaneously entertained. If the amount of available data is small, this can often result in situations where a particular trend or contrast may fail if tested along with other contrasts but might have proved significant if tested in isolation. This problem is not acute if there is an abundance of data. The routine use of multiple hypotheses and *post hoc* hypotheses should be restricted to such situations.

Beginners may feel that the system is open to abuse by unscrupulous researchers who might refuse to acknowledge that their tests were, in fact, *post hoc* and prefer to present their results as single comparisons. In practice, this form of cheating is often easily spotted since researchers must use hypotheses which are consistent with their own theories or previous work or with the theories and work of others. They are not usually at liberty to 'pull hypotheses out of a hat'. Where they succeed in this dastardly manoeuvre they run the risk of vilification by other researchers, who will certainly report any failures to replicate the result. In any case, researchers who are intent on cheating do much better to simply forge their data, since this is much less easily detected than confusing *post hoc* and *a priori* techniques.

12.2   Comparison with a control – example

The data for this example are given in section 10.4, and concern four different methods of making a motorcyclist more conspicuous. Sample 1 is a control condition. We wish to compare each method with the control to ascertain whether or not it is effective in making the motorcyclist more visible. This involves four different comparisons. Each comparison involves computing the value of Z using the method appropriate for that design. Table 12.1 shows the computations and the four resulting values of Z.

Before consulting table C in the appendix, we must decide how many hypotheses are actually under test. If the hypotheses are bidirectional (i.e. positive or negative differences are *both* of interest), then count two for every comparison. (In this case, all values of Z should be treated as positive when consulting table C.) If the hypotheses specify only one acceptable direction of difference between the treatment and the control, then count only one per comparison. In our example, we expect every method to increase the visibility of the motorcyclist. We therefore count only one per comparison, giving a count of four hypotheses.

The appropriate row in table C is therefore

Q	5%	2.5%	1%	0.1%
4	$\cdot z \geqslant 2.24$	2.50	2.81	3.48

Table 12.1   Comparisons with a control for an example taken from Fulton (1983) (section 10.4). The conditions were (1) control (2) headlight on (3) single running light (4) double running light (5) fluorescent jacket

Condition	1	2	3	4	5	$\sum\limits_{j=1}^{k} \lambda_j S_{\cdot j}$	$\sum\limits_{j=1}^{k} \lambda_j^2$
$S_{\cdot j}$	28	51	42	45	43		
λ_{1j}	-1	$+1$	0	0	0	23	2
λ_{2j}	-1	0	$+1$	0	0	14	2
λ_{3j}	-1	0	0	$+1$	0	17	2
λ_{4j}	-1	0	0	0	$+1$	15	2

From section 10.4 we have $\Sigma S_i. = 209$ and $\Sigma S_i^2 = 795$. Using the expression (taken from section 10.4)

$$Z = \sqrt{\left\{ \frac{k(k-1)\left(\sum\limits_{j=1}^{k} \lambda_j S_{\cdot j} \right)^2}{k \sum\limits_{i=1}^{m} S_i. - \sum\limits_{i=1}^{m} S_i^2 \; (\Sigma\lambda_j^2)} \right\}} = \sqrt{\left\{ \frac{5(4)(\Sigma\lambda_j S_{\cdot j})^2}{[5(209) - 795]2} \right\}}$$

$$= 0.2\Sigma\lambda_j S_{\cdot j}$$

1 vs 2 $Z = 0.2(23) = 4.6$

1 vs 3 $Z = 0.2(14) = 2.8$

1 vs 4 $Z = 0.2(17) = 3.4$

1 vs 5 $Z = 0.2(15) = 3.0$

To be significant our value of Z must be greater than or equal to the critical value for the appropriate level. We may, therefore, summarize our results as follows:

Comparison	Z	Significance level
(1) control vs (2) headlight	4.6	$<0.1\%$
(1) control vs (3) single running light	2.8	$<2.5\%$ (almost 1%)
(1) control vs (4) double running light	3.4	$<1\%$ (almost 0.1%)
(1) control vs (5) fluorescent jacket	3.0	$<1\%$

All of the comparisons are significant, which supports the idea that all of the visibility aids yielded higher detection rates than the control condition.

12.3   Pairwise contrasts – example

The data for this example are taken from section 9.6, where Southall and Stone (1982) give glare thresholds for (a) partially sighted children (b) normally sighted children and (c) normally sighted adults. In section 9.6 we performed an overall test which was not significant. Here we shall compare each sample with both of the other samples. Each comparison involves computing the value of Z appropriate for that design. In our case we use the methods of section 9.3. The computations are given in table 12.2.

Before consulting table C in the appendix, we must decide how many hypotheses are actually under test. If the hypotheses are bidirectional (i.e. positive or negative differences would both be accepted), then count two for each comparison. In this case, all values of Z should be treated as positive when consulting table C. In our example, we have no specific alternative hypotheses to predict the direction of difference. There are three comparisons and we therefore count six hypotheses altogether. If we do have specific alternative hypotheses which indicate only one acceptable direction of difference, then we need count only one hypothesis per comparison.

The appropriate row in table C is therefore

Q	5%	2.5%	1%	0.1%
6	$z \geqslant 2.39$	2.64	2.94	3.59

To be significant our value of Z must be greater than or equal to the critical value for the appropriate level. We may therefore summarize our results as follows:

Comparison	Z	*Significance level*
Partially sighted vs normal children	0.70	not sig ($p > 0.05$)
Partially sighted children vs adults	1.79	not sig ($p > 0.05$)
Normal children vs adults	1.32	not sig ($p > 0.05$)

None of the comparisons is statistically significant.

12.4   *Post hoc* analysis – example

Specific alternative hypotheses which are proposed *post hoc* (i.e. after the data have been collected and examined) are to be analysed in the same way as any other specific alternative hypothesis. The difference arises when the time comes to evaluate the statistical significance of the result. First a conversion is

Table 12.2 Pairwise comparisons for an example taken from Southall and Stone (1982) (section 9.6). The conditions were (a) partially sighted children, (b) normally sighted children and (c) normally sighted adults

Comparison		(a)	(b)	(c)	
	R_j	598	234	158	$T = 0.9998$
	n_j	24	10	10	$N = 44$
(a) vs (b)	λ_{1j}	+1	−1	0	$L = 364$
			$\Sigma n_j \lambda_j^2 = 34$		$\Sigma n_j \lambda_j = 14$
(a) vs (c)	λ_{2j}	+1	0	−1	$L = 440$
			$\Sigma n_j \lambda_j^2 = 34$		$\Sigma n_j \lambda_j = 14$
(b) vs (c)	λ_{3j}	0	+1	−1	$L = 76$
			$\Sigma n_j \lambda_j^2 = 20$		$\Sigma n_j \lambda_j = 0$

Using the expression

$$Z = \frac{L - (N + 1)(\Sigma n_j \lambda_j)/2}{\sqrt{\{(N + 1)[N \Sigma n_j \lambda_j^2 - (\Sigma n_j \lambda_j)^2] T/12\}}}$$

(a) vs (b)

$$Z = \frac{364 - (45)(14)/2}{\sqrt{\{(45)[44(34) - 14^2](0.9998)/12\}}} = 0.70$$

(a) vs (c)

$$Z = \frac{440 - (45)(14)/2}{\sqrt{\{(45)[44(34) - 14^2](0.9998)/12\}}} = 1.79$$

(b) vs (c)

$$Z = \frac{76 - (45)(0)/2}{\sqrt{\{(45)[44(20) - 0^2](0.9998)/12\}}} = 1.32$$

made from Z to H:

$$H = Z^2$$

and then H is referred to tables of chi-square with $k - 1$ degrees of freedom. This gives the researcher total freedom to invent *post hoc* hypotheses, but at the price of requiring a much higher value of Z to reach significance.

By way of illustration we shall use the data in section 10.9 to illustrate both *post hoc* analysis and an important principle. Parkes (1982) compared measures of student nurses' mood before the investigation then during and after exposure to two wards. The sample rank sums for the five groups are given in table 12.3. If we wished to obtain the most statistically significant result possible, we could do no better than use the sample rank means themselves as coefficients. These will be perfectly correlated with the actual sample rank means and optimally correlated with the individual ranks in the corresponding samples. At first sight

Table 12.3 Computation of *post hoc* analysis based on section 10.9 using sample rank means as coefficients

Example taken from table 10.8

		Before ward allocation	*Male medical*		*Female surgical*	
			Mid	*End*	*Mid*	*End*
rank sum	R_j	37	51.5	48	38.5	35
	λ_j	2.65	3.68	3.43	2.75	2.5

Note that the λ_j, on this occasion, are taken from the sample rank means $(\bar{R}_j)$.

Other values taken from section 10.9

$$m = 14; \quad k = 5; \quad \Sigma T_i = 11.85$$

Compute, using the method given in section 10.3

$$L = \Sigma \lambda_j R_j = 645.215$$

$$\Sigma \lambda_j = 15$$

$$\Sigma \lambda_j^2 = 46.09$$

$$E(L) = \tfrac{1}{2} m(k+1) \, \Sigma \lambda_j = \tfrac{1}{2}(14)(6)(15) = 630$$

$$\text{var}(L) = \tfrac{1}{12}(k+1)[k \Sigma \lambda_j^2 - (\Sigma \lambda_j)^2] \, \Sigma T_i$$

$$= \tfrac{1}{12}(6)[(6)(46.09) - (15)^2](11.85)$$

$$= 32.273$$

$$Z = \frac{L - E(L)}{\sqrt{\text{var}(L)}} = \frac{645.215 - 630}{\sqrt{32.273}}$$

$$= 2.678$$

this may seem to be an unscrupulous method for achieving reportable results but, as we shall see, it is perfectly meaningful within our hypothesis testing system. Table 12.3 shows the computations for this operation, with a resulting value of $Z = 2.678$. For a *post hoc* analysis, which this most certainly is, we must convert Z to H:

$$H = Z^2 = 2.678^2 = 7.17$$

which should be treated as a chi-square value with four degrees of freedom. It is not statistically significant.

Of course, *any* coefficients can be used in a *post hoc* test. The sample rank means were only used by way of illustration of the freedom given to the test user.

A link between specific and nonspecific tests

The value of H obtained above is precisely the same as that obtained in section 10.9 using a nonspecific test (i.e. $H = 7.17$ with four degrees of freedom). It is generally the case that a nonspecific test will give the same result as a *post hoc* specific test which uses the sample rank means as coefficients. Indeed we might use this as an operational definition of what we mean by a nonspecific test. Put in simple terms, a nonspecific test asks whether the observed pattern of sample rank means can be a reliable guide to the pattern of effects which exist in the world from which the data were sampled. In this case the hypothesis to be tested can only be formulated after the data have been inspected and is, therefore, genuinely *post hoc*. The idea that nonspecific tests are merely optimum *post hoc* specific tests is a powerful explanatory concept which can act as a useful unifying principle across a wide range of hypothesis testing methods.

There is more to be extracted from this example, however. It is asserted, without proof, that no set of coefficients will yield a higher value of Z than the sample rank means. We have also illustrated in this section the principle that this maximum value of Z corresponds (after squaring) to the value of H obtained from a nonspecific analysis, which we know is distributed approximately as chi-square with $k - 1$ degrees of freedom. The critical values of chi-square for $k - 1$ degrees of freedom are, therefore, appropriate criteria for Z^2 when using sample rank means as coefficients. These critical values must, therefore, be conservative (i.e. too high) for Z^2 derived using *any other* set of coefficients (which must necessarily produce a lower value of the statistic). The words 'any other' are crucial here since they permit the researcher to generate his *post hoc* specific hypothesis on any basis he likes. *The above logic guarantees that non-specific tests will always be relatively conservative* and that, in most cases, it will be a low power test so that often the null hypothesis will not be rejected even though it is false. This power loss becomes increasingly strong as the number of samples increases. It is also very marked where there are small numbers of scores in the sample. In conclusion *post hoc* procedures are very safe but very insensitive and are typically used only as a last resort.

13 Factorial designs

13.1 Introduction

Factorial designs permit the simultaneous investigation of the effects of two or more independent variables and their mutually interactive effects upon a dependent variable. In the preceding chapters, attention has been focused on designs involving a *single* independent variable and its simple effect upon a dependent variable. In the examples given in the following we examine designs with more than one independent variable.

Designs involving multiple independent variables are common and result from two distinct motives. The first motive is one of economy; many investigators wish to examine the effects of a range of variables within the bounds of a single experiment. The second motive is more subtle – a wish to examine the mutual interaction of variables, where the term 'interaction' is to be understood as 'a modifying influence of one variable upon the effect of a second variable on the dependent variable'.

Complexity of design is usually associated with a variety of hazards and penalties. These include (a) problems of interpretation, (b) increases in the minimum number of observations needed, (c) restrictions in the cell frequencies (which must always be equal) and (d) increased complexity of computation. None of these complications has prevented factorial analysis from becoming very popular in the domain of parametric analysis. We might reasonably assume that the same popularity will one day extend to factorial analysis in the nonparametric domain.

Very few new concepts are required to understand the methods of factorial analysis. These mainly involve repeated application of the simpler methods of chapters 6-10. Because we are asking a number of simultaneous questions, the principles discussed in chapter 12 are also relevant. The basic scheme of non-specific factorial analysis of variance by ranks has been modelled closely on accepted procedures currently used in parametric analysis of variance (see e.g. Winer, 1971). This should be of benefit to those who have already mastered that particular art. However, we will place greater stress on the possibilities of factorial analysis using *specific hypotheses*, which are typically given less emphasis in the parametric analogue.

Factorial analysis using specific hypotheses

To avoid too many abstractions, the basic features of factorial analysis will be illustrated using the example in section 13.2. This involves 16 students whose arithmetic competence was assessed either in the very hot conditions of a heated environmental chamber or at a moderate ambient temperature. Although we might reasonably expect performance to be worse at high temperatures than at low temperatures, it was known that many investigators had found little difference. One possibility was that subjects were highly motivated by the challenge of the 'hot room' and, by extra effort, compensated for any of the performance depressing effects of the physical environment. Gledhill (1976) sought to investigate this by lowering the motivation of half of the students by treating them in an offhand way and generally giving them the impression that the experimenter was incompetent. He expected that the depressing effect of heat would be more obvious in the low motivation group. Gledhill's data are given in table 13.1. The pattern of his expectations is summarized in figure 13.1. In brief he expects that:

(a) Low motivation subjects will perform worse than high motivation subjects.

Table 13.1　Number of correct addition sums for 16 subjects under two conditions of environmental temperature and two levels of motivation (example in section 13.2)

Motivation	Temperature	
	Normal	*Hot*
High (interested)	83	91
	83	91
	116	97
	73	87
	(sample 1)	(sample 2)
Low (irritated)	81	81
	102	69
	95	90
	79	65
	(sample 3)	(sample 4)

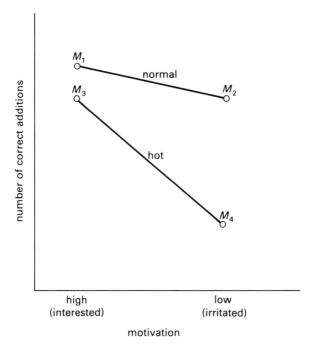

Figure 13.1   Graphical representation of Gledhill's (1976) three main expectations (see text). The circles represent the medians of scores in the four samples

   (b) Subjects working in high ambient temperatures will perform worse than those in moderate ambient temperatures.

   (c) The performance depressing effect of raised temperatures will be more marked in low motivation subjects.

To evaluate these three specific hypotheses we need to recast the data in table 13.1 into a more familiar format. We can line up the four samples to form a 2 × 2 hierarchical pattern:

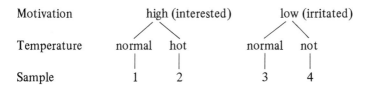

This recasting is fully illustrated in table 13.2. In our example, these are four independent samples and we can proceed using the methods given in chapter 9. If the samples were related, then the techniques of chapter 10 would be

Table 13.2 Computational layout for example in section 13.2

Motivation	Interested				Irritated			
Temperature	Normal		Hot		Normal		Hot	
Sample	1		2		3		4	
	Score	Rank	Score	Rank	Score	Rank	Score	Rank
	73	3	87	9	79	4	65	1
	83	7.5	91	11.5	81	5.5	69	2
	83	7.5	91	11.5	95	13	81	5.5
	116	16	97	14	102	15	90	10
rank sums R_j	34		46		37.5		18.5	
sample size n	4		4		4		4	
rank means $\bar{R}_j$	8.5		11.5		9.37		4.625	
					N	16		

Correction factor for ties

Tied score	Rank	Tie length t	$t^3 - t$
81	5.5	2	6
83	7.5	2	6
91	11.5	2	6

$$\Sigma t^3 - t = 18$$

$$T = 1 - \frac{\Sigma t^3 - t}{N^3 - N} = 1 - \frac{18}{16^3 - 16} = 1 - \frac{18}{4080}$$

$$= 0.9956$$

appropriate. A factorial analysis using related samples is illustrated in section 13.3.

We have three specific hypotheses to test, and these must be evaluated using three separate analyses. Each analysis requires a separate set of coefficients. Hypothesis (a) implies that high temperature motivation subjects (samples 1 and 2) will perform better than low motivation subjects (samples 3 and 4). Using the methods of section 2.10, we can express this hypothesis in terms of the four sample medians:

Hypothesis (a) $M_1 + M_2 > M_3 + M_4$

where M_j is the median of the *j*th sample.

Using normal algebraic conventions, rearrange the inequality so that all medians are to the left hand side of the expression, and make all signs explicit:

$$+M_1 + M_2 - M_3 - M_4 > 0$$

We can use the signs attached to each median to form the λ coefficients for this hypothesis:

Sample	1	2	3	4
λ_j	$+1$	$+1$	-1	-1

Similarly, for hypothesis (b), we expect the high temperature condition (samples 2 and 4) to yield a poorer performance than the low temperature condition (samples 1 and 3). Using the same procedure:

Hypothesis (b) $M_1 + M_3 > M_2 + M_4$

When rearranged, this yields the appropriate coefficients:

$$+M_1 \quad -M_2 \quad +M_3 \quad -M_4 > 0$$

Sample	1	2	3	4
λ_j	$+1$	-1	$+1$	-1

This is an elaborate method of arriving at coefficients which could have been devised very quickly using an intuitive approach. However, it is very useful for making explicit the meaning of the coefficients used for hypothesis (c). This hypothesis asserts that the depressing effect of heat on performance will be greater for the less motivated group. In other words, the difference due to heat in the low motivation samples (i.e. between samples 2 and 4) will be greater than the difference due to heat in the high motivation group (i.e. between samples 1 and 3).

Hypothesis (c) $M_2 - M_4 > M_1 - M_3$

On rearranging, this becomes:

	$-M_1$	$+M_2$	$+M_3$	$-M_4 > 0$
Sample	1	2	3	4
λ_j	-1	$+1$	$+1$	-1

which are the coefficients to be used in section 13.2.

In summary, we can analyse the three specific hypotheses in this example using three separate analyses using the following lambda coefficients:

hypothesis (a)	λ_j	$+1$	$+1$	-1	-1
hypothesis (b)	λ_j	$+1$	-1	$+1$	-1
hypothesis (c)	λ_j	-1	$+1$	$+1$	-1

For more complex situations, the discovery of appropriate λ coefficients can prove more difficult. As long as these reflect the essence of the relevant hypothesis, however, there should be few problems.

Hypotheses (a) and (b) are referred to as *main effects* because they concern the simple effects of two independent variables. Hypothesis (c) is called an *interaction effect* because it refers to the moderating effect of one variable upon another (i.e. the effect of motivation upon the effect of heat upon performance). If the interaction effect is statistically significant, our interpretation of the main effects must be qualified since it implies that the main effects vary in strength according to the level of the other variable. For example, a significant interaction in our example might imply that heat impairs performance but only in low motivational situations. As a result, a simple generalization that heat impairs performance (i.e. a main effect) would be unjustified since it tells only half the story.

Nonspecific factorial analysis

Parametric analysis of variance has tended to emphasize nonspecific methods, so that the experienced ANOVA practitioner will find this section more familiar than the last. The advantage of nonspecific hypothesis testing is that it lends itself well to a mindless and mechanical approach to the computational procedures. The intellectual effort is not required until after the data have been analysed, when the researcher must sit down and work out what it all means!

Basically, the procedure is simple. For each independent variable we separately evaluate a 'main effect'. For each pair of independent variables we separately evaluate a 'first-order interaction effect'. For each triplet of independent variables we evaluate a 'second-order interaction effect' and so on. In the example introduced in this section (and analysed in section 13.2) we have only two main effects: (a) motivation and (b) temperature, one for each independent variable. Since we have only two independent variables, we can have only one possible interaction effect, that between motivation and temperature. As the number of independent variables increases, the number of effects to be evaluated can increase alarmingly. This is the price to pay for complex experiments.

The computational procedures are based on the methods given in chapters 9 and 10 for evaluating nonspecific hypotheses according to whether the samples are unrelated (see chapter 9) or related (see chapter 10). Some preliminary rearrangement of the data table is usually necessary to get the samples into a hierarchically ordered set. This procedure has already been illustrated in this section. Once this has been done and the data ranked, we need to find a new set of rank sums for each effect being studied. Thus for the 'motivation' effect we have two rank sums, one for each level of motivation. These two rank sums are found by combining all the rank sums represented at the corresponding level of the effect. Once the rank sums have been found the computation of the statistic H proceeds as for simpler designs.

When computing interaction effects the value of H must be adjusted downwards by subtracting the values of H corresponding to all lower order effects. Thus a first-order interaction effect is adjusted downwards by subtracting the H values for the two relevant main effects. For example, when computing the interaction effect for motivation and temperature we compute H using the four rank sums which correspond to each combination of both levels of motivation and temperature. Then we must subtract the two main effects for motivation and temperature to get the true interaction:

$$H_{M \times T} = H - H_M - H_T$$

This technique ensures that our estimate of the interaction of two main effects is not contaminated by the presence of the main effects. The procedures are illustrated in detail in the sections which follow.

13.2   2 × 2 factorial analysis

Example

Gledhill (1976) investigated the effect of high temperatures on performance of a task which involved repeated simple (four-number, two-digit) addition sums.

Eight students performed the task in hot ambient temperatures (effective temperature 87 °F at relative humidity of 50 per cent), and a further eight students performed the same task at normal temperatures (effective temperature 73 °F at a relative humidity of 57 per cent). Many previous experiments had shown no performance change at high temperatures, and some others had even shown a performance improvement at high temperatures. Gledhill speculated that these effects, which run contrary to regular expectations, might be explained by an increase in motivation which occurs in an experimental situation where students respond to the challenge of a novel stress. This 'challenge' effect might not be expected to occur in an industrial situation, where workers are likely to resent the high temperatures. Gledhill investigated this effect by subjecting half of the subjects in both groups to a motivation depressing treatment. This simply involved treating the subjects in an offhand manner and giving them the impression that he was not a very competent experimenter. For convenience we shall call these two levels of motivation 'interested' and 'irritated'. The dependent variable was the number of correct additions in a 30 min period. All subjects were debriefed at the end of the experiment, when the purpose of the procedures were fully explained. The results have already been given in Table 13.1.

The method to be presented has been partly discussed in section 13.1. Because we are dealing with four independent samples, the methods of section 9.3 are used. For the nonspecific tests we compare two samples using the methods of sections 7.7 and 9.6. Chapter 6 contains general advice on procedure in rank sum analysis. To avoid repetition, it will be assumed that the reader is familiar with this material.

Analysis using specific alternative hypotheses

Step 1 – formulate the problem
  (a) Specify the hypotheses:
       Null hypothesis (H_0): arithmetic ability is unaffected by environmental temperature, level of motivation or any interaction of these two effects.

       There are three specific alternative hypotheses:

       H_1: low motivation (irritation) degrades arithmetic ability.
       H_2: heat degrades arithmetic ability
       H_3: heat and low motivation combine to degrade arithmetic ability to an extent which is in excess of the combined simple effects of the two.
  (b) Specify the variables:
       Independent variables: (a) heat, normal/hot (b) motivation, irritated/ interested.
       Dependent variable: arithmetic ability expressed in terms of the number of addition sums correctly completed within a 30 minute period.

(c) Choose the coefficients:
These have been discussed in section 13.1:

H_1: λ_j $+1$ $+1$ -1 -1

H_2: λ_j $+1$ -1 $+1$ -1

H_3: λ_j -1 $+1$ $+1$ -1

Step 2 – find the sample rank sums and means

(a) Rearrange the data to form columns with the scores arranged in ascending order for ease of ranking. See table 13.2.
(b) Rank the scores, irrespective of sample, remembering to give low ranks to low scores. Write the ranks alongside the original scores.
(c) Find the sample rank sums (R_j).
(d) Check that the ranking has been carried out correctly:

$$N(N + 1)/2 = \Sigma R_j$$

$$16 \times 17/2 = 34 + 46 + 37.5 + 18.5$$

$$136 = 136$$

(e) Find the sample rank means $(\bar{R}_j = R_j/n_j)$. In our case these are 8.5, 11.5, 9.37 and 4.625.

Note that all of the procedures for step 2 are taken from section 9.3 because the samples are independent. For related samples use the procedures given in section 10.3.

Step 3 – interpret the rank means

As expected, the worst performance is found in the 'irritated' group working at high temperature. Their rank mean (4.625) is the lowest of all. Among both 'interested' groups, the highest sample rank mean comes from the high temperature group (11.5) compared with the normal temperature group (8.5). As expected, the high temperature appears to have created a 'challenge' condition to which the subjects have responded.

Step 4 – compute and evaluate our key statistics (L and Z)

Because we have independent samples, we use the following expressions from section 9.3:

$$Z = [L - E(L)]/\sqrt{\mathrm{var}(L)}$$

$$E(L) = (N + 1) \sum_{j=1}^{k} n_j\lambda_j/2$$

$$\mathrm{var}(L) = (N + 1)\left[N\Sigma n_j\lambda_j^2 - \left(\sum_{j=1}^{k} n_j\lambda_j \right)^2 \right]T/12$$

Since all of our samples are the same size, we set

$$n = n_j$$

The coefficients have also been chosen, so that

$$\Sigma n\lambda_j = 0$$

We may therefore simplify the above expressions:

$$E(L) = 0$$

$$\text{var}(L) = (N + 1)\, NnT\Sigma\lambda_j^2/12$$

So a single expression for Z can be used:

$$Z = L/\sqrt{[(N + 1)\, NnT\Sigma\lambda_j^2/12]}$$

Note that if we were dealing with related samples we would use expressions derived from section 10.3. These are illustrated in section 13.3.

We need to compute three values for Z, one for each hypothesis. The following values remain constant across all three computations:

$$N = 16, \quad n = 4, \quad \Sigma\lambda_j^2 = 4 \quad \text{and} \quad T = 0.9956$$

(The computation of T is given in table 13.3.)

We can enter these constants into the above formula to obtain

$$Z = L/\sqrt{[(17)(16)(4)(0.9956)(4)/12]}$$

$$= L/19.00$$

(a) Find Z:

Applying this to our three hypotheses, we obtain

$$L = \Sigma\lambda_j R_j$$

H_1:

$$L_1 = (+1)(34) + (+1)(46) + (-1)(37.5) + (-1)(18.5)$$

$$= 24$$

$$Z_1 = 24/19.00 = 1.26$$

H_2:

$$L_2 = (+1)(34) + (-1)(46) + (+1)(37.5) + (-1)(18.5)$$

$$= 7$$

$$Z_2 = 7/19.00 = 0.37$$

H_3:

$$L_3 = (-1)(34) + (+1)(46) + (+1)(37.5) + (-1)(18.5)$$

$$Z_3 = 31/19.00 = 1.63$$

(b) Evaluate Z by consulting a table of the normal distribution (table A in the appendix). The relevant critical values are given here:

	5%	2.5%	1%	0.1%
$z \geqslant$	1.64	1.96	2.33	3.10

To be significant at any of the four levels, Z must be equal to or greater than the corresponding value. We have three values for Z and none of them is greater than even the lowest significance level, although Z_3 is very close.

Step 5 – draw conclusions
None of our three results is statistically significant ($p > 0.05$) and we are, therefore, not justified in rejecting H_0 in favour of any of the three hypotheses. We must note that hypothesis 3 came very close to the 5 per cent significance level, at $Z_3 = 1.63$. It is quite possible that only slightly larger samples might have given rise to a significant effect. This was the interaction hypothesis which implied that heat and irritation combined to depress performance even more than would be expected from the action of the two main effects working in isolation.

Nonspecific analysis

Step 1 – formulate the problem
(a) Specify the hypothesis:
    The null hypothesis is expressed in terms of three different null hypotheses:

$H_{0/1}$: heat does not affect arithmetic ability.
$H_{0/2}$: low motivation (irritation) does not affect arithmetic ability.

$H_{0/3}$: the combined effects of heat and low motivation do not affect arithmetic ability over and above the combined simple effect of both variables working together.

(b) Specify the variables:
Independent variables: (a) heat, normal/hot (b) motivation, irritated/interested.
Dependent variable: arithmetic ability expressed in terms of the number of addition sums correctly completed within a 30 minute period.

Step 2 - find the sample rank sums and means
See the specific analysis in table 13.2.

Step 3 - interpret the rank means
Figure 13.2 shows the four sample rank means in graphical form. It is clear that under the hot condition the 'irritated' group performed much worse than the 'interested' group. On the other hand, there was little difference between the two groups for the normal temperature condition.

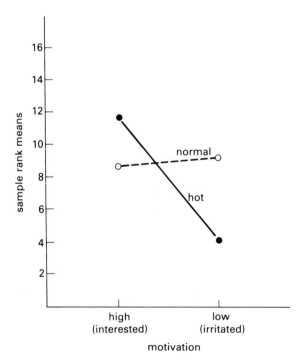

Figure 13.2   Sample rank means for example in section 13.2

Step 4 – compute and evaluate our key statistic (H)
(a) Find H:
We need to compute three different values for H corresponding to our three null hypotheses. See section 7.7.

Heat main effect
Find the two rank sums for the two heat conditions by pairing samples $1 + 3$ and $2 + 3$. These are

$$R_{\text{normal}} = 34 + 37.5 = 71.5$$

$$R_{\text{hot}} = 46 + 18.5 = 64.5$$

Compute H using the two-group formula from section 7.7:

$$H = \left[\frac{12(R_1^2 + R_2^2)}{nN(N + 1)} - 3(N + 1) \right] \Big/ T$$

Using $n = 8$ (because eight scores contribute to each rank sum), $N = 16$ and $T = 0.9956$ (see table 13.2):

$$H_1 = \left[\frac{12(71.5^2 + 64.5^2)}{8(16)(17)} - 3(17) \right] \Big/ 0.9956 = 0.14$$

Motivation main effect
Find the two rank sums for the two motivation conditions by pairing samples $1 + 2$ and $3 + 4$. There are

$$R_{\text{interested}} = 34 + 46 = 80$$

$$R_{\text{irritated}} = 37.5 + 18.5 = 56$$

Compute H using the formula for two unrelated samples from section 7.7:

$$H = \left[\frac{12(R_1^2 + R_2^2)}{nN(N + 1)} - 3(N + 1) \right] \Big/ T$$

Using $n = 8$ (because eight scores contribute to each rank sum), $N = 16$ and $T = 0.9956$ (see table 13.2):

$$H_2 = \left[\frac{12(80^2 + 56^2)}{8(16)(17)} - 3(17) \right] \Big/ 0.9956$$

$$= 1.59$$

Interaction effect (temperatures × motivation)

Compute the overall H score and then subtract the two main effects. To find the overall H use all four rank sums in the formula for independent samples:

$$H = \left[\frac{12}{nN(N+1)} \Sigma R_j^2 - 3(N+1) \right] \Big/ T$$

Using $n = 4$ (because only four scores contribute to each rank sum), $N = 16$ and $T = 0.9956$ (see table 13.2):

$$H_{\text{overall}} = \left[\frac{12}{4(16)(17)} (34^2 + 46^2 + 37.5^2 + 18.5^2) - 3(17) \right] \Big/ 0.9956$$

$$= 4.39$$

We may now compute the interaction effect:

$$H_{\text{interaction}} = H_{\text{overall}} - H_1 - H_2$$

$$= 4.39 - 0.14 - 1.59$$

$$= 2.66$$

Find the degrees of freedom associated with each H value. For *main effects*, the degrees of freedom are simply one less than the number of rank sums appearing in the computations. Only two rank sums were used for both motivation' and 'temperature' effects. Both values of H, therefore, are associated with only one degree of freedom. For interaction effects we compute the degrees of freedom in the same way but then subtract the degrees of freedom associated with each component main effect. The interaction effect computation was based on four rank sums and therefore begins with three degrees of freedom. From this value subtract the degrees of freedom associated with each component effect thus:

$$\text{df}_{\text{motivation × temperature}} = \text{df}_{\text{overall}} - \text{df}_{\text{motivation}} - \text{df}_{\text{temperature}}$$

$$= \quad 3 \quad - \quad 1 \quad - \quad 1$$

$$= 1$$

As a consequence, all of our effects have only one degree of freedom.

(b) Evaluate H by consulting tables of chi-square (table B in the appendix. In the case of the 2×2 design, all three values of H have only one degree of freedom. The appropriate row of table B is therefore

Degrees of freedom	5%	2.5%	1%	0.1%
1	$\chi^2 \geqslant 3.8$	5.0	6.6	10.8

None of our three values ($H_1 = 0.14$, $H_2 = 1.59$ and $H_3 = 2.66$) approaches even the least significant value of χ^2.

Step 5 - draw conclusions
Since none of our results is significant ($p > 0.05$ in all three cases) we are unable to reject any of our three null hypotheses. Although not strictly necessary for this relatively simple factorial design, the results have been tabulated in summary form in table 13.3.

13.3 3 × 2 factorial analysis

The move from a 2 × 2 to a 3 × 2 analysis may seem a small one but it introduces a world of difference between the specific and nonspecific procedures for analysing the same data set. The added complications when dealing with main effects are, of course, no more than the complications introduced in chapters 9 and 10 when more than two samples were being analysed. It is the interaction effects which present difficulties of conceptualization. In the case of the specific tests, the researcher is required to think particularly clearly and to represent in advance, using coefficients, precisely which aspect of interaction between a number of alternatives is to be considered. In the case of the nonspecific test a purely mechanical procedure is available for computing the interaction effect and evaluating its statistical significance. The researcher is still left with the problem of visualizing the interaction effect and grasping the implications for the research in hand. Both procedures, therefore, call for very clear thinking.

In section 13.2 we were able to compute the interaction by computing the overall H value (based on all the sample rank sums) and subtracting the two H

Table 13.3 Summary table for the 2 × 2 factorial problem used in example in section 13.2

Source	H	df	p
Heat	0.14	1	>0.05
Motivation	1.59	1	>0.05
Heat × motivation	2.66	1	>0.05
Overall	4.39	3	

values attributable to the main effect.

$$H_{\text{interaction}} = H_{\text{overall}} - H_{\text{main effect A}} - H_{\text{main effect B}}$$

The same line of reasoning applies to the degrees of freedom to be associated with each H value:

$$df_{\text{interaction}} = df_{\text{overall}} - df_{\text{main effect A}} - df_{\text{main effect B}}$$

It is a simple matter to determine the degrees of freedom to be associated with the overall and main effects. We simply subtract one from the number of rank sums which were used in the computation of the particular H value. A general rule, which becomes more useful as the designs become more complex, is that the degrees of freedom associated with any H value (*excepting the overall value*) can be found using the product of the degrees of freedom of the main effects which are named in the effect. Thus the A × B interaction term degrees of freedom can be found thus:

$$df_{A \times B} = df_A \times df_B$$

The results of this computation should be the same as that using the alternative expression just given.

Example

As part of a much larger experiment, Meudell, Mayes and Neary (1980) investigated the recognition memory for pictures in six amnesic patients diagnosed as alcoholic Korsakoff. The patients were shown 108 cartoons with captions removed which they were expected to recognize 24 hours later. They were tested using 108 pairs of cartoons where one member of each pair had been shown on the previous day but the other was new. The patient's task was to decide which was new and which he had seen the day before. The experimenters counted the number of correct choices. Half of the cartoons were 'funny' (even without the captions) and half were 'neutral'. The maximum number of correct choices in each category was therefore 54.

The experimenters believed that a failure to remember among amnesic patients may be caused by a failure to interpret the full significance of their experiences when they are occurring. To explore this possibility, they repeated the experiment three times (on three successive days) using the same six patients but giving them different instructions on each occasion:

(1) to *describe* each cartoon as it was presented
(2) to look at the pictures and remember them (*learn*)
(3) to *spot the difference* between each cartoon and a similar cartoon with minor differences which was presented alongside it.

Table 13.4 Number of pictures correctly recognized by a group of amnesic patients following various learning conditions. Example in section 13.3 from Meudell, Mayes and Neary (1982)

Orienting task	Describe		Learn		Spot the difference	
Humour	Funny	Neutral	Funny	Neutral	Funny	Neutral
Sample	1	2	3	4	5	6
Patient						
1	16	13	17	13	10	10
2	15	6	10	9	5	7
3	16	11	10	9	8	13
4	14	13	12	11	14	10
5	17	15	18	12	14	11
6	15	14	13	14	11	15

The results are given in table 13.4.

The design is basically six samples organized into six blocks (i.e. six *related* samples). The computational formulae therefore must be suited to related samples. Appropriate formulae can be found in chapters 8 and 10.

Analysis using specific alternative hypotheses

Step 1 - formulate the problem
The experimenters make a number of clear predictions about their results which can be used as specific alternative hypotheses. Firstly, they predict that amnesics, like normal people, will find funny cartoons more memorable. Secondly, they predict that better recognition will be obtained when the patients are required to attend to the meaning of the pictures by being asked to *describe* them. On the other hand, they expect worst recall in the *spot the difference* conditions when their attention is focused on irrelevant details. Thirdly, and most importantly, they predict that humour will improve the amnesics' recall most when the patient's are required to *describe* the cartoons. This crucial prediction is based on the idea that amnesics do not spontaneously appreciate the full significance of their experiences unless some effort is made to do so. As a further consequence, the experimenters would expect least benefit from humour in the *spot the difference* condition where the task should distract the patient from noticing that some of the cartoons were funny. The first two hypotheses can be tested using a straightforward application of appropriate coefficients to the sample rank sums. The third hypothesis, which involves the interaction of two

L

factors (humour and orienting task), requires the evasive manoeuvre of computing a humour-neutral difference score for each patient in each orienting task. The first two hypotheses will be tackled now, the third later.

(a) Specify the hypotheses:

Specific alternative hypothesis (H_1): humour improves recall for pictures.

Specific alternative hypothesis (H_2): describing pictures will yield better subsequent recognition than simply being asked to remember them. Both of these conditions will be better than a *spot the difference* condition.

(b) Specify the variables:

Independent variables: (i) humour, funny/neutral; (ii) orienting task, describe/learn/spot the difference.

Dependence variable: number of correct recognitions out of 108 (in a forced, two-choice situation).

(c) Choose the coefficients:

H_1: funny/neutral

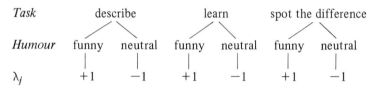

Task	describe		learn		spot the difference	
Humour	funny	neutral	funny	neutral	funny	neutral
λ_j	$+1$	-1	$+1$	-1	$+1$	-1

H_2: effect of orienting task

λ_j	$+1$	$+1$	0	0	-1	-1

These coefficients have been chosen to sum to zero. This is not essential but it does simplify the computations in situations where the sample sizes are equal.

Step 2 – find the sample rank sums and means

(a) Create table 13.5. To avoid confusion only the ranks will be entered here.

(b) Rank the scores *within each block separately*, remembering to assign low ranks to low scores.

(c) Find the sample rank sums (R_j).

(d) Check that the ranking has been carried out correctly. In the related samples case with one score per cell we expect

$$mk(k + 1)/2 = \Sigma R_j$$

$$6(6)(7)/2 = 33 + 21 + 25 + 17 + 13 + 17$$

$$126 = 126$$

(e) Find the sample rank means ($\bar{R}_j = R_j/n_j$). See table 13.5.

Table 13.5 Computational layout for example in section 13.3. Values in the tables are ranks

Orienting task	Describe		Learn		Spot the difference	
Humour	Funny	Neutral	Funny	Neutral	Funny	Neutral
Patient						
1	5	3.5	6	3.5	1.5	1.5
2	6	2	5	4	1	3
3	6	4	3	2	1	5
4	5.5	4	3	2	5.5	1
5	5	4	6	2	3	1
6	5.5	3.5	2	3.5	1	5.5
rank sum R_j	33	21	25	17	13	17
sample size n	6	6	6	6	6	6
rank mean $\bar{R}_j$	5.5	3.5	4.2	2.8	2.2	2.8

(f) It is convenient to compute the tie correction factor (ΣT_i) at this stage to avoid repeat calculations later:

Block	Tie lengths		
1	2, 2	$1 - 2(2^3 - 2)/(6^3 - 6)$	0.943
2	–		1
3	–		1
4	2	$1 - (2^3 - 2)/(6^3 - 6)$	0.971
5	–		1
6	2, 2	$1 - 2(2^3 - 2)/(6^3 - 6)$	0.943
		ΣT_i	5.857

Step 3 - interpret the rank means

The sample rank means have been plotted in figure 13.3. There is a clear recognition advantage for funny over neutral pictures in both the 'describe' and 'learn' conditions but not for the 'spot the difference' condition. Among funny pictures there is also a trend in the expected direction across the three orienting tasks. This trend is not evident for neutral pictures except that the 'describe' condition was superior to the other two.

Step 4 - compute and evaluate our key statistics (L and Z)

(a) Find L and Z:

Because we have related samples, we must use the methods of section 10.3. The appropriate formulae can be simplified because we have only one

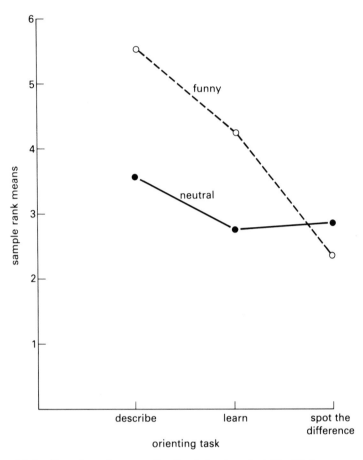

Figure 13.3   Sample rank means for the data given in table 13.4

score per cell and the coefficients sum to zero.

$$L = \Sigma \lambda_j R_j$$
$$E(L) = 0 \quad (\text{because } \Sigma \lambda_j = 0)$$
$$\text{var}(L) = \tfrac{1}{12} k(k+1)(\Sigma \lambda_j^2) \Sigma T_i$$
$$Z = [L - E(L)]/\sqrt{\text{var}(L)}$$

Specific alternative hypothesis H_1

Using the coefficients $\lambda_j = +1, -1, +1, -1, +1, -1$, we obtain

$$L_1 = (+1)(33) + (-1)(21) + (+1)(25) + (-1)(17) + (+1)(13)$$
$$+ (-1)(17)$$
$$= 33 - 21 + 25 - 17 + 13 - 17$$
$$= 16$$
$$\Sigma\lambda_j^2 = (+1)^2 + (-1)^2 + (+1)^2 + (-1)^2 + (+1)^2 + (-1)^2 = 6$$
$$\text{var}(L_1) = \tfrac{1}{12}(6)(7)(6)\ 5.857 = 123$$
$$Z_1 = L_1\sqrt{\text{var}(L_1)} = 16/\sqrt{123} = 1.44$$

Specific alternative hypothesis H_2

Using the coefficients $\lambda_j = +1, +1, 0, 0, -1, -1$, we obtain

$$L_2 = (+1)(33) + (+1)(21) + (0)(25) + (0)(17) + (-1)(13)$$
$$+ (-1)(17)$$
$$= 33 + 21 - 13 - 17 = 24$$
$$\Sigma\lambda_j^2 = (+1)^2 + (+1)^2 + (0)^2 + (0)^2 + (-1)^2 + (-1)^2 = 4$$
$$\text{var}(L_2) = \tfrac{1}{12}(6)(7)(4)\ 5.857 = 81.82$$
$$Z_2 = L_2/\sqrt{\text{var}(L_2)} = 24/\sqrt{82} = 2.65$$

(b) Evaluate Z by consulting tables of the normal distribution (table A in the appendix). The relevant values are

5%	2.5%	1%	0.1%
$z \geqslant 1.64$	1.96	2.33	3.10

To be significant at any of the four levels, Z must be greater than or equal to the corresponding critical value. Z_1 (1.44) is not significant at the 5 per cent level ($p < 0.05$) and Z_2 (2.65) is significant at the 1 per cent level ($p < 0.01$).

Step 5 – draw conclusions

Because Z_2 is statistically significant, we may reject H_0 in favour of the hypothesis that the orienting task affects recall in the manner predicted. However, figure 13.3 shows that the situation is not quite as simple as this. If this hypothesis was the only one needed to explain the results, we would expect two parallel lines in the graph. The crossover remains unexplained. We must consider, now, hypothesis 3.

Interaction effect H_3

The third hypothesis predicts that the effect of humour will be most marked in the 'describe' condition where the patients are most likely to notice that the pictures are funny. It should be least marked in the 'spot the difference' condition where the patients are distracted from the content of the picture by the task which requires that they focus on irrelevant details. We can tackle this issue by using *difference scores* (for each patient in each condition) for the funny and neutral pictures. These difference scores are given in table 13.6. We can now proceed using regular procedures for assessing a trend across three related samples. See section 10.3.

Step 1 - formulate the problem

(a) Specify the hypotheses:
Null hypothesis (H_0): orienting task does not affect the benefit attributable to humour in pictures.
Specific alternative hypothesis (H_1): humour is least beneficial in the 'spot the difference' condition and most beneficial in the 'describe' condition.

(b) Specify the variables:
Independent variable: orienting task, describe/learn/spot the difference.
Dependent variable: difference between reocgnition score for 'funny' and 'neutral' pictures.

Table 13.6   Difference scores obtained by subtracting recognition scores for 'neutral' pictures from those for 'funny' pictures

Patient	Orienting task		
	Describe	Learn	Spot the difference
1	3	4	0
2	9	1	−2
3	5	1	−5
4	1	1	4
5	2	6	3
6	1	−1	−4

(c) Choose the coefficients:

The following coefficients represent a simple trend across the three samples. They have been chosen to sum to zero to simplify the computations:

$$\lambda_j \qquad +1 \qquad 0 \qquad -1$$

Step 2 – find the sample rank sums and means

(a) Rank the scores (see table 13.7) separately within each block.
(b) Find the sample rank sums. In our case they are $R_1 = 13.5$, $R_2 = 13.5$, $R_3 = 9$.
(c) Check that the ranking has been carried out correctly. In the case of related samples with only one score per cell we expect

$$mk(k + 1)/2 = \Sigma R_j$$

$$6(3)(4)/2 = 13.5 + 13.5 + 9$$

$$36 = 36$$

(d) Find the sample rank means ($\bar{R}_j = R_j/n_j$). In our case these are $\bar{R}_1 = 2.25$, $\bar{R}_2 = 2.25$, $\bar{R}_3 = 1.5$.

Table 13.7   Computational layout for trend analysis of difference scores given in table 13.5

Orienting task	Describe		Learn		Spot the difference	
	Score	Rank	Score	Rank	Score	Rank
Patient						
1	3	2	4	3	0	1
2	9	3	1	2	−1	1
3	5	3	1	2	−5	1
4	1	1.5	1	1.5	4	3
5	2	1	6	3	3	2
6	1	3	−1	2	−4	1
rank sum R_j		13.5		13.5		9
sample size n		6		6		6
rank mean $\bar{R}_j$		2.25		2.25		1.5

(e) Find the correction factor for ties. Patient 4 is the only one with a tie (of length 2):

$$T_i = 1 - \frac{\Sigma(t^3 - t)}{k^3 - k} = 1 - \frac{2^3 - 2}{3^3 - 3} = 1 - \frac{6}{24} = 0.75$$

The remaining patients have $T_i = 1$, therefore $\Sigma T_i = 5.75$.

Step 3 - interpret the rank means
'Spot the difference' produced a smaller 'funny-neutral' difference score than the other two conditions. However, the 'describe' condition showed no superiority over the learn condition. This does not rule out the possibility of a trend across the groups, of course, and it is worth proceeding with the calculations.

Step 4 - compute and evaluate our key statistics (L and Z)
(a) Find L and Z:
Because this is a related samples design with a single score in each cell, we must use the procedures given in section 10.3. The coefficients have been chosen to sum to zero so that we may compute Z directly using

$$Z = L / \sqrt{[\tfrac{1}{12}k(k + 1)(\Sigma\lambda_j^2)(\Sigma T_i)]}$$

following arguments given above. First of all, find L:

$$L = \Sigma\lambda_j R_j = (+1)(13.5) + (0)(13.5) + (-1)(9) = 13.5 - 9 = 4.5$$
$$\Sigma\lambda_j^2 = (+1)^2 + (0)^2 + (-1)^2 = 2$$

so that

$$Z = 4.5 / \sqrt{[\tfrac{1}{12}(3)(4)(2)(5.75)]}$$
$$= 1.33$$

(b) Evaluate Z using the critical values for the normal distribution given above. Clearly our value of Z is nonsignificant ($p > 0.05$).

Step 5 - draw conclusions
Even though the results were broadly in line with our expectations, the differences between the samples were not strong enough to cause us to reject H_0.

We may summarize our analysis using specific alternative hypotheses as follows. There is a strong evidence that the orienting tasks affect the recognition task in the way we might expect. It also appears to be true that amnesic patients benefit from humour in the pictures when it comes to later recognition. Although the patients showed improved recognition for the 'describe' and 'learn' conditions but no improvement for the 'spot the difference' condition, this interaction effect was not strong enough to be statistically significant.

Interaction effect, alternative method

The difference score method given, has the advantage that it is (relatively) easy to follow the logic of the computation. It has the disadvantage that it involves a different set of procedures from those used for computing the main effects. Moreover, it would be impossible to use that method in the case of unrelated samples. If the scores in the two samples are not matched, how could we ever decide which scores need to be subtracted from which? Ideally, we need a method for computing specific interaction effects which uses the same procedure as that for computing the main effects.

This can be achieved simply enough, if we can devise a set of λ coefficients which express our hypothesis faithfully enough. In analysing the 2×2 design in section 13.2 we were able to do this using some logic given in section 13.1. However, the situation was much simpler there. Figure 13.4a shows a graphical attempt to express the hypothesis in terms of λ coefficients. It certainly captures the idea that the difference between funny and neutral conditions is greatest for 'describe' and least for 'spot the difference' conditions. However, it also implies the truth of the first two hypotheses: (a) that funny cartoons are more memorable than neutral and (b) that there is a downward trend in memorability from 'describe' to 'learn' to 'spot the difference'. Ideally, we should devise a set of coefficients which look at our interaction hypothesis in isolation.

Figure 13.4b offers an alternative solution which is better suited to our needs. Firstly, it does not imply any difference between the 'funny' and 'neutral' conditions because the sums of the three coefficients for both conditions are the same (i.e. zero in both cases). Secondly, it does not imply any difference between the three orientation conditions because the sums of the two coefficients for the three conditions are all the same (i.e. zero in all three cases). Thirdly, it does express our expectation that the 'funny-neutral' difference will be largest for the 'describe' condition and least for the 'spot the difference' condition. We can check this by finding the difference between the 'funny' and 'neutral' coefficients for the three conditions

describe	$(+1) - (-1) = +2$
learn	$(0) - (0) = 0$
spot the difference	$(-1) - (+1) = -2$

This set of coefficients also has the useful computational property of having a zero sum, which keeps the calculations simple. The set is also independent of the two previous sets. We can check this by finding the cross-product of pairs of sets of coefficients. If the cross-product is zero, then the coefficients are inde-

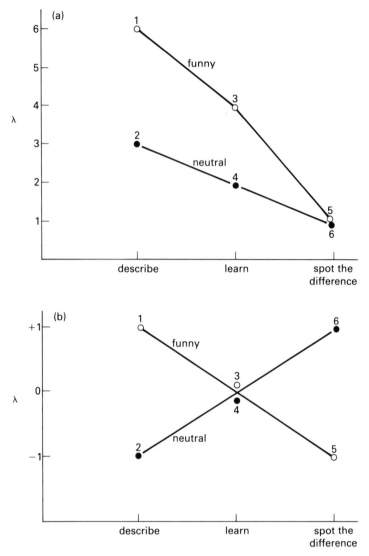

Figure 13.4   Two attempts to find appropriate λ coefficients to express the interaction hypothesis (a) $\lambda_j = 6, 3, 4, 2, 1, 1$ (b) $\lambda_j = +1, -1, 0, 0, -1, +1$

pendent. Check this first against the first main effect:

hypothesis (a)	+1	−1	+1	−1	+1	−1
hypothesis (c)	+1	−1	0	0	−1	+1
cross-product	+1	+1	0	0	−1	−1

$$\Sigma \lambda_a \lambda_c = 0$$

Then check it against the second main effect:

hypothesis (b)	+1	+1	0	0	−1	−1
hypothesis (c)	+1	−1	0	0	−1	+1
cross-product	+1	−1	0	0	+1	−1

$$\Sigma \lambda_b \lambda_c = 0$$

In both cases the independence check is satisfied.

Now that we have a satisfactory set of coefficients we can slot our computations into step 4 of our original calculations:

$$E(L) = 0 \quad (\text{because } \Sigma \lambda = 0)$$

$$L = \Sigma \lambda_j R$$

$$= (+1)(33) + (-1)(21) + (0)(25) + (0)(17) + (-1)(13)$$

$$+ (+1)(17)$$

$$= 16$$

$$\Sigma \lambda_j^2 = (+1)^2 + (-1)^2 + (0)^2 + (0)^2 + (-1)^2 + (+1)^2 = 4$$

$$\text{var}(L) = \tfrac{1}{12}(6)(7)(4)(5.857) = 81.66$$

$$Z = [L - E(L)]/\sqrt{\text{var}(L)}$$

$$= 16/\sqrt{82} = 1.76$$

Our value of Z is significant at the 5 per cent level, which provides some support for the hypothesis that humour contributes to the memorability of cartoons more when the patient is required to analyse the significance of what he is looking at. It is unfortunate that the conclusion drawn here is different from that using the difference score method given. This is a sobering reaffirmation of the general truth in statistics that when the question is rephrased the answer is often slightly different.

It is not surprising, perhaps, that users are discouraged from using the more precise tool of specific hypothesis testing by the difficulty in formulating their hypotheses in terms of appropriate sets of λ coefficients. The more mechanical and simple procedures using nonspecific hypotheses will inevitably appear more attractive. However, these are not without their problems too, as we shall now see.

Analysis using nonspecific hypotheses

Step 1 - formulate the problem
(a) Specify the hypotheses:
     Null hypotheses:
     $H_{0/1}$: humour does not affect picture recall in amnesics.

$H_{0/2}$: orientating task does not affect picture recall in amnesics.

$H_{0/3}$: there is no interaction effect for humour and orientating task.

(b) Specify the variables:

Independent variables: (i) humour, funny/neutral; (ii) orientating task, describe/learn/spot the difference.

Dependent variable: number of pictures subsequently correctly recognized (maximum 20)

Step 2 – find the sample rank sums and means

See specific analysis given earlier in text and table 13.5.

Step 3 – interpret the rank means

See specific analysis given earlier.

Step 4 – compute and evaluate our key statistic (H)

In what follows,

$$m \text{ (number of blocks)} = 6$$
$$nk \text{ (number of scores per block)} = b = 6$$

NB   The values of n and k will vary according to how the samples are combined but the product nk should remain the same.

Tie correction factor $\Sigma T_i = 5.875$ (see table 13.5).

(a) Find H:

Humour main effect

Find the two rank sums for the two humour conditions using table 13.5 by combining samples $1 + 3 + 5$ and $2 + 4 + 6$:

$$R_{\text{funny}} = 33 + 25 + 13 = 71$$

$$R_{\text{neutral}} = 21 + 17 + 17 = 55$$

Note the cell frequency (i.e. number of scores per block in each subgroup) $n = 3$, and number of subgroups $k = 2$. Check that nk is the number of scores per block:

$$nk = 3(2) = 6$$

Compute H, using methods given in chapter 10:

$$H = \left[\frac{12\Sigma R_j^2}{n^2 k(nk + 1)} - 3m^2(nk + 1) \right] \Big/ \Sigma T_i$$

$$= \left[\frac{12(71^2 + 55^2)}{3^2(2)(7)} - 3(6^2)(7) \right] \Big/ 5.875$$

$$= 2.07$$

Degrees of freedom:

$$df = k - 1 = 1$$

Note that the methods of section 9.6 would be used if the data formed independent samples.

Orientating task main effect

Find the three rank sums for the three orienting tasks using table 13.5 by combining samples $1 + 2, 3 + 4$ and $5 + 6$:

$$R_{describe} = 33 + 21 = 54$$

$$R_{learn} = 25 + 17 = 42$$

$$R_{spot\ the\ difference} = 13 + 17 = 30$$

Note the cell frequency $n = 2$, and the number of subgroups $k = 3$. Check that nk is the number of scores per block:

$$nk = 2(3) = 6$$

Compute H:

$$H = \left[\frac{12\Sigma R_j^2}{n^2 k(nk + 1)} - 3m^2(nk + 1) \right] \Big/ \Sigma T_i$$

$$= \left[\frac{12(54^2 + 42^2 + 30^2)}{2^2(3)(7)} - 3(6^2)(7) \right] \Big/ 5.875$$

$$= 7.00$$

Degrees of freedom:

$$df = k - 1 = 2$$

Interaction, humour × orienting task

We must first compute the overall value of H across all six samples, so that $k = 6, n = 1$:

$$\Sigma R_j^2 = 33^2 + 21^2 + 25^2 + 17^2 + 13^2 + 17^2$$

$$= 2902$$

$$H_{overall} = \left[\frac{12\Sigma R_j^2}{n^2 k(nk + 1)} - 3m^2(nk + 1) \right] \Big/ \Sigma T_i$$

$$= \left[\frac{12(2902)}{1^2(6)(7)} - 3(6^2)(7) \right] \Big/ 5.875$$

$$= 12.45$$

Degrees of freedom:

$$\mathrm{df} = k - 1 = 5$$

We can now compute the interaction effect:

$$H_{\mathrm{humour} \times \mathrm{task}} = H_{\mathrm{overall}} - H_{\mathrm{humour}} - H_{\mathrm{task}}$$

$$= 12.45 - 2.07 - 7.00$$

$$= 3.38$$

Degrees of freedom:

$$\mathrm{df}_{\mathrm{humour} \times \mathrm{task}} = \mathrm{df}_{\mathrm{overall}} - \mathrm{df}_{\mathrm{humour}} - \mathrm{df}_{\mathrm{task}}$$

$$= 5 - 1 - 2$$

$$= 2$$

Alternatively:

$$\mathrm{df}_{\mathrm{humour} \times \mathrm{task}} = \mathrm{df}_{\mathrm{humour}} \times \mathrm{df}_{\mathrm{task}}$$

$$= 1 \times 2 = 2$$

(b) Evaluate H by consulting tables of chi-square (table B in the appendix). We need the critical values for 1 and 2 degrees of freedom:

Degrees of freedom	5%	2.5%	1%	0.1%
1	$\chi^2 \geqslant 3.8$	5.0	6.6	10.8
2		6.0	7.4	9.2	13.8

The humour main effect ($H = 2.07$, df $= 1$) is not significant ($p > 0.05$). The orienting task main effect ($H = 7.00$, df $= 2$) is significant at the 5 per cent level ($p < 0.05$). The task $\times$ humour interaction effect ($H = 3.38$, df $= 2$) is also not significant ($p > 0.05$).

Step 5 – draw conclusions
We are able to reject the null hypothesis only in the case of the differences between orienting tasks. Clearly, choice of orienting task does influence memory

for pictures among amnesic patients. The humour effect is, however, not statistically significant.

The interaction effect (humour × orienting task) is also clearly nonsignificant here even though it was marginally significant in the specific analysis performed earlier. This is, however, not just a consequence of the greater power of specific tests. The two tests are asking quite different interaction questions. The question posed by the specific test has been discussed at some length, but we need to look at the question of just what is meant by the idea of a nonspecific interaction effect. To help us, figure 13.5a represents the situation where there is no interaction effect (i.e. the situation where $H_{0/3}$ is true). The key feature of this graph is that the two lines are parallel. They can be slid up or down or tilted in any direction (reflecting variations in the strength of the two main effects) but they must remain parallel if there is no interaction.

We can now define an interaction effect (in the nonspecific sense) as any reliable deviation from this situation. Obviously, there are a variety of different configurations which could be thought of as nonparallel. Figures 13.5b and c represent two of these. Any reliable nonparallel configuration will give rise to a statistically significant result using a nonspecific analysis, whereas only one configuration can give a significant result using a specific analysis. Strictly speaking we must interpret a nonspecific significant interaction simply in terms of the nonparallel aspect of the graphical representation of the data. One way of expressing this in words is to say that there are some nonrandom factors at work in addition to the main effects (if any). This is not a very informative conclusion to draw, especially in complex designs, but the researcher is free to point to his diagram and imply that a similar overall pattern of results might be expected in future, if ever the experiment or survey were repeated.

13.4   3 × 2 factorial analysis for frequency tables

Frequency tables can be analysed in the same way as raw scores. The formulae in this chapter, have, however, been tailored for designs with equal cell frequencies, and it is important to check this point before embarking on the calculations. Analysis using *specific* alternative hypotheses can be attempted even when cell frequencies are unequal, but the appropriate formulae must be used from sections 9.4 and 10.7. The following example is taken from an experiment where the sample sizes were fixed by the researcher.

Example

Zeichner and Phil (1979) compared inebriated (1.32 ml alcohol/kg body weight) male social drinkers with sober controls in terms of their willingness to admini-

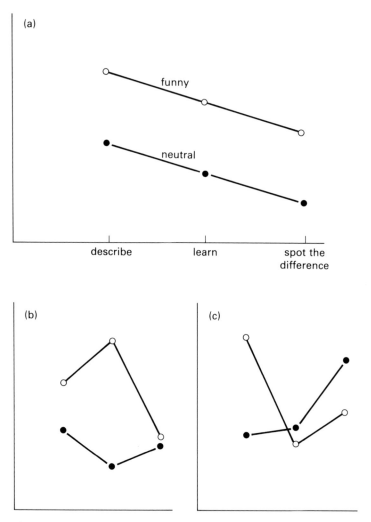

Figure 13.5 Possible configurations of results for the data used in the example in section 13.3. (a) represents the null hypothesis condition of 'no interaction' between the two main effects. (b) and (c) represent situations where an interaction effect is probably present

ster electric shock to a third party. The sober controls were divided into two further groups, a placebo group given orange juice which they believed contained alcohol, and a control group given nothing to drink. The experiment was so arranged (using the Buss technique) that the subject, who believed that he was assessing the pain threshold of another, received aversive loud tone stimuli immediately after delivering each shock to the third party. For half of the subjects this aversive tone was linked in intensity to the strength of the electric shock given (correlated group). For the remainder of subjects the tones were equally aversive but unrelated to the strength of the electric shock (unrelated). In the correlated group the subject was, in fact, punishing himself. The experiment involved considerable deception and, in reality, no electric shocks were given to the third party. There were six subjects in each subgroup, making a total of 72 subjects altogether. The electric shocks could be administered at five intensities (rated 1-5). The data in table 13.8 are based on the average shock intensity over 25 trials for each subject.

Zeichner and Phil (1979) analysed their data using parametric analysis of variance of the raw scores. Table 13.8 is a fabricated frequency table using a crude four-point scale derived from their published data. The conversion to a frequency table and the associated loss of information seemed acceptable since the samples were quite widely separated. However, readers interested in the original analysis should consult the authors' own account.

Analyses using specific alternative hypotheses

It is obvious that Zeichner and Phil expected (a) alcohol to make their subjects more aggressive and (b) correlated aversive consequences to reduce the intensity of shocks meted out to the third partner. Thus we already have two hypotheses to test. The authors were, however, most interested in the possibility that alcohol would lessen the restraining effects of linked punishment. It is clear from figure 13.6 that the inebriated subjects were largely unaffected by receiving (in return for their electric shocks) painful noises which were linked in intensity. The two sober groups, by contrast, were quick to reduce the intensity of shocks given when their own noise stimuli were linked in aversiveness.

Table 13.8   Mean intensity of electric shock delivered by 72 subjects to a third party

Mean shock strength	Alcohol		Placebo		Control	
	Correlated	Unrelated	Correlated	Unrelated	Correlated	Unrelated
1-2	0	0	11	2	12	3
2-3	0	0	1	8	0	7
3-4	3	1	0	2	0	2
4-5	9	11	0	0	0	0
total	12	12	12	12	12	12

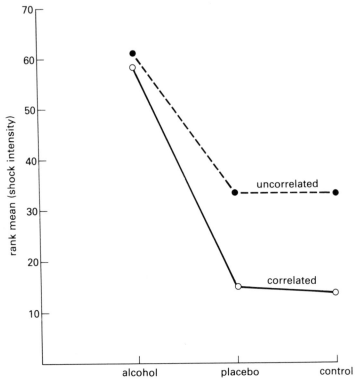

Figure 13.6   Men shock intensity ratings derived from table 13.8

The authors took the precaution of introducing a placebo group to control for the 'expectancy effects' of alcohol consumption. Judging by figure 13.6 the control and placebo group appear to have responded indentically to the treatments. The analysis of the differences between the two groups will be postponed until exercise 13.1, but for our purposes it will be convenient to think of them as similar groups. It will be particularly convenient when considering coefficients for the interaction hypothesis which are much easier to generate in a 2 × 2 situation. By effectively combining placebo and control samples into a single 'sober' sample, we reduce the alcohol consumption variable from three to two levels.

Step 1 - formulate the problem
(a) Specify the hypotheses:
    Specific alternative hypotheses H_1: alcohol consumption increases the intensity of electric shocks given to a third party.

Specific alternative hypothesis H_2: linked aversive consequences reduce the intensity of electric shocks given.

Specific alternative hypothesis H_3: linked aversive consequences will have a smaller effect in persons following alcohol consumption than in sober persons.

(b) Specify the variables:

Independent variables: (a) alcohol consumption, alcohol/placebo/control; (b) aversive consequences of shock administration, linked/unlinked to intensity of electric shock.

Dependent variable: intensity of shock administered, mean value per subject over 25 trials – reduced to four categories.

(c) Choose the coefficients:

H_1: *alcohol/sober*

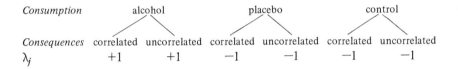

Consumption	alcohol		placebo		control	
Consequences	correlated	uncorrelated	correlated	uncorrelated	correlated	uncorrelated
λ_j	+1	+1	−1	−1	−1	−1

To simplify the calculations later we should adjust the coefficients to sum to zero:

λ_j	+2	+2	−1	−1	−1	−1

The two sets of coefficients are linearly dependent (see section 2.10) and will give the same result.

H_2: *correlated/uncorrelated*

λ_j	−1	+1	−1	+1	−1	+1

H_3: *interaction (consumption × consequences)*

To find the coefficients we can use a device used in section 13.1. First of all set up an inequality of the following form:

$$R_2 - R_1 < \tfrac{1}{2}(R_4 - R_3) + \tfrac{1}{2}(R_6 - R_5)$$

This says that the reduction in shock intensity caused by correlated aversive consequences in the alcohol consumption group $(R_2 - R_1)$ will be less than

($<$) the average of the differences for the two other groups. We can rewrite this as

$$R_4 - R_3 + R_6 - R_5 > 2R_2 - 2R_1$$

and then

$$+2R_1 - 2R_2 - R_3 + R_4 - R_5 + R_6 > 0$$

When we have >0 to the right hand of the expression we can use the left hand to specify the coefficients:

λ_j $+2$ -2 -1 $+1$ $|$ -1 $+1$

Figure 13.7 shows the pattern of results implied by this set of λ coefficients. Note that the coefficients do not imply any main effect due to alcohol or to correlation of tone stimuli with shocks since, within these categories, the coefficients sum to zero. The interaction coefficients are also linearly independent of the coefficients for the two main effects, since the cross-products ($\Sigma\lambda_a\lambda_c$ and $\Sigma\lambda_b\lambda_c$) are zero. These points are illustrated at greater length in section 13.3.

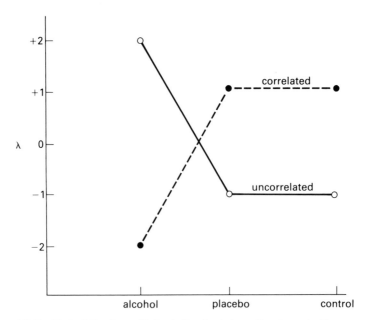

Figure 13.7   Plot of the λ coefficients for the interaction term in the example in section 13.4. $\lambda_j = +2, -2, -1, +1, -1, +1$

Step 2 - find the sample rank sums and means
(a) Draw up the frequency table as in table 13.9. Find the category totals (t_q) and the cumulative category total (Σt_q).
(b) Find the shared ranks for each category using the expression

$$r_q = \Sigma t_q - t_q/2 + 0.5$$

(c) Find the sample rank sums by multiplying the shared rank for each category by the number of scores in that sample and adding, e.g.:

$$R_1 = 0(14.5) + 0(36.5) + 3(48.5) + 9(62.5) = 708$$

(d) Check that the ranking has been carried out correctly. For an independent samples design (see section 9.7) we expect:

$$N(N + 1)/2 = \Sigma R_j$$
$$72(73)/2 = 708 + 736 + 196 + 418 + 174 + 396$$
$$2628 = 2628$$

(e) Note the sample sizes (in this case, the cell frequencies) and check that they are all equal. In our case $n = 12$. Compute the sample rank means $(\bar{R}_j = R_j/n_j)$ - see table 13.9.
(f) Compute the tie correction factor T. The values of $t_q^3 - t_q$ are given in the end column of table 13.9. Since we have only one block,

$$T = 1 - \Sigma(t^3 - t)/(N^3 - N)$$
$$= 1 - 34488/(72^3 - 72)$$
$$= 1 - 34\,488/373\,176 = 0.9076$$

Step 3 - interpret the rank means
Figure 13.6 summarizes the results in graphical form. The alcohol consumption group administered more intense shocks than the other two groups, who were very similar. When the aversive tone stimuli (delivered to the subject) were linked to the intensity of shock administered to the third party, both of the sober groups showed a reduced average intensity of shocks given. However, this was not so for the inebriated subjects, who were apparently unaffected by the unpleasant consequences of their actions.

Step 4 - compute and evaluate our key statistics (L and Z)
(a) Find L and Z using the formulae

$$L = \Sigma \lambda_j R_j$$

$$E(L) = 0 \quad (\text{because } \Sigma n_j \lambda_j = 0)$$

$$\text{var}(L) = \tfrac{1}{12} m(kn + 1)\, kn^2 \Sigma \lambda_j^2 \Sigma T_i$$

Note that $k = 6$, $m = 1$, $n = 12$ and $\Sigma T_i = T = 0.9076$.

H_1: *alcohol/sober*

$$\Sigma \lambda_j^2 = 2^2 + 2^2 + (-1)^2 + (-1)^2 + (-1)^2 + (-1)^2 = 12$$

$$L_1 = +2(708) + 2(736) - 1(196) - 1(418) - 1(174) - 1(396)$$

$$= 1704$$

$$\text{var}(L_1) = \tfrac{1}{12}(73)(6)(12^2)(12)(0.9076)$$

$$= 57\,244$$

$$Z_1 = L_1 / \sqrt{\text{var}(L_1)} = 1704 / \sqrt{57\,244} = 7.12$$

H_2: *correlated/uncorrelated consequences*

$$\Sigma \lambda_j^2 = (-1)^2 + 1^2 + (-1)^2 + 1^2 + (-1)^2 + 1^2 = 6$$

$$L_1 = -1(708) + 1(736) - 1(196) + 1(418) - 1(174) + 1(396)$$

$$= 472$$

$$\text{var}(L_2) = \tfrac{1}{12}(73)(6)(12^2)(6)(0.9076)$$

$$= 28\,622$$

$$Z_2 = L_2 / \sqrt{\text{var}(L_2)} = 472 / \sqrt{28\,622} = 2.79$$

H_3: *interaction (consumption × consequences)*

$$\Sigma \lambda_j^2 = 2^2 + (-2)^2 + (-1)^2 + 1^2 + (-1)^2 + 1^2 = 12$$

$$L_3 = +2(708) - 2(736) - 1(196) + 1(418) - 1(174) + 1(396)$$

$$= 388$$

$$\text{var}(L_3) = \tfrac{1}{12}(73)(6)(12^2)(12)(0.9076)$$

$$= 57\,244$$

$$Z_3 = L_3 / \sqrt{\text{var}(L_3)} = 388 / \sqrt{57\,244} = 1.62$$

(b) Evaluate Z by consulting tables of the normal distribution (table A in the appendix). The relevant values are

	5%	2.5%	1%	0.1%
$z \geqslant$	1.64	1.96	2.33	3.10

Z_1 (7.12) is significant at better than 0.1 per cent level ($p < 0.001$).
Z_2 (2.79) is significant at better than the 1 per cent level ($p < 0.01$).
Z_3 (1.62) is almost (but not quite) significant at the 5 per cent level
($p \simeq 0.05$)

Step 5 - draw conclusions
Alcohol is clearly increasing the intensity of the shocks administered ($p < 0.001$). Correlated aversive consequences also reduce the intensity of the shocks ($p < 0.01$). However, the interaction effect is almost significant. This qualifies the main effect to the more limited conclusion that aversive consequences reduce shock intensities but only in the sober subjects.

Analysis using nonspecific hypotheses

Step 1 - formulate the problem
(a) Specify the hypotheses:
    Null hypotheses:
    $H_{0/1}$: alcohol consumption or the belief that one has consumed alcohol does not affect shock administration behaviour (alcohol consumption main effect).
    $H_{0/2}$: linking punishing consequences to the intensity of shock administered does not affect shock administration behaviour (aversive consequences main effect).
    $H_{0/3}$: none of the three consumption groups is *differentially* affected by linking punishing consequences to the administration of shock (interaction effect).
(b) Specify the variables:
    Independent variables: (a) alcohol consumption, alcohol/placebo/control; (b) aversive consequences of shock administration, linked/unlinked with intensity of electric shock.
    Dependent variables: intensity of shock administered, mean value per subject over 25 trials - reduced to four categories.

Step 2 - find the sample rank sums and means
See above table 13.9.

Step 3 - interpret the rank means
See earlier.

Step 4 - compute and evaluate our key statistic (H)
(a) Find H:
    In our design we have only one block and a total of 72 scores in that block, therefore $m = 1$ and $nk = 72$. Because we have only one block, $\Sigma T_i = T = 0.09076$.

Table 13.9 Computational layout for example in section 13.4

Shock strength	Alcohol		Placebo		Control		t_q	Σt_q	r_q	$t_q^3 - t_q$
	1 Correlated	2 Uncorrelated	3 Correlated	4 Uncorrelated	5 Correlated	6 Uncorrelated				
1-2	0	0	11	2	12	3	28	28	14.5	21 924
2-3	0	0	1	8	0	7	16	44	36.5	4 080
3-4	3	1	0	2	0	2	8	52	48.5	504
4-5	9	11	0	0	0	0	20	72	62.5	7 980
										34 488
rank sum R_j	708	736	196	418	174	396				
sample size n_j	12	12	12	12	12	12 $\quad N = 72$				
rank mean $\bar{R}_j$	59	61.3	16.3	34.8	14.5	33				

Alcohol consumption, main effect
Find the three rank sums for the three consumption conditions using table
13.9 by combining samples $1 + 2, 3 + 4$ and $5 + 6$:

$$R_{\text{alcohol}} = 708 + 736 = 1444$$

$$R_{\text{placebo}} = 196 + 418 = 614$$

$$R_{\text{control}} = 174 + 396 = 570$$

Note the cell frequency (i.e. the number of scores per block in each subgroup) $n = 24$, and the number of subgroups $k = 3$. Check that nk is the
number of scores per block:

$$nk = 24(3) = 72$$

Compute H:

$$H = \left[\frac{12\Sigma R_j^2}{n^2 k(nk + 1)} - 3m^2(nk + 1) \right] \Big/ T$$

$$= \left[\frac{12(1444^2 + 614^2 + 570^2)}{24^2(3)(73)} - 3(73) \right] \Big/ 0.9076$$

$$= 50.824$$

Degrees of freedom:

$$df = k - 1 = 3 - 1 = 2$$

Aversive consequences, main effect
Find the two rank sums for the correlated and uncorrelated conditions
using table 13.9 by combining samples $1 + 3 + 5$ and $2 + 4 + 6$:

$$R_{\text{correlated}} = 708 + 196 + 174 = 1078$$

$$R_{\text{uncorrelated}} = 736 + 418 + 396 = 1550$$

Note the cell frequency (i.e. the number of scores per block in each subgroup) $n = 36$, and the number of subgroups $k = 2$. Check that nk is the
number of scores per block:

$$nk = 32(2) = 72$$

Compute H:

$$H = \left[\frac{\Sigma 12 R_j^2}{n^2 k(nk + 1)} - 3m^2(nk + 1) \right] \Big/ T$$

$$= \left[\frac{12(1078^2 + 1550^2)}{36^2(2)(73)} - 3(73) \right] \Big/ 0.9076$$

$$= 7.78$$

Degrees of freedom:

$$df = k - 1 = 2 - 1 = 1$$

Interaction, alcohol × aversive consequences
We must first compute the overall value of H across all six samples so that $n = 12$ and $k = 6$:

$$\Sigma R_j^2 = 708^2 + 736^2 + 196^2 + 418^2 + 174^2 + 396^2$$

$$= 1\,443\,192$$

$$H_{\text{overall}} = \left[\frac{12 \Sigma R_j^2}{n^2 k(nk + 1)} - 3m^2(nk + 1) \right] \Big/ T$$

$$H = \left[\frac{12(1\,443\,192)}{12^2(6)(73)} - 3(73) \right] \Big/ 0.9076$$

$$= 61.238$$

Degrees of freedom:

$$df = k - 1 = 6 - 1 = 5$$

We may now compute the interaction effect:

$$H_{\text{alcohol × consequences}} = H_{\text{overall}} - H_{\text{alcohol}} - H_{\text{consequences}}$$

$$= 61.238 - 50.824 - 7.78$$

$$= 2.63$$

Degrees of freedom:

$$df_{\text{alcohol × consequences}} = df_{\text{overall}} - df_{\text{alcohol}} - df_{\text{consequences}}$$

$$= 5 - 2 - 1$$

$$= 2$$

(b) Evaluate *H* by consulting tables of chi-square (table B in the appendix). We need the critical ranks for 1 and 2 degrees of freedom:

Degrees of freedom	5%	2.5%	1%	0.1%
1	$\chi^2 \geqslant 3.8$	5.0	6.6	10.8
2		6.0	7.4	9.2	13.8

The alcohol consumption main effect ($H = 50.8$, df $= 2$) is significant at better than the 0.1 per cent level ($p < 0.001$). The aversive consequences main effect ($H = 7.78$, df $= 1$) is significant at better than the 2.5 per cent level ($p < 0.025$). The interaction effect ($H = 2.63$, df $= 2$) is not significant ($p > 0.05$). Table 13.10 summarizes the results.

Step 5 - draw conclusions
We are able to reject the null hypothesis both in the case of alcohol consumption and the effect of correlated aversive consequences of administering the shock. Surprisingly the interaction effect is not significant despite a clear indication in figure 13.6 that aversive consequences may depress shock intensity in the sober group but not in the inebriated group. In the original study using parametric analysis a significant interaction was found, although it was small in comparison to the main effects. Considerable loss of information may have occurred in the transfer from raw scores to a crude four-point category scale.

13.5 Discussion

The aim of this chapter was to introduce the reader to the ability of rank sum analysis to deal with factorial designs. Such analysis can be confusing for the beginner but it is easily mastered because very few new principles are involved. The complications arise from the attempt to evaluate a number of hypotheses

Table 13.10 Summary table for example in section 13.4 for an analysis using nonspecific hypotheses

Source	*H*	*df*	*p*
Alcohol	50.8	2	<0.001
Aversive consequences	7.78	1	<0.025
Alcohol × aversive consequences	2.63	3	not significant
Overall	61.24	5	

simultaneously when previously only one hypothesis was tackled per data set. However, the reader should not be too irritated to discover that complex designs entail complex analysis. In general, however, the analyses required will not be as complex as those given earlier in the chapter. The analyses looked bulky because both specific and nonspecific analyses were attempted, when normally a researcher will choose only one or the other. The reader who has come this far should have little difficulty in deciding which one he needs.

The key to understanding factorial analysis by ranks is the ability to create a table which represents each combination of the levels of the independent variables as a separate column, each with its own rank sum. In this way a 2×3 design has six rank sums. Once this is done, the methods of chapters 9 and 10 can be applied quite straightforwardly. Use chapter 9 if the design is based on unrelated scores. Use chapter 10 if the design is based on related scores. The particular method used depends on whther a specific or nonspecific analysis is required. It may also depend on certain features of the design such as single scores per cell.

It should go without saying that the test user must be able to identify his independent variables quite positively at the outset. A common mistake is to confuse independent variables with blocking variables. For example, parametric analysis of variance would treat the example in section 13.3 as a $2 \times 3 \times 6$ factorial design, i.e. a 2 (funny/neutral) $\times$ 3 (describe/learn/spot the difference) $\times$ 6 (patients) design. However, rank sum analysis treats the patients as a blocking variable and classifies the example as a 2×3 *related* scores design. It so happens that this also makes it easier to compute and less confusing.

Nonspecific factorial analysis requires that the sample sizes (i.e. column frequencies) are equal. This restriction does not apply to a specific analysis. The reason for the restriction can be traced to the computation of the interaction term, using the following expression:

$$H_{interaction} = H_{overall} - H_{main\ effect\ A} - H_{main\ effect\ B}$$

This method only works when the column frequencies are equal. If they are unequal then considerably more advanced techniques are required, and these lie beyond the scope of this book.

The difficulties associated with factorial analysis (either parametric or nonparametric) are not necessarily associated with computational procedures. The problem usually lies with a researcher's inability to frame his question clearly in a manner which is amenable to statistical analysis. To frame such questions he needs to have clear reasons for collecting the data as well as a knowledge of the kind of questions which statistics can help answer. If either is deficient, problems follow. The benefit of using a specific analysis is that the researcher must frame his question explicitly in the form of a set of coefficients. This is usually quite simple for the main effects (i.e. effects associated with a single independent variable) but becomes difficult when dealing with interaction effects. This is to be expected because the idea of an interaction effect which is statistically independent of its associated main effects is very slippery indeed. Looking for appropriate coefficients concentrates the mind wonderfully.

Some of these problems are avoided in parametric analysis of variance by concentrating almost exclusively on nonspecific tests, and these are commonly used even when researchers could easily specify directional alternative hypotheses for their main effects at least. Use of nonspecific analysis makes interaction effects easier to describe (e.g. any deviation from the pattern of results predicted from a knowledge of the relevant main effects). Unfortunately, good researchers are not normally interested in 'any deviation'. They should, and usually do, have a clear idea of precisely how the scores will deviate from the pattern established by the main effects. This is certainly true of all the examples quoted in this chapter. If this is the case then a specific analysis is called for and an effort should be made to find the appropriate coefficients.

This chapter represents only a start in the search for ergonomic computing frames for analysing factorial designs using rank sum techniques. All kinds of extensions can be envisaged. Parametric analysis of variance has blazed the trail which rank sum analysis must follow. What seems important at this stage is to bear in mind the essential difference between the two - that rank sum analysis concerns rank means and not estimated population means. Often the distinction makes little difference to the final conclusions, but sometimes it can be crucial. Vigilance will be essential.

Exercises

13.1 In the example in section 13.4 it was assumed for the sake of convenience that placebo and control groups did not differ in any way, but this needs to be tested explicitly. Remove the data for the alcohol group from table 13.8 then attempt a 2×2 nonspecific factorial analysis on the remaining data

13.2 In the example in section 10.9 Parkes (1982) studied depression ratings of student nurses during and at the end of a training period on two wards in the hospital. If the control data are removed from table 10.17 the problem could be recast as a 2×2 factorial analysis for related scores. For hypotheses we might use the following:

H_1: medical wards are more depressing than surgical wards.

H_2: medical wards become more depressing the longer a nurse works there, whereas the surgical ward becomes less depressing.

13.3 In the following experiment three subjects were woken once per night over 12 nights. On each occasion they were asked to rate the depth of sleep that they were experiencing immediately prior to waking. This rating was made on a 5 point scale (5 = deeply asleep, 1 = not asleep). Four awaken-

Table 13.11 Depth of sleep ratings (5 = deep) for three subjects woken from three different sleep stages both early and late in the night. See exercise 13.3

Sleep type	Awakening time	Subject		
		1	*2*	*3*
DQS	early	3	5	4
		2	2	3
	late	5	4	4
		4	3	4
LQS	early	1	3	3
		1	2	4
	late	2	2	5
		3	3	3
AS	early	2	4	5
		2	2	5
	late	3	3	5
		5	4	4

ings were made from each of three different kinds of sleep: (a) deep quiet sleep (DQS); (b) light quiet sleep (LQS); and (c) active sleep (AS). Half of these awakenings were made early in the night while the remainder were made late in the night.

The experiment was abandoned after running three subjects because the results were so unpromising. Was this a justifiable decision? The experimenter had expected reports of lightest sleep from LQS and deepest sleep from DQS. He had also expected reports of deeper sleep early in the night. He did not expect any interaction effect.

Part III

14 Mathematical considerations

14.1 Some properties of ranks

Analysis of variance by ranks procedures require that all scores are converted to ranks before any computation is begun. We are, therefore, exclusively concerned with the mathematical properties of these ranks. We are particularly concerned with the situation where a set of N ranks is partitioned in a random fashion into k samples. First of all, however, we shall look at the basic properties of a set of N ranks.

Mean and variance of a set of N ranks

Let r_i be the ith rank in a set of N ranks without ties. The sum of N ranks is

$$\sum_{i=1}^{N} r_i = N(N+1)/2 \tag{14.1}$$

and the mean rank is

$$\bar{r} = \frac{\Sigma r_i}{N} = (N+1)/2 \tag{14.2}$$

The sum of squares of N ranks is

$$\sum_{i=1}^{N} r_i^2 = N(N+1)(2N+1)/6 \tag{14.3}$$

We define the variance of a set of N ranks as the mean squared difference between each score and the mean rank:

$$\text{var}(r) = \sum_{i=1}^{N} \frac{(r_i - \bar{r})^2}{N} = \frac{\sum_{i=1}^{N} r_i^2}{N} - \bar{r}^2 \tag{14.4}$$

If we substitute for Σr^2 and $\bar{r}$, using (14.2) and (14.3) in (14.4), we obtain

$$\text{var}(r) = \frac{(N+1)(2N+1)}{6} - \left(\frac{N+1}{2}\right)^2$$

$$= \frac{4N^2 + 6N + 2 - (3N^2 - 6N - 3)}{12}$$

$$= \frac{N^2 - 1}{12} \quad \text{or} \quad \frac{(N+1)(N-1)}{12} \tag{14.5}$$

Table 14.1 illustrates these formulae with a simple example.

Table 14.1 Illustration of the use of formulae for computing various statistics of a set of five ranks with no ties

r_i	r_i^2	$r_i - \bar{r}$	$(r_i - \bar{r})^2$
1	1	-2	4
2	4	-1	1
3	9	0	0
4	16	1	1
5	25	2	4
			N 5
Σr_i 15	Σr_i^2 55		$\Sigma(r_i - \bar{r})^2$ 10

(1) Sum of ranks (Σr_i) is 15.
Using (14.1), $\Sigma r_i = N(N+1)/2 = 5(6)/2 = 15$.

(2) Mean rank ($\bar{r}_i$) is $\Sigma r_i/N = 15/5 = 3$.
Using (14.2), $\bar{r}_i = (N+1)/2 = 6/2 = 3$.

(3) Sum of the squares of the ranks (Σr_i^2) is 55.
Using (14.3), $\Sigma r_i^2 = N(N+1)(2N+1)/6 = 5(6)(11)/6 = 55$.

(4) Variance of ranks is $\Sigma(r_i - \bar{r})^2/N = 10/5 = 2$.
Using (14.5), $\text{var}(r) = (N^2 - 1)/12 = 24/12 = 2$.

M

The formulae are more complicated when we have ties. Let Q be the number of *distinct* values of the scores to be ranked (where $Q \leqslant N$). If we deal with tied scores, using the method of tied ranks described in chapter 6, we will have Q different values for the ranks. Some of these will occur only once, whereas others will have a higher frequency of occurrence. Let t_q be the frequency of the qth distinct assigned ranking. We call t_q the length of the qth tie, noting that tie lengths of 1 are both possible and common. The following correction value will be found useful:

$$T = 1 - \frac{\sum\limits_{q=1}^{Q} (t_q^3 - t_q)}{N^3 - N} \tag{14.6}$$

where $0 \leqslant T \leqslant 1$. It is convenient that, in this expression, ties of length 1 (i.e. $t_q = 1$) can be ignored since they do not affect the value of T at all. This is because $(t_q^3 - t_q) = 0$ when $t_q = 1$.

Ties do not affect the expression for the sum of ranks (Σr) or the mean rank $(\bar{r})$ but they do reduce the variance:

$$\text{var}(r) = \frac{N^2 - 1}{12} T \quad \text{or} \quad \frac{(N+1)(N-1)}{12} T \tag{14.7}$$

When there are no ties, $T = 1$ and can be omitted from (14.7). Table 14.2 illustrates the computation of the mean and variance of a set of ranks including ties.

The method of assigning shared ranks is not the only technique for dealing with tied scores. There are others (see Bradley, 1968, pp. 48–54) but it is the only method used here and is probably the only practical method for routine laboratory computation. The problem of coping with ties in a set of ranks should not be confused with the difficulty of deciding how to deal with a set of ranks which are completely tied and therefore contribute nothing to the differences between rank sums. Our correction procedure effectively eliminates a completely tied set from the computations by reducing the variance to zero. With all scores tied, we have only one tie of length N, so that

$$T = 1 - \frac{\Sigma(t_q^3 - t)}{N^3 - N} = 1 - \frac{N^3 - N}{N^3 - N} = 0$$

and

$$\text{var}(r) = \frac{N^2 - 1}{12} T = 0$$

Table 14.2   Illustration of the use of formulae for computing various statistics of a set of five ranks with ties

r_i	r_i^2	$r_i - \bar{r}$	$(r_i - \bar{r})^2$
1	1	-2	4
3	9	0	0
3	9	0	0
3	9	0	0
5	25	2	4

N 5

Σr_i 15 Σr_i^2 53 $\Sigma(r_i - \bar{r})^2$ 8

(1) Sum of ranks (Σr_i) is 15.
     Using (14.1), $\Sigma r_i = N(N+1)/2 = 15$.

(2) Mean rank ($\bar{r}$) is $\Sigma r_i/N = 15/5 = 3$.
     Using (14.2), $\bar{r}_i = (N+1)/2 = 6/2 = 3$.

(3) Variance of the ranks is $\Sigma(r_i - \bar{r})^2/N = 8/5 = 1.6$.
     Using (14.6),

$$T = 1 - \Sigma(t^3 - t)/(N^3 - N)$$

$$= 1 - (3^3 - 3)/(5^3 - 5) = 1 - 24/120 = 0.8$$

Note that we have only one tie (t) of length 3.
Using (14.7), $\text{var}(r) = (N^2 - 1)T/12 = 24(0.8)/12 = 1.6$.

This convenience applies only to approximate tests, however, and not to exact tests. The recommended procedure for dealing with completely tied sets of ranks in an exact test is to delete them from the data set altogether. Bradley (1968, pp. 48–54) should again be consulted for a discussion of the advantages and disadvantages of alternative procedures.

Check sums. Expression (14.1) is the basis for the various check sum expressions to be found throughout the exposition of the test procedure. If we have m sets of ranks with b_i ranks in the ith set, then the grand total sum of ranks will be

$$\sum_{i=1}^{m} b_i(b_i + 1)/2 \qquad (14.8)$$

and this should equal the sum of the sample rank sums ($R_1 + R_2 + \ldots + R_k$). Expression (14.8) simplifies for different designs. For example, in the matched pair test (see section 8.3) we know that $b_i = 2$:

$$\sum_{i=1}^{m} b_i(b_i + 1)/2 = \sum_{i=1}^{m} 2(3)/2 = 3m \qquad (14.9)$$

Since the derivation of each checksum is simple, they will not be enumerated here.

Sampling from a set of ranks

If we take a sample of n_j ranks from a set of N ranks, we may compute the sum of ranks of that sample R_j. If the sample is repeatedly drawn at random from the complete set of N ranks, R_j will vary. In rank sum analysis, we are interested in the relative frequencies of occurrence of different values of R_j following repeated sampling. In particular, we want to know the mean or expected value,

Table 14.3   Computation of $E(R_j)$ and var(R_j) by enumeration and direct computation in the absence of ties

The population of ranks in this example is $(1, 2, 3, 4)$ and the population size is four $(N = 4)$. To examine the effects of taking samples of size two $(n_j = 2)$, we must draw up the following list of all possible combinations of two ranks:

Number	Sample	R_j	$R_j - \bar{R}_j$ $(R_j - 5)$	$(R_j - \bar{R})^2$
1	1, 2	3	-2	4
2	1, 3	4	-1	1
3	1, 4	5	0	0
4	2, 3	5	0	0
5	2, 4	6	1	1
6	3, 4	7	2	4
	ΣR_j	30	$\Sigma(R_j - \bar{R}_j)^2$	10

(1) Expected value of R_j is $E(R_j) = \bar{R}_j = \Sigma R_j/6 = 30/6 = 5$.
    Using (14.10) $E(R_j) = n_j(N + 1)/2 = 2(5)/2 = 5$.

(2) Variance of R_j is var$(R_j) = \Sigma(R_j - \bar{R}_j)^2/6 = 10/6 = 1.67$.
    Using (14.11) and noting that $T = 1$ because there are no ties,

$$\text{var}(R_j) = n_j(N + 1)(N - n_j)T/12$$
$$= 2(5)(2)(1)/12 = 1.67$$

$E(R_j)$, and then its variance, $\mathrm{var}(R_j)$. To visualize the meaning of these two expressions, we have to imagine an infinite number of independent random samples of equal size (n_j). Our estimates of $E(R_j)$ and $\mathrm{var}(R_j)$ would need to be based on this very large number of samples. Fortunately, we do not need to actually carry out this labour. We can specify $E(R_j)$ and $\mathrm{var}(R_j)$ by drawing up a *representative* set of samples. This representative set is a collection of all possible combinations of n_j ranks taken from N ranks. Tables 14.3 and 14.4 illustrate this enumeration in the case of samples of two $(n_j = 2)$ taken from a population of four $(N = 4)$ ranks.

The use of all combinations of n_j ranks from N is justified by the assumption that each combination is equally likely to occur as a random sample. This

Table 14.4 Computation of $E(R_j)$ and $\mathrm{var}(R_j)$ by enumeration and direct computation with ties

The population of ranks in this example contains a tie of length two $(t = 2)$ and is $(1, 2.5, 2.5, 4)$. To examine the effects of taking samples of size two $(n_j = 2)$, we must draw up the following list of all possible combinations of two ranks:

Number	Sample	R_j	$(R_j - \bar{R}_j)$ $(R_j - 5)$	$(R_j - \bar{R}_j)^2$
1	1, 2.5	3.5	−1.5	2.25
2	1, 2.5	3.5	−1.5	2.25
3	1, 4	5	0	0
4	2.5, 2.5	5	0	0
5	2.5, 4	6.5	1.5	2.25
6	2.5, 4	6.5	1.5	2.25
		ΣR_j 30		$\Sigma(R_j - \bar{R}_j)^2$ 9

(1) Expected value of R_j is $E(R_j) = \bar{R}_j = \Sigma R_j/6 = 30/6 = 5$.
   Using (14.10) $E(R_j) = n_j(N + 1)/2 = 2(5)/2 = 5$.

(2) Variance of R_j is $\mathrm{var}(R_j) = \Sigma(R_j - \bar{R}_j)^2/6 = 9/6 = 1.5$.
   There is one tie of length 2, so that

$$T = 1 - \Sigma(t^3 - t)/(N^3 - N) = 1 - (2^3 - 2)/(4^3 - 4)$$

$$= 1 - 6/60 = 0.9$$

Using (14.11),

$$\mathrm{var}(R_j) = n_j(N + 1)(N - n_j)T/12$$

$$= 2(5)(2)(0.9)/12 = 1.5$$

assumption is crucial to hypothesis testing using analysis of variance by ranks and is the heart of the null hypothesis (H_0). No matter how H_0 is expressed in the context of a particular analysis, its deepest meaning is simply that all combinations of ranks are equally likely to occur as our random sample. The above method of finding $E(R_j)$ and $\mathrm{var}(R_j)$ by the process of enumerating combinations is, therefore, appropriate to the situation where H_0 is true.

Rather than enumerate all combinations, we can compute $E(R_j)$ and $\mathrm{var}(R_j)$ directly:

$$E(R_j) = n_j(N+1)/2 \tag{14.10}$$

$$\mathrm{var}(R_j) = n_j(N+1)(N-n_j)\,T/12 \tag{14.11}$$

These computations and comparisons with the enumerative method are illustrated in table 14.3 without ties and table 14.4 with ties. When there are no ties T can be omitted because it is equal to one. The derivation of (14.10) and (14.11) is as follows:

The expected value of R_j is simply the sum of the expected values of the individual ranks:

$$E(R_j) = \sum_{p=1}^{n_j} E(r_p)$$

but since these expected ranks are all the same,

$$E(r_p) = \bar{r} = (N+1)/2$$

i.e. the mean rank (2), we obtain

$$E(R_j) = n_j E(r_p) = n_j(N+1)/2$$

The variance of R_j is generated using the equality

$$\mathrm{var}(R_j) = \sum_{p=1}^{n_j} \mathrm{var}(r_p) + \sum_{p \neq q}^{n_j} \sum^{n_j} \mathrm{covar}(r_p, r_q)$$

The variances and covariances are all the same, so that

$$\mathrm{var}(R_j) = n_j\,\mathrm{var}(r) + n_j(n_j - 1)\,\mathrm{covar}(r, r) \tag{14.12}$$

The covariance of r can be found by the simple expedient of considering the case when $n_j = N$, i.e. when the sample is the same size as the population. In this case R_j is fixed and has no variance:

$$\mathrm{var}(R_j) = 0 = N\,\mathrm{var}(r) + N(N-1)\,\mathrm{covar}(r, r)$$

So

$$\text{covar}(r, r) = -\frac{\text{var}(r)}{N-1}$$

Substituting for var(r), using (14.7),

$$\text{covar}(r, r) = -\frac{(N+1)(N-1)}{12(N-1)} T = -\frac{N+1}{12} T$$

Substituting in (14.12) for covar(r, r) and var(r),

$$\text{var}(R_j) = n_j(N+1)(N-1)\, T/12 - n_j(n_j - 1)(N+1)\, T/12$$
$$= n_j(N+1)(N-n_j)\, T/12$$

These results assume that there is only one block ($m = 1$). When there are many blocks, we have m independent sets of ranks. We can find $E(R_j)$ and var(R_j) simply by summing across the blocks. Notice that results (14.10) and (14.11) now refer to cell rank sums (R_{ij}). We must, therefore, adjust the sample size (n_j) to the cell frequency (n_{ij}) and the total sample size (N) to the block frequency (b_i) before summing. The correction term (T) also becomes the block correction term (T_i). We now have, using (14.10),

$$E(R_j) = \sum_{i=1}^{m} E(R_{ij}) = \sum_{i=1}^{m} n_{ij}(N+1)/2$$

$$= \tfrac{1}{2}(N+1) \sum_{i=1}^{m} n_{ij} \tag{14.13}$$

and, using (14.11),

$$\text{var}(R_j) = \sum_{i=1}^{m} \text{var}(R_{ij})$$

$$= \sum_{i=1}^{m} n_{ij}(b_i + 1)(b_i - n_{ij})\, T_i/12 \tag{14.14}$$

Sample rank sum as a deviation from the expected value

It is important not to confuse the expected rank sum, $E(R_j)$, with R_j, the rank sum of an actual sample. The former is the average over all possible samplings, whereas the latter is the sum of ranks in a particular sample. The phrase 'expected rank sum' is slightly misleading, since only a small number of the possible samples are expected to have this value as their rank sum. However, this value is as likely or more likely to occur than any other possible value of R_j. What is much more important, the majority of possible values of R_j will cluster closely around $E(R_j)$. Just how tightly they cluster is measured by $\text{var}(R_j)$. A smaller value of $\text{var}(R_j)$ indicates a tighter clustering.

A central preoccupation of rank sum analysis is the extent to which a sample rank sum in an experimental investigation deviates from the kind of value which would be expected if H_0 were true. A useful statistic to measure this is the standard score:

$$Z_j = \frac{R_j - E(R_j)}{\sqrt{\text{var}(R_j)}} \tag{14.15}$$

Large values of Z_j (positive and negative) indicate large deviations from the expected value of R_j, relative to the other possible values of R_j. Z_j has some useful properties, namely

$$E(Z_j) = 0$$

and

$$\text{var}(Z_j) = 1$$

These properties are illustrated in table 14.5 using the six possible values of R_j when two ranks are sampled from the ranks 1, 2, 3, 4. Because of these properties, Z is called a standard score.

Distribution of the sample rank sum. When the number of ranks (N) is increased, the distribution of all the possible values of R_j becomes increasingly similar to the normal distribution (see section 3.2). This is a consequence of the *central limit theorem*, which can be paraphrased, for our purposes, to say (1) that any score which is an aggregate of other scores which are (a) free to vary at random and (b) at least partly independent of one another, will be distributed approximately normally, and (2) that the approximation to the normal distribution will become increasingly close as the number of scores in the aggregate increases. Our sample rank sum (R_j) meets these requirements since (a) it is an aggregate of other scores (the ranks in the sample) (b) the ranks in the sample are chosen at random and (c) the sample ranks are at least partly independent. It follows, then, that as the sample size (n_j) increases, the distribution of the rank sum (R_j) will be increasingly 'normal'.

If R_j is approximately normally distributed, it follows that Z_j will also be approximately normally distributed since it is a simple linear transform of R_j. Because Z_j has $E(Z_j) = 0$ and $\text{var}(Z_j) = 1$, its distribution will approximate that

Table 14.5 Demonstration of two important properties of standard scores using the sample rank sums for samples of size two from a set of four ranks

The population of ranks in this example is (1, 2, 3, 4), and we know from table 14.3 that samples of size two ($n_j = 2$) from this set of ranks have the following properties:

$$E(R_j) = 5 \quad \text{and} \quad \text{var}(R_j) = 1.67$$

Using these values we can compute Z_j for each sample using (14.14). For example, for sample 1,

$$Z = [R_j - E(R_j)]/\sqrt{[\text{var}(R_j)]} = (3 - 5)/\sqrt{1.67} = -1.55$$

Number	Sample	R_j	Z_j	$(Z_j - \bar{Z}_j)$ $(Z_j - 0)$	$(Z_j - \bar{Z}_j)^2$
1	1, 2	3	−1.55	−1.55	2.4
2	1, 3	4	−0.77	−0.77	0.6
3	1, 4	5	0	0	0
4	2, 3	5	0	0	0
5	2, 4	6	0.77	0.77	0.6
6	3, 4	7	1.55	1.55	2.4
			ΣZ_j 0	$\Sigma(Z_j - \bar{Z}_j)^2$	6.0

$$E(Z) = \bar{Z} = \Sigma Z/6 = 0/6 = 0$$

$$\text{var}(Z) = \Sigma(Z - \bar{Z})^2/6 = 6/6 = 1$$

of the standard normal distribution, which has the same properties and is tabulated in table A in the appendix. Whereas the distribution of R_j changes dramatically with both the sample size (n_j) and the total number of ranks sampled (N), the distribution of Z_j changes very little and remains close to the standard normal distribution. For this reason, we routinely convert R_j to Z_j in order to decide just how far our observed value of R_j has deviated from the value expected, $E(R_j)$, in the situation where H_0 is true. The table of the standard normal distribution given in the appendix can then be used to put a figure to the probability of obtaining as extreme or more extreme a value of Z_j when H_0 is true.

A simple statistical test. If we substitute the expressions for computing $E(R_j)$ and $\text{var}(R_j)$ in our formula for computing Z_j, we find (when there are no ties)

$$Z_j = \frac{R_j - E(R_j)}{\sqrt{\text{var}(R_j)}} = \frac{R_j - n_j(N + 1)/2}{\sqrt{[n_j(N + 1)(N - n_j)/12]}} \qquad (14.16)$$

If we take only one sample of size n_1 from the N ranks, we can think of the remaining $N-n_1$ ranks as a second sample of size n_2. Substituting these new values in (14.16) and multiplying the numerator and denominator by 2, we obtain

$$Z_1 = \frac{2R_1 - n_1(N+1)}{\sqrt{[n_1 n_2(N+1)/3]}}$$

which can be used in conjunction with table A in the appendix to discover how often such a large (or small) value as R_1 would occur if H_0 were true.

We could alternatively have used the rank sum of the second sample as the basis of our calculations, but we would have obtained the same result except for a change of sign:

$$Z_2 = \frac{2R_2 - n_2(N+1)}{\sqrt{[n_1 n_2(N+1)/3]}}$$

From (14.1), we know that

$$R_1 + R_2 = N(N+1)/2$$

and

$$n_2 = N - n_1$$

Substituting for R_2 and n_2 in the numerator, we have

$$Z_2 = \frac{2[N(N+1)/2 - R_1] - (N - n_1)(N+1)}{\sqrt{[n_1 n_2(N+1)/3]}}$$

$$= \frac{-[2R_1 - n_1(N+1)]}{\sqrt{[n_1 n_2(N+1)/3]}} = -Z_1$$

Since the situation is symmetrical with respect to the sample used, we can look upon (14.16) as a two-sample test even though only one sample rank sum is used in the calculations. In fact, this is the test used in section 7.3 for two unrelated samples. The only difference is that the statistic L is used to replace R_1 in section 7.3. Since $R_1 = L$, the two methods are identical.

14.2 Specific tests: derivation of working formulae

We concluded the last section with a formula for evaluating the results of a two-sample experiment. The formula had been derived from first principles, and the way is now clear to proceed using similar techniques to derive the appropriate working formulae for the full range of designs used in previous chapters. It is more economical, however, to look for a single set of formulae which will cope with the most complex design of interest. The working formulae for the simpler designs will then be special cases of the general formulae and will be derived quite quickly using a little algebra.

The general case

For our purposes, the most complex design of interest is illustrated in section 10.6. For this design we expect any number of samples (k), any number of blocks (m) and the pattern of cell frequencies (n_{ij}) to be irregular. If we have working formulae for this design, we can obviously cope with the simpler situations where we may have either only two samples ($k = 2$) or only one block ($m = 1$) or regular cell frequencies (e.g. $n_{ij} = 1$), and the working formulae can be simplified accordingly. These formulae have been given for the case of the specific tests in Meddis (1980).

$$L = \sum_{j=1}^{k} \lambda_j R_j \tag{14.17}$$

$$E(L) = \tfrac{1}{2} \sum_{i=1}^{m} \left[(b_i + 1) \sum_{j=1}^{k} n_{ij} \lambda_j \right] \tag{14.18}$$

$$\mathrm{var}\,(L) = \tfrac{1}{12} \sum_{i=1}^{m} \left\{ (b_i + 1) \left[b_i \sum_{j=1}^{k} n_{ij} \lambda_j^2 - \left(\sum_{j=1}^{k} n_{ij} \lambda_j \right)^2 \right] T_i \right\} \tag{14.19}$$

They are not the easiest computational formulae to work with, and the benefits of simplification for specific designs are immediately obvious! Section 10.6 does illustrate their application, however, and a computer program for performing the calculations is given in chapter 15. The logic of using the statistic L is explained in section 3.2, together with the reasons for converting L to a standard score (Z):

$$Z = \frac{L - E(L)}{\sqrt{\mathrm{var}\,(L)}} \tag{14.20}$$

Further discussion of standard scores will be found in section 10.1. None of these matters will be rehearsed here.

The derivation of expressions (14.17), (14.18) and (14.19) will now be given for the sake of completeness. Begin by considering only a single block of scores (the ith block) containing b_i ranks. Let r_{ip} be the pth rank in the ith block. Also let λ_{ip} be the expected value of r_{ip} given that the specific alternative hypothesis is true. Define

$$l_i = \sum_{p=1}^{b_i} \lambda_{ip} r_{ip} \tag{14.21}$$

We first need to find $E(l_i)$:

$$E(l_i) = b_i E(r_{ip}) E(\lambda_{ip}) \tag{14.22}$$

since l_i is based on b_i ranks. We know from (14.2) that

$$E(r_{ip}) = (b_i + 1)/2 \tag{14.23}$$

and

$$E(\lambda_{ip}) = \left(\sum_{p=1}^{b_i} \lambda_{ip} \right) \Big/ b_i \tag{14.24}$$

Substituting for $E(r_{ip})$ and $E(\lambda_{ip})$ in (14.22), we obtain

$$E(l_i) = \tfrac{1}{2}(b_i + 1) \sum_{p=1}^{b_i} \lambda_{ip} \tag{14.25}$$

Normally, we will use the same value of λ_{ij} for all ranks in the same sample. We can therefore express (14.25) in terms of k samples rather than b_i ranks:

$$E(l_i) = \tfrac{1}{2}(b_i + 1) \sum_{j=1}^{k} n_{ij} \lambda_{ij} \tag{14.26}$$

If there are m blocks,

$$E(L) = \sum_{i=1}^{m} E(l_i) = \tfrac{1}{2} \sum_{i=1}^{m} \left[(b_i + 1) \sum_{j=1}^{k} n_{ij} \lambda_{ij} \right] \tag{14.27}$$

It is customary, but not obligatory, to use the same value λ_j for all ranks in the same sample, irrespective of block. We can therefore substitute λ_j for λ_{ij} in (14.27) to obtain our final computational formula for $E(L)$:

$$E(L) = \tfrac{1}{2} \sum_{i=1}^{m} (b_i + 1) \sum_{j=1}^{k} n_{ij} \lambda_j \tag{14.28}$$

Following a similar process, beginning with (14.21), we obtain

$$L = \sum_{i=1}^{m} \sum_{p=1}^{b_i} \lambda_{ip} r_{ip}$$

$$= \sum_{j=1}^{k} \lambda_j R_j \tag{14.29}$$

It is worth pausing to note that the λ_j coefficients are normally applied to the sample rank sums R_j, but the underlying theory requires predictions (λ_{ij}) for each cell and even (λ_{ip}) for each rank. It is merely a computational convenience that we treat all λ_{ip} in the same sample as having the same value. In certain situations it might be necessary to be more precise, to use the predicted cell mean (λ_{ij}) or even the predicted value of individual ranks λ_{ip}. This is most likely to occur when cell frequencies vary very widely, since the expected value of a rank is at least partly a function of the number of scores being ranked.

The derivation of var(L) is also more complex. Begin with a single block of scores (see (14.21)):

$$\text{var}(l_i) = \text{var}\left(\sum_{p=1}^{b_i} \lambda_{ip} r_{ip}\right) \tag{14.30}$$

Using a result given in Kendall and Stuart (1967, vol. 2, 31.19), we have

$$\text{var}\left(\sum_{p=1}^{b_i} \lambda_{ip} r_{ip}\right) = \sum_{p=1}^{b_i} (\lambda_{ip} - \bar{\lambda})^2 \sum_{p=1}^{b_i} (r_{ip} - \bar{r})^2 \Big/ (b_i - 1) \tag{14.31}$$

By the definition in (14.4) we have

$$\sum_{p=1}^{b_i} (r_{ip} - \bar{r})^2 = b_i \, \text{var}(r) \tag{14.32}$$

and using (14.7) to substitute for var(r),

$$\sum_{p=1}^{b_i} (r_{ip} - \bar{r})^2 = b_i (b_i^2 - 1) \, T_i / 12$$

$$= b_i (b_i + 1)(b_i - 1) \, T_i / 12 \tag{14.33}$$

Again, using the definition in (14.4), we have

$$\sum_{p=1}^{b_i} (\lambda_{ip} - \bar{\lambda})^2 = b_i \, \text{var}(\lambda)$$

$$= b_i \left[\sum_{p=1}^{b_i} \lambda_{ip}^2 \bigg/ b_i - \left(\sum_{p=1}^{b_i} \lambda_{ip} \bigg/ b_i \right)^2 \right]$$

$$= \sum_{p=1}^{b_i} \lambda_{ip}^2 - \left(\sum_{p=1}^{b_i} \lambda_{ip} \right)^2 \bigg/ b_i \qquad (14.34)$$

Using (14.33) and (14.34) we can rewrite (14.30):

$$\mathrm{var}(l_i) = \mathrm{var}\left(\sum_{p=1}^{b_i} \lambda_{ip} r_{ip} \right) = \tfrac{1}{12} b_i (b_i + 1) \left[\sum_{p=1}^{b_i} \lambda_{ip}^2 - \left(\sum_{p=1}^{b_i} \lambda_{ip} \right)^2 \bigg/ b_i \right] T_i$$

When all λ_{ip} in a given cell are the same (λ_{ij}),

$$\mathrm{var}(l_i) = \tfrac{1}{12} b_i (b_i + 1) \left[\sum_{j=1}^{k} n_{ij} \lambda_{ij}^2 - \left(\sum_{j=1}^{k} n_{ij} \lambda_{ij} \right)^2 \bigg/ b_i \right] T_i$$

$$= \tfrac{1}{12}(b_i + 1) \left[b_i \sum_{j=1}^{k} n_{ij} \lambda_{ij}^2 - \left(\sum_{j=1}^{k} n_{ij} \lambda_{ij} \right)^2 \right] T_i \qquad (14.35)$$

Since the blocks are statistically independent, we can combine the variances in the following way:

$$\mathrm{var}(L) = \sum_{i=1}^{m} \mathrm{var}(l_i)$$

$$= \tfrac{1}{12} \sum_{i=1}^{m} (b_i + 1) \left[b_i \sum_{j=1}^{k} n_{ij} \lambda_{ij}^2 - \left(\sum_{j=1}^{k} n_{ij} \lambda_{ij} \right)^2 \right] T_i \qquad (14.36)$$

If the λ_{ij} are all the same within a given sample, we may substitute λ_j for λ_{ij} in (14.36) and obtain the computational formula given in (14.17).

The λ coefficients

Our statistic Z in expression (14.20) is, essentially, a scale-free measure of linear correlation between the individual ranks (r_{ip}) and the values predicted for them (λ_{ip}) by the specific alternative hypothesis. We do not, of course, go to the elaborate lengths of computing the actual predicted values for each rank. We make use of the fact that all ranks in the same cell will have the same predicted value, so that one prediction will do for all in that cell. Nor do we specify exactly the estimated value of the ranks but merely choose a set of λ coefficients which reflect the pattern of expected differences between the cells. We can do this because Z is a scale-free statistic, and any set of coefficients which are perfectly linearly correlated will yield the same numerical value of Z. If the correlation is good but not perfect between two sets of coefficients, we still expect the results to be very similar. This gives us considerable latitude in choosing values for λ_j without changing the outcome of the analysis very much.

The simplest way of deciding whether two sets of coefficients are linearly correlated is to plot corresponding pairs of coefficients on graph paper. If a straight line can be drawn through all of the points, then they should yield exactly the same value for Z. In the special case where we have only two samples, and hence only two λ coefficients, any set of coefficients will be equivalent to any other. This is because our graph has only two points and a straight line can always be drawn through these two points wherever they are. As a result we always choose the most convenient pair ($\lambda_j = 1, 0$), which greatly simplifies the computational formulae.

Single-block designs

When we have only one block, expressions (14.18) and (14.19) can be simplified immediately. In this case $m = 1$, $b_i = N$, and $n_{ij} = n_j$, $T_i = T$. Since there is only one block, we can drop the summation over blocks ($\Sigma_{i=1}^{m}$). Substituting appropriately, we obtain

$$E(L) = \tfrac{1}{2}(N + 1) \sum_{j=1}^{k} n_j \lambda_j \tag{14.37}$$

$$\mathrm{var}\,(L) = \tfrac{1}{12}(N + 1) \left(N \sum_{j=1}^{k} n_j \lambda_j^2 - \left(\sum_{j=1}^{k} n_j \lambda_j \right)^2 \right) T \tag{14.38}$$

which are the expressions used in section 9.3. If there are no ties, $T = 1$ and can be omitted from (14.38). Marascuilo and McSweeney (1967) presented the case with equal cell frequencies ($n_j = n$) and where the coefficients were chosen to sum to zero ($\Sigma \lambda_j = 0$). In this case we have

$$E(L) = 0 \tag{14.39}$$

$$\mathrm{var}\,(L) = \tfrac{1}{12} n N(N + 1) \, \Sigma \lambda_j^2 \, T \tag{14.40}$$

Dunn (1964) uses similar expressions to evaluate contrasts.

Equal cell frequency designs

If all cells contain the same number of scores (n), we can simplify expressions (14.18) and (14.19) considerably. In this case $n_{ij} = n$, $b_i = kn$, $N = mkn$ and $n_j = mn$.

$$E(L) = \tfrac{1}{2} \sum_{i=1}^{m} \left[(b_i + 1) \sum_{j=1}^{k} n_{ij}\lambda_j \right]$$

$$= \tfrac{1}{2} m \left[(kn + 1) \, n \, \Sigma\lambda_j \right]$$

$$= \tfrac{1}{2} mn(kn + 1) \, \Sigma\lambda_j \tag{14.41}$$

$$\text{var}(L) = \tfrac{1}{12} \sum_{i=1}^{m} \left[(b_i + 1) \left(b_i \sum_{j=1}^{k} n_{ij}\lambda_j^2 - \left(\sum_{j=1}^{k} n_{ij}\lambda_j \right)^2 \right) \right] T_i$$

$$= \tfrac{1}{12} (kn + 1) [kn^2 \Sigma\lambda_j^2 - n^2 (\Sigma\lambda_j)^2] \sum_{i=1}^{m} T_i \tag{14.42}$$

If there are no ties, $\Sigma T_i = m$ and we have

$$\text{var}(L) = \tfrac{1}{12} mn^2 (kn + 1) [k \Sigma\lambda_j^2 - (\Sigma\lambda_j)^2] \tag{14.43}$$

The most popular instance of this design occurs when we have only one score per cell ($n = 1$), in which case (14.41) and (14.42) simplify further:

$$E(L) = \tfrac{1}{2} m (k + 1) \, \Sigma\lambda_j \tag{14.44}$$

$$\text{var}(L) = \tfrac{1}{12} (k + 1) [k \Sigma\lambda_j^2 - (\Sigma\lambda_j)^2] \sum_{i=1}^{m} T_i \tag{14.45}$$

If there are no ties, $\Sigma T_i = m$ and we have

$$\text{var}(L) = \tfrac{1}{12} m (k + 1) [k \Sigma\lambda_j^2 - (\Sigma\lambda_j)^2] \tag{14.46}$$

Page (1963) presented a trend test which restricted itself to the case which used the coefficients 1 to k (i.e. $\lambda_j = j$). In this case we have, from (14.1),

$$\Sigma\lambda_j = \sum_{j=1}^{k} j = k(k + 1)/2$$

$$E(L) = \tfrac{1}{4} mk(k + 1)^2 \tag{14.47}$$

Similarly, we have, from (14.4) and (14.5),

$$k\Sigma\lambda_j^2 - (\Sigma\lambda_j)^2 = k^2 \, \text{var}(\lambda)$$

$$= k^2 (k^2 - 1)/12$$

Substituting in (14.45) we obtain

$$\text{var}(L) = \tfrac{1}{144} k^2 (k^2 - 1)(k + 1) \Sigma T_i \tag{14.48}$$

When we have no ties, $\Sigma T_i = m$:

$$\text{var}(L) = \tfrac{1}{144} mk^2 (k^2 - 1)(k + 1) \tag{14.49}$$

which is the expression given by Page and presented as a computationally economical alternative in section 10.3. These formulae apply only when the specific alternative hypothesis indicates a monotonic ascending trend across the samples 1 to k.

Marascuilo and McSweeney (1967) allowed for any kind of trend or contrast for the single score per cell design but restricted the coefficients so that they summed to zero:

$$\Sigma \lambda_j = 0$$

This permits the following simplifications by substituting for $\Sigma \lambda_j$ in (14.44) and (14.45):

$$E(L) = 0 \tag{14.50}$$

$$\text{var}(L) = \tfrac{1}{12} k(k + 1)(\Sigma \lambda_j^2) \Sigma T_i \tag{14.51}$$

When there are no ties, $\Sigma T_i = m$ and we have

$$\text{var}(L) = \tfrac{1}{12} mk(k + 1)(\Sigma \lambda_j^2) \tag{14.52}$$

which is the expression given by Marascuilo and McSweeney and illustrated in section 10.3.

Dichotomous data often occur in designs with a single score per cell (e.g. when individuals answer questionnaires 'yes/no' or are rated 'success/failure' following various treatments). In such cases we can avoid the need to rank the scores if we note that there are only two types of score, which we shall call successes and failures. Let us adopt the following notation for our purposes:

S_{ij}: the number of successes in the i, jth cell ($S_{ij} = 1$ or 0).

$S_{i.}$: the number of successes in the ith block:

$$S_{i.} = \sum_{j=1}^{k} S_{ij}$$

$S._j$: the number of successes in the jth sample:

$$S._j = \sum_{i=1}^{m} S_{ij}$$

We can now assign a rank to a 'success' in the ith block because all successes will share the same rank, $(S_{i.} + 1)/2$. The shared rank for 'failures' will be $(k + S_{i.} + 1)/2$. This allows us to find the rank sum for the jth sample using this expression:

$$R_j = \tfrac{1}{2} \sum_{i=1}^{m} [S_{ij}(S_{i.} + 1) + (1 - S_{ij})(k + S_{i.} + 1)]$$

$$= \tfrac{1}{2} \left[\sum^{m} S_{ij}S_{i.} + \sum^{m} S_{ij} + mk + \sum^{m} S_{i.} + m - k \sum^{m} S_{ij} - \sum^{m} S_{ij}S_{i.} - \sum^{m} S_{ij} \right]$$

$$= \tfrac{1}{2} \left[\sum^{m} S_{i.} - k \sum^{m} S_{ij} + m(k + 1) \right]$$

But

$$\sum^{m} S_{ij} = S._j, \quad \text{so}$$

$$R_j = \tfrac{1}{2} \left[\sum^{m} S_{i.} - kS._j + m(k + 1) \right] \tag{14.53}$$

To simplify further we must accept the restriction that the coefficients sum to zero ($\Sigma\lambda_j = 0$), so that

$$L = \Sigma\lambda_j R_j = \tfrac{1}{2} \left[\sum_{j=1}^{k} \lambda_j \sum_{i=1}^{m} S_{i.} - k \sum_{j=1}^{k} \lambda_j S._j + m(k + 1) \sum_{j=1}^{k} \lambda_j \right]$$

But $\Sigma\lambda_j = 0$, so

$$L = \tfrac{1}{2} k \sum_{j=1}^{k} \lambda_j S._j \tag{14.54}$$

From (14.50) we also have

$$E(L) = 0 \tag{14.55}$$

The expression for var(L), given in (14.43) for this design, can be rewritten in terms of the $S_{i.}$. First note that, from (14.6),

$$T_i = 1 - \frac{\Sigma(t^3 - t)}{b^3 - b_i} = \frac{k^3 - \Sigma t^3}{k(k+1)(k-1)} \tag{14.56}$$

since, for this design, $\Sigma t = k$ and $b_i = k$. We can rewrite Σt^3 because there are only two tie lengths (for 0 and 1). These lengths are $S_{i.}$ and $k - S_{i.}$:

$$\Sigma t^3 = S_{i.}^3 + (k - S_{i.})^3$$
$$= k^3 + 3k(kS_{i.} - S_{i.}^2)$$

Substituting for Σt^3 in (14.56) and summing:

$$\sum_{i=1}^{m} T_i = \frac{3}{(k+1)(k-1)} \left(k \sum_{i=1}^{m} S_{i.} - \sum_{i=1}^{m} S_{i.}^2 \right) \tag{14.57}$$

Expression (14.53) can now be written:

$$\text{var}(L) = \tfrac{1}{12} k(k+1)(\Sigma\lambda_j^2) \Sigma T_i$$

$$= \tfrac{1}{4} k(\Sigma\lambda_j^2) \left(k \sum_{i=1}^{m} S_{i.} - \sum_{i=1}^{m} S_{i.}^2 \right) \Big/ (k-1) \tag{14.58}$$

The computation of Z can now be attempted directly in term ᵒf the $S_{i.}$ and $S_{.j}$ using (14.20), (14.54), (14.55) and (14.58):

$$Z = \frac{L - E(L)}{\sqrt{\text{var}(L)}} = \sqrt{\frac{L^2}{\text{var}(L)}}$$

$$= \sqrt{\left[\frac{k(k-1) \sum_{j=1}^{k} (\lambda_j S_{.j})^2}{\left(\sum_{j=1}^{k} \lambda_j^2 \right) \left(k \sum_{i=1}^{m} S_{i.} - \sum_{i=1}^{m} S_{i.}^2 \right)} \right]} \tag{14.59}$$

Two-sample designs

When we have only two samples ($k = 2$), the choice of coefficients is arbitrary and does not influence the result of the computations, except possibly for sign. This is because any two pairs of coefficients are necessarily perfectly linearly correlated (see note on λ coefficients earlier in this section). We are, therefore, free to fix the coefficients at the computationally convenient values

$$\lambda_1 = 1 \quad \text{and} \quad \lambda_2 = 0 \tag{14.60}$$

which imply that the rank mean of sample 1 will be greater than the rank mean of sample 2. It also means that, for all two-sample designs,

$$L = \Sigma \lambda_j R_j = (1)R_1 + (0)R_2 = R_1 \qquad (14.61)$$

It is important, therefore, to make quite sure that sample 1 is, in fact, the sample which is expected to have the higher rank mean.

From (14.18), we have

$$E(L) = \tfrac{1}{2} \sum_{i=1}^{m} \{(b_i + 1)[n_{i1}(1) + n_{i2}(0)]\}$$

$$= \tfrac{1}{2} \sum_{i=1}^{m} n_{i1}(b_i + 1)$$

or

$$E(L) = \tfrac{1}{2} \sum_{i=1}^{m} n_{i1}(n_{i1} + n_{i2} + 1) \qquad (14.62)$$

From (14.19), we have

$$\mathrm{var}(L) = \tfrac{1}{12} \sum_{i=1}^{m} (b_i + 1)(b_i n_{i1} - n_{i1}^2) T_i$$

$$= \tfrac{1}{12} \sum_{i=1}^{m} n_{i1} n_{i2}(n_{i1} + n_{i2} + 1) T_i \qquad (14.63)$$

which are the expressions used in section 8.4. For further discussion of this design see Wilcoxon (1947), Bradley (1968, p. 116) and Meddis (1975).

If we have only *one block*, we can omit the summation operators and use the fact that $b_i = N = n_{i1} + n_{i2}$ to obtain the working formulae

$$E(L) = \tfrac{1}{2} n_1(n_2 + n_2 + 1) = \tfrac{1}{2} n_1(N + 1) \qquad (14.64)$$

$$\mathrm{var}(L) = \tfrac{1}{12} n_1 n_2(N + 1) T \qquad (14.65)$$

which are the expressions used in section 7.3. They were first presented by Wilcoxon (1945, 1947) and are equivalent to the test presented by Mann and Whitney (1947). When there are no ties, $T = 1$ and can be omitted from (14.67). We can find Z using (14.20), which can be rewritten as

$$Z = \frac{2L - n_1(N + 1)}{\sqrt{[n_1 n_2(N + 1) T/3]}} \qquad (14.66)$$

which is a convenient computational form. This expression was derived by a different method in section 14.1 (see expression (14.16)).

In a two-sample multiple block design with only one score per cell (*matched pairs*), we can rewrite (14.62) and (14.63) using the following equivalence:

$$n_{i1} = n_{i2} = 1$$

Thus

$$E(L) = \tfrac{1}{2} \sum_{}^{m} (1)(3)$$

$$= 3m/2 \tag{14.67}$$

$$\text{var}(L) = \tfrac{1}{12} \sum_{}^{m} (1)(1)(3) \, T_i$$

$$= \tfrac{1}{4} \sum_{}^{m} T_i \tag{14.68}$$

When there are no ties $\Sigma T_i = m$ and we can rewrite (14.68) as

$$\text{var}(L) = m/4$$

When ties do occur, $T_i = 0$, so that $\Sigma T_i = m'$, where m' is the number of untied pairs of scores. We can now find Z using (14.15), which can be rewritten as

$$Z = \frac{2L - 3m}{\sqrt{m'}} \tag{14.69}$$

which is the convenient computational form used in section 8.2.

This matched pairs test was originally presented by Dixon and Mood (1946) in a form known as the sign test. Instead of ranking the scores in each pair, they simply count x, the number of pairs in which the sample 1 score is greater than the corresponding score in sample 2. However, when there are no ties,

$$L = R_1 = x + m \tag{14.70}$$

$$\Sigma T_i = m \tag{14.71}$$

Substituting in (14.69), we obtain

$$Z = \frac{2x - m}{\sqrt{m}} = \frac{x - m/2}{\tfrac{1}{2}\sqrt{m}} \tag{14.72}$$

which is the normal approximation to the binomial distribution, typically used in this context. The only difference between the two tests is the explicit use of ranks in the former case. There is little difference in computation convenience between them.

2 × 2 frequency table quick test. A 2 × 2 frequency table is an alternative method of presenting data from a two-sample single-block design. The appropriate computational formula is therefore (14.66):

$$Z = \frac{2L - n_1(N + 1)}{\sqrt{[n_1 n_2(N + 1) T/3]}} \tag{14.73}$$

Table 14.6 illustrates the notation used. There are only two sets of shared ranks for this problem, which we find using the method outlined in section 6.4:

$$r_q = \Sigma t_q - t_q/2 + 0.5 \tag{14.74}$$

For the first row, we have $\Sigma t_q = A + B$, $t_q = A + B$, so

$$r_1 = A + B - (A + B)/2 + 0.5 = (A + B + 1)/2 \tag{14.75}$$

For the second row, we have $\Sigma t_q = A + B + C + D$, $t_q = C + D$, so

$$r_2 = A + B + C + D - (C + D)/2 + 0.5 = (2A + 2B + C + D + 1)/2 \tag{14.76}$$

Using these shared ranks we can now find $2L$:

$$\begin{aligned} 2L = 2R_1 &= Ar_1 + Cr_2 \\ &= A(A + B + 1) + C(2A + 2B + C + D + 1) \end{aligned} \tag{14.77}$$

Table 14.6   Notation used for 2 × 2 frequency tables. A, B, C and D are frequencies

	Sample 1	Sample 2	
Category 1	A	B	$t_1 = A + B$
Category 2	C	D	$t_2 = C + D$
	$n_1 = A + C$	$n_2 = B + D$	$N = A + B + C + D$

Since $n_1 = A + C$ and $N = A + B + C + D$, we have

$$n_1(N + 1) = (A + C)(A + B + C + D + 1) \qquad (14.78)$$

So that, using (14.77) and (14.78), the numerator of (14.73) is

$$2L - n_1(N + 1) = A^2 + AB + A + 2AC + 2BC + C^2 + CD + C$$
$$-(A^2 + AB + AC + AD + AC + BC + C^2 + CD + A + C)$$
$$= BC - AD \qquad (14.79)$$

The denominator of (14.73) is

$$n_1 n_2 (N + 1)\, T/3 = (A + C)(B + D)(N + 1)\, T/3 \qquad (14.80)$$

From (14.6), noting that $b_i = N$ and $\Sigma t = N$,

$$T = 1 - \frac{\Sigma(t_q^3 - t)}{N^3 - N}$$
$$= \frac{N^3 - \Sigma t_q^3}{N(N^2 - 1)} \qquad (14.81)$$

We have only two ties, of length $t_1 = A + B$ and $t_2 = C + D$. Let

$$X = A + B$$
$$N - X = C + D$$

From (14.81),

$$T = \frac{N^3 - X^3 - (N - X)^3}{N(N^2 - 1)}$$
$$= \frac{N^3 - X^3 - N^3 + 3N^2 X - 3NX^2 + X^3}{N(N^2 - 1)}$$
$$= \frac{3NX(N - X)}{N(N + 1)(N - 1)}$$
$$= \frac{3(A + B)(C + D)}{(N + 1)(N - 1)} \qquad (14.82)$$

From (14.80) and (14.82) we can find the denominator:

$$n_1 n_2 (N+1) \, T/3 = \frac{(A+C)(B+D)(N+1)3(A+B)(C+D)}{3(N+1)(N-1)}$$

$$= \frac{(A+C)(B+D)(A+B)(C+D)}{N-1} \qquad (14.83)$$

From (14.73), (14.79) and (14.83) we arrive at the computational formula

$$Z = \frac{BC-AD}{\sqrt{\left[\dfrac{(A+C)(B+D)(A+B)(C+D)}{N-1}\right]}} \qquad (14.84)$$

which was used in chapter 7. This formulation emerged as the most highly recommended from a recent survey of alternative procedures for analysing data from a 2 × 2 comparative trial (Upton, 1982).

14.3   Nonspecific tests: derivation of working formulae

The general case

In a nonspecific test, we seek merely to evaluate the null hypothesis (H_0) that all the samples have the same expected rank mean. An obvious approach to a solution is to consider each individual rank sum and ask whether it deviates from its expected value, $E(R_j)$, to an extent greater than is normally found under conditions of purely random fluctuation. In section 14.1 expression (14.15), a method was proposed for computing a standard index which measures the extent of deviation:

$$Z_j = \frac{R_j - E(R_j)}{\sqrt{\mathrm{var}(R_j)}} \qquad (14.85)$$

The statistic Z_j is approximately normally distributed (mean $= 0$, variance $= 1$) with large sample sizes. By consulting tables of the normal distribution, we can estimate how often such a deviation from expectation would be equalled or exceeded if H_0 were true.

This method is satisfactory for a single sample but we wish to take all of the samples into account. A simple statistic which aggregates the deviations of all R_j from $E(R_j)$ might be

$$H^* = \sum_{j=1}^{k} Z_j^2 \qquad (14.86)$$

It is intuitively clear that H^* will be large when the sample rank means are widely separated. If we knew the distribution of H^* (given H_0 true), we would immediately be able to estimate the likelihood of obtaining such a wide spread of sample means, and the problem of generating computational formulae would be easily solved. Unfortunately, it is not so simple. Expression (14.86) does, however, bring us close to the solution we seek, since H^* would be distributed as chi-square with k degrees of freedom if, and only if, the Z_j were *statistically independent*. The Z_j are, however, mutually dependent to a moderate degree. This can be best appreciated by considering the sample rank sums, whose total is constrained by the grand total of ranks in the data set. If one rank sum goes up, at least one of the others must come down, so they cannot be wholly independent. The complications which attend the derivation of nonspecific computational procedures are mainly the consequence of this correlation which exists between sample rank sums.

The eventual solution takes us briefly into the realms of matrix algebra. We define u_j, the 'reduced rank sum', as

$$u_j = R_j - E(R_j) \tag{14.87}$$

where, from (14.13),

$$E(R_j) = \tfrac{1}{2} \sum_{}^{m} n_{ij}(n_i + 1) \tag{14.88}$$

We define also σ_{jj}^2, the variance of u_j, from (14.14):

$$\sigma_{jj}^2 = \tfrac{1}{12} \sum_{i=1}^{m} n_{ij}(b_i - n_{ij})(b_i + 1)\,T_i \tag{14.89}$$

which is $\text{var}(R_j)$.

The covariance of two reduced rank sums $(R_j, R_{j'})$ is given by

$$\sigma_{jj'}^2 = -\tfrac{1}{12} \sum_{i=1}^{m} n_{ij}n_{ij}(b_i + 1)\,T_i \tag{14.90}$$

We can picture the covariance as the average product of u_j and $u_{j'}$ over all permutations of ranks; it is a measure of the correlation which necessarily exists between the rank sums of two samples. We now define

$$H = UV^{-1}U' \tag{14.91}$$

where U is a vector of the k reduced rank sums, u_j, and V^{-1} is the *generalized* inverse of the $k \times k$ variance–covariance matrix with elements σ_{jj}^2 and $\sigma_{jj'}^2$.

When H is calculated in this way, we obtain a value which is similar to H^* in (14.87) but which has been corrected for the statistical dependence which exists among the sample rank sums. It is now the case that H is distributed approximately as chi-square with $(k-1)$ degrees of freedom (Brunden and Mohberg, 1976), and (14.91) can be used as the seed for deriving all of our computational formulae for nonspecific tests.

Instead of using the generalized inverse of the matrix V, Benard and Van Elteren (1953) recommended using the regular inverse of a reduced version of V in conjunction with reduced vectors U and U'. Because of the statistical interdependence of the sample rank sums, any one row or column of the matrix V can be inferred from a knowledge of the other $(k-1)$ rows or columns. In other words, the 'rank' of the matrix is $k-1$. They suggest deleting any row (and its corresponding column) from the matrix V and deleting the corresponding elements from the vectors U and U'. If we name these reduced vectors as V^* and U^* we obtain

$$H = U^* V^{*-1} U^{*'} \tag{14.92}$$

where V^{*-1} is the *regular* inverse of the reduced matrix V. Brunden and Mohberg (1976) have shown that (14.92) and (14.91) are equivalent procedures.

Expression (14.92) is of special interest in the *two-sample* situation where the reduction procedure reveals a very simple structure. By deleting the last elements, we obtain

$$H = (u_1 \ u_2) \begin{vmatrix} \sigma_{11}^2 & \sigma_{12}^2 \\ \sigma_{21}^2 & \sigma_{22}^2 \end{vmatrix}^{-1} (u_1 \ u_2)' \tag{14.93}$$

$$= u_1 \sigma_{11}^{-2} u_1' = u_1^2 / \sqrt{\sigma_{11}^2}$$

$$= \frac{[R_1 - E(R_1)]^2}{\text{var}(R_1)} \tag{14.94}$$

This is a result we shall need later when considering two-sample designs.

When a design has more than two samples, more than one block and unequal cell frequencies, the test user has no alternative to the use of either (14.92) or (14.91). Since few researchers will have the necessary skill to invert a variance-covariance matrix, this will best be done using a computer program. Such a procedure is illustrated in chapter 15. No attempt can be made here to teach matrix arithmetic to the requisite level of ability. If the researcher does not have a computer available he should consider whether or not he really needs to carry out a nonspecific test on the data. A specific test is considerably more powerful and informative. He may find, on reflection, that the result may not warrant the effort in pursuing the goal.

If the cell frequencies are unequal but regular, (14.92) can be simplified. For example if, for all n_{ij}, the following condition is met:

$$n_{ij} = n_j b_i / N \tag{14.95}$$

we can use the following expression given by Benard and Van Elteren (1953):

$$H = \frac{12N \sum\limits_{j=1}^{k} \left[R_j - \frac{1}{2} \sum\limits_{i=1}^{m} n_{ij}(b_i + 1) \right]^2 \Big/ n_j}{\sum\limits_{i=1}^{m} b_i^2 (b_i + 1) T_i} \tag{14.96}$$

Many of the derivations in the following use this expression, which was given in section 10.2.

Equal cell frequency designs

The only design in this class which interests us is the case where all cells contain only one score. In this case $n_{ij} = 1$, $b_i = k$, $n_j = m$ and $N = mk$. First we expand (14.96) and then simplify:

$$H = \frac{12N \sum\limits_{j=1}^{k} \left\{ R_j^2 - R_j \sum\limits_{i=1}^{m} n_{ij}(b_i + 1) + \frac{1}{4} \left[\sum\limits_{i=1}^{m} n_{ij}(b_i + 1) \right]^2 \right\} \Big/ n_j}{k^2(k + 1) \sum\limits_{i=1}^{m} T_i}$$

$$= \frac{12k \left[\sum\limits_{i=1}^{k} R_j^2 - m(k + 1) \sum\limits_{j=1}^{k} R_j + \frac{1}{4} mk(k + 1)^2 \right]}{k^2(k + 1) \Sigma T_i}$$

Noting that $\Sigma R_j = mk(k + 1)/2$, we obtain

$$H \sum\limits_{i=1}^{m} T_i = \frac{12}{k(k + 1)} \sum\limits_{j=1}^{k} R_j^2 - \frac{12}{k(k + 1)} [m^2 k(k + 1)^2/2 - m^2 k(k + 1)^2/4]$$

$$= \frac{12}{k(k + 1)} \sum\limits_{j=1}^{k} R_j^2 - 3m^2(k + 1)$$

$$H = \left[\frac{12}{k(k + 1)} \Sigma R_j^2 - 3m^2(k + 1) \right] \Big/ \sum\limits_{i=1}^{m} T_i \tag{14.97}$$

When there are no ties, $\Sigma T_j = m$ and

$$H = \frac{12}{N(k+1)} \Sigma R_j^2 - 3m(k+1) \tag{14.98}$$

which is the expression originally given by Friedman (1937), and which is used in section 10.9. In this case K, our 'statistic of disagreement', is ΣR_j^2 and can be readily substituted in (14.98).

Cochran (1950) examined the special case of this design where each cell contains only one score on a binary scale (dichotomous data). To avoid repetition, we shall use results (14.53) and (14.57) and the associated notation from section 14.2. Thus

$$R_j = \tfrac{1}{2}\left[\sum_{i=1}^{m} S_{i.} - kS_{.j} + m(k+1) \right] \tag{14.99}$$

and

$$\sum_{i=1}^{m} T_i = \frac{3}{(k+1)(k-1)}\left(k \sum_{i=1}^{m} S_{i.} - \sum_{i=1}^{m} S_{i.}^2 \right) \tag{14.100}$$

Note that

$$\sum_{i=1}^{m} S_{i.} = \sum_{j=1}^{k} S_{.j}$$

so that, from (14.99),

$$\sum_{j=1}^{k} R_j^2 = \tfrac{1}{4} \sum_{j=1}^{k}\left[\sum_{j=1}^{k} S_{.j} - kS_{.j} + m(k+1) \right]^2$$

$$= \tfrac{1}{4}\left[k\left(\sum_{j=1}^{k} S_{.j} \right)^2 - 2k\left(\sum_{j=1}^{k} S_{.j} \right)^2 + 2mk(k+1)\left(\sum_{j=1}^{k} S_{.j} \right) \right.$$

$$+ k^2 \sum_{j=1}^{k} S_{.j}^2 - 2m(k+1)k\left(\sum_{j=1}^{k} S_{.j} \right) + km^2(k+1) \Bigg]$$

$$= \tfrac{1}{4}\left[k^2 \sum_{j=1}^{k} S_{.j}^2 - k\left(\sum_{j=1}^{k} S_{.j} \right)^2 + km^2(k+1)^2 \right] \tag{14.101}$$

We can now recast expression (14.97) in terms of the new notation by substituting for ΣR_j^2 and ΣT_i using (14.101) and (14.100). After some elementary manipulation we obtain

$$H = \frac{(k-1)\left[k \sum\limits_{j=1}^{k} S_{\cdot j}^2 - \left(\sum\limits_{j=1}^{k} S_{\cdot j}^2\right)^2\right]}{k \sum\limits_{i=1}^{m} S_{i\cdot} - \sum\limits_{i=1}^{m} S_{i\cdot}^2} \tag{14.102}$$

which is equivalent to the expression given by Cochran (1950). The use of this procedure, which is a particularly effort-free method of analysing large quantities of data by hand, is illustrated in section 10.10.

Single-block designs

When we have only one block, expression (14.96) can be considerably simplified. In this case $m = 1$, $b_i = N$, $n_{ij} = n_j$ and $T_i = T$. Since there is only one block, we can drop the summation over blocks (Σ^m). Substituting appropriately, we obtain

$$H = \frac{12N \sum\limits_{j=1}^{k} [R_j - \tfrac{1}{2}n_j(N+1)]^2 / n_j}{N^2(N+1)T}$$

$$= \frac{12N\left(\sum\limits_{j=1}^{k} R_j^2 / n_j - (N+1)\,\Sigma R_j + \tfrac{1}{4}(N+1)^2 \Sigma n_j\right)}{N^2(N+1)T}$$

Noting that $\Sigma R_j = N(N+1)/2$ and $\Sigma n_j = N$, we have

$$H = \left[\frac{12 \sum\limits_{j=1}^{k} R_j^2 / n_j}{N(N+1)} - 3(N+1)\right] \Big/ T \tag{14.103}$$

When there are no ties, $T = 1$ and can be omitted:

$$H = \frac{12 \sum\limits_{j=1}^{k} R_j^2 / n_j}{N(N+1)} - 3(N+1) \tag{14.104}$$

This expression was first given by Kruskal (1952), and its use is illustrated in section 9.6. When we are using K, the 'statistic of disagreement',

$$K = \sum_{j=1}^{k} R_j^2/n_j$$

and

$$H = \frac{12K}{N(N+1)} - 3(N+1) \tag{14.105}$$

Two-sample designs

The derivation of computational formulae for two-sample problems is greatly simplified by the use of expression (14.94). Substituting for $E(R_1)$ and $\text{var}(R_1)$ using expressions (14.88) and (14.89), we obtain

$$H = \frac{\left[R_1 - \frac{1}{2}\sum_{i=1}^{m} n_{i1}(b_i + 1)\right]^2}{\sum_{i=1}^{m} n_{i1}n_{i2}(b_i + 1)\,T_i\Big/12}$$

$$= \frac{3\left[2K - \sum_{i=1}^{m} n_{i1}(b_i + 1)\right]^2}{\sum_{i=1}^{m} n_{i1}n_{i2}(b_i + 1)\,T_i} \tag{14.106}$$

where $K = R_1$, the rank sum of the sample with the higher observed rank mean. When there are no ties, all $T_i = 1$ and can be omitted. See section 8.8 for an illustration of the use of this expression.

When the cell frequences are equal ($n_{ij} = n$), we can simplify further:

$$H = \frac{3[2K - mn(2n + 1)]^2}{n^2(2n + 1)\sum_{i=1}^{m} T_i} \tag{14.107}$$

When there are no ties, $\Sigma T_i = m$:

$$H = \frac{3[2K - mn(2n + 1)]^2}{mn^2(2n + 1)} \tag{14.108}$$

In the simplest design with only one score per cell (i.e. matched pairs),

$$H = \frac{(2K - 3m)^2}{\sum\limits_{i=1}^{m} T_i} \tag{14.109}$$

Let m' be the number of untied pairs. Then

$$m' = \sum_{i=1}^{m} T_i$$

since $T_i = 1$ for an untied pair and $T_i = 0$ for tied pairs. In which case

$$H = \frac{(2K - 3m)^2}{m'} \tag{14.110}$$

The application of this formula to the matched pairs problem is illustrated in section 8.6.

When we have only one block, $m = 1$, and expression (14.106) simplifies immediately to

$$H = \frac{3[2K - n_1(N + 1)]^2}{n_1 n_2 (N + 1) T} \tag{14.111}$$

where $K = R_1$. If there are no ties, $T = 1$ and can be omitted. The use of this expression is illustrated in chapter 7.

In the two-sample case, a close relationship exists between H and Z, as might be expected from (14.86):

$$H = Z^2 \tag{14.112}$$

It is instructive to note that in (14.112), a large value of H could represent either a large negative or a large positive value of Z. Thus, although Z would register statistically significant only when large *and* positive, H can also be statistically significant when Z is large and negative. Thus when H_0 is true, a value of H is twice as likely to occur as the equivalent positive value of Z. Statistical tables take this into account. For example, $H = 3.9$ is rated as being significant at the 5 per cent level in table B in the appendix, whereas $Z = \sqrt{3.9} = 1.97$ is rated as significant at the 2.5 per cent level in table A in the appendix. Because H can be thought of as representing both positive and negative tails of the distribution of Z, it is sometimes referred to as a 'two-tail test'. This applies, of course, only to tests for designs with two samples.

2 × 2 frequency table quick test. A 2 × 2 frequency table is an alternative method for presenting data for a two-sample single-block design. Rather than derive the computational formula from (14.111) we can note, with gratitude, the relationship (14.112) between specific and nonspecific tests for two-sample designs. We may, therefore, immediately assert from (14.84) that

$$H = Z^2 = \frac{(N-1)(BC - AD)^2}{(A + C)(B + D)(A + B)(C + D)} \tag{14.113}$$

The use of this expression with 2 × 2 frequency tables is illustrated in chapter 7.

14.4   Correlation and concordance

Correlation

Correlation using rank sum methods concerns k pairs of ranks (X_j, Y_j) and is based on the statistic

$$\rho = \frac{\sum\limits_{j=1}^{k} (X_j - \bar{X})(Y_j - \bar{Y})}{\sqrt{\left[\sum\limits_{j=1}^{k} (X_j - \bar{X})^2 \sum\limits_{j=1}^{k} (Y_j - \bar{Y})^2 \right]}} \tag{14.114}$$

which has a range from $+1$ to -1. Values close to $+1$ or -1 indicate a strong correlation between the variables. If we define $x_j = (X_j - \bar{X})$ and $y_j = (Y_j - \bar{Y})$ we have

$$\rho = \frac{\Sigma xy}{\sqrt{(\Sigma x^2 \Sigma y^2)}} \tag{14.115}$$

We wish to express ρ in terms of the rank sums R_j where

$$R_j = X_j + Y_j \tag{14.116}$$

To this end we also define

$$r_j = x_j + y_j = R_j - (\bar{X} + \bar{Y}) \tag{14.117}$$

When there are no ties, we have from (14.4) and (14.5)

$$\Sigma x_j^2 = \Sigma y_j^2 = k \, \text{var}(x) = k(k^2 - 1)/12$$
$$= (k^3 - k)/12 \tag{14.118}$$

We can find Σxy from the following equality:

$$\Sigma r_j^2 = \Sigma(x_j + y_j)^2 = \Sigma x_j^2 + \Sigma y_j^2 + 2\Sigma xy$$

Whence, from (14.118),

$$\Sigma xy = (\Sigma r_j^2 - \Sigma x_j^2 - \Sigma y_j^2)/2$$
$$= [\Sigma r^2 - (k^2 - k)/6]/2 \tag{14.119}$$

Substituting for Σx_j^2, Σy_j^2 (14.118) and xy (14.119) in (14.115), we obtain

$$\rho = \frac{\Sigma r_j^2 - \dfrac{k^3 - k}{6}}{2\sqrt{\left[\left(\dfrac{k^3 - k}{12}\right)^2\right]}}$$

$$= \frac{6\Sigma r_j^2}{k^3 - k} - 1 \tag{14.120}$$

From (14.2) we know that

$$\bar{X} = \bar{Y} = (k + 1)/2 \tag{14.121}$$

So that, from (14.117),

$$r_j = R_j - (k + 1) \tag{14.122}$$

Noting that $\Sigma R_j = k(k + 1)$, we find that

$$\sum_{j=1}^{k} r_j^2 = \sum_{j=1}^{k} R_j^2 - 2(k + 1) \sum_{j=1}^{k} R_j + \sum_{j=1}^{k} (k + 1)^2$$
$$= \Sigma R_j^2 - 2k(k + 1)^2 + k(k + 1)^2$$
$$= \Sigma R_j^2 - k(k + 1)^2 \tag{14.123}$$

Substituting for r_j in (14.120), we conclude that

$$\rho = \frac{6[\Sigma R_j^2 - k(k + 1)^2]}{k^3 - k} - 1$$

$$= \frac{6\Sigma R_j^2}{k^3 - k} - 6\frac{(k + 1)}{k - 1} - 1 \tag{14.124}$$

N

Spearman (1906) originally proposed the rank correlation statistic in terms of D_j, the difference between the scores in each pair $(D_j = X_j - Y_j)$:

$$\rho = 1 - \frac{6 \sum\limits_{j=1}^{k} D_j^2}{k^3 - k} \tag{14.125}$$

The two formulations are identical, and this can be readily shown using the identity

$$\Sigma R_j^2 = 2\Sigma X_j^2 + 2\Sigma Y_j^2 - \Sigma D_j^2$$
$$= 4k(k+1)(2k+1)/6 - \Sigma D_j^2$$

So that, from (14.124),

$$\rho = \frac{6[4k(k+1)(2k+1)/6 - \Sigma D_j^2]}{k^3 - k} - \frac{6(k+1)}{k-1} - 1$$
$$= \frac{4(2k+1)}{k-1} - \frac{6\Sigma D_j^2}{k^3 - k} - \frac{6(k+1)}{k-1} - 1$$
$$= 1 - \frac{6\Sigma D_j^2}{k^3 - k}$$

The design is, in essence, a k-sample, two-block, single-score-per-cell design. Therefore, from (14.98), noting that $m = 2$ and $N = 2k$,

$$H = \frac{12}{N(k+1)} \Sigma R_j^2 - 3m(k+1)$$
$$= \frac{6}{k(k+1)} \Sigma R_j^2 - 6(k+1) \tag{14.126}$$

From which we see, using (14.124), that

$$\rho = H/(k-1) - 1 \tag{14.127}$$

and

$$H = (\rho + 1)(k - 1) \tag{14.128}$$

We might use H as a large-sample approximation to the chi-square distribution with $(k-1)$ degrees of freedom. However, the approximation can be improved by the following adjustment:

$$H = \rho^2(k-1) \tag{14.129}$$

where H is distributed as chi-square with *only one degree of freedom*. This test is nonspecific in the sense that it can reject H_0 given either large positive or large negative values of ρ. If the observed correlation is in the direction predicted by a specific alternative hypothesis, the following expression holds:

$$Z = \sqrt{H} = \sqrt{[\rho^2(k-1)]} \tag{14.130}$$

where Z is approximately normally distributed. Note that Z will always be positive using (14.130) because of the effect of squaring ρ before taking the square root. It is important, therefore, to make quite sure that the correlation is in the expected direction before using (14.130).

The effect of ties is to complicate expression (14.125) to a degree which discourages its use. Instead, compute H using

$$H = \left[\frac{12}{k(k+1)} \sum_{j=1}^{k} R_j^2 - 12(k+1) \right] \bigg/ (T_x + T_y) \tag{14.131}$$

where T_x and T_y are the correction factors for the X and Y variables:

$$T_x = 1 - \frac{\displaystyle\sum_{q=1}^{Q} (t_{xq}^3 - t_{xq})}{k^3 - k} \tag{14.132}$$

and then compute ρ using (14.127).

Correcting for ties normally has the effect of increasing H (or Z) and makes the result slightly more significant. Failure to correct for ties is, therefore, a conservative strategem. The tie correction procedure given here, however, has an unsymmetrical effect on ρ: it increases positive values of ρ (i.e. makes them more positive) but also increases negative values of ρ (i.e. makes them less negative). Since ρ is purely a descriptive measure, the test user must decide how important accuracy is in determining ρ. If no correction is applied, the error in determining ρ is unsymmetrical with respect to its sign.

Concordance

When we have a number of variables which need testing for concordance, we may use the many-block design with a single score per cell. Each block contains the scores of one variable and each sample represents a single individual assessed on each variable. Kendall (1945) proposed the statistic

$$W = \frac{S}{S_{max}} \qquad (14.133)$$

where

$$S = \sum_{j=1}^{k} (R_j - \bar{R})^2$$

and

$$S_{max} = \tfrac{1}{12} m^2 (k^3 - k)$$

W has a maximum value of $+1$ and reflects the degree of concordance among the m variables across the k individuals.

We may expand S using (14.4):

$$S = \sum_{j=1}^{k} (R_j - \bar{R})^2 = k \, \mathrm{var}(R_j) = \Sigma R_j^2 - (\Sigma R_j)^2 / k$$

Noting that

$$\Sigma R_j = mk(k + 1)/2$$

we find that

$$S = \Sigma R_j^2 - m^2 k^2 (k + 1)^2 / 4k$$

from which

$$W = \frac{S}{S_{max}} = \frac{12\Sigma R_j^2}{m^2 k(k + 1)(k - 1)} - \frac{12m^2 k(k + 1)^2}{4m^2 k(k - 1)(k - 1)}$$

$$= \frac{12\Sigma R_j^2}{m^2 k(k + 1)(k - 1)} - \frac{3(k + 1)}{(k - 1)} \qquad (14.134)$$

But we know from (14.98) that

$$H = \frac{12 \sum_{j=1}^{k} R_j^2}{mk(k + 1)} - 3m(k + 1)$$

From which we can see that

$$W = H/m(k-1) \qquad\qquad (14.135)$$

In this design it is clearly expedient to compute H first, in the normal way (correcting for ties if necessary). H can then be used both for computing W and for performing the test of statistical significance.

15 Computer program

15.1 Introduction to the program

The listing given in section 15.3 is a BASIC program suitable for running on almost any kind of computer which is equipped with a BASIC interpreter. The program has been kept as simple as possible so that it can be easily understood and so that it will be compatible with almost all dialects of BASIC. The program will handle *all* of the examples given in the book except for the correlation example using frequency tables (section 11.3). The printout is designed to give the user access to all of the detailed computations involved in each exercise. If these are not needed, simply delete the relevant PRINT statements.

Data input is achieved by using numbered DATA statements which are added to the program before runtime. When the DATA statements have been entered, the program will execute following a RUN statement. DATA statements have two major advantages for this kind of application. Firstly, they are common to all dialects of BASIC, which means that the program will 'travel' well. Secondly, they remain in the program and can be easily edited after a run that fails for some reason. As a result, the data do not need to be re-entered in full after every crash. Users who prefer immediate keyboard input can simply replace READ statements with INPUT statements. Disk filing and retrieval subroutines can also be easily added to the program.

The data input sequence always begins with the number of samples and blocks, e.g.

    10000   DATA   3, 1

and is followed by a letter, in quotes, indicating raw score (R) or frequency table (F) input, e.g.

    10001   DATA   'R'

or

    10001   DATA   'F'

If the data are in the form of raw scores then the next value indicates the number of scores per cell, e.g.

    10002   DATA   5

Then enter the data values. All scores from the first block should precede scores from the second block etc. Within each block, put the scores in sample 1 before sample 2 etc. If we observe these strict rules, the program has no difficulty in handling the scores correctly. If the cell frequencies vary then no single 'score per cell' value can be entered and a zero should be registered, e.g.

    10002   DATA   0

When this happens, the data for each cell should be preceded by a value which specifies the number of scores in that cell. For example, if we have two samples with five scores in the first sample and six in the second the data should look like this:

    10000   DATA   2, 1
    10001   DATA   'R'
    10002   DATA   0
    10003   DATA   5
    10004   DATA   20, 22, 19, 17, 18
    10005   DATA   6
    10008   DATA   17, 23, 26, 29, 14, 17

For frequency tables, input is quite straightforward. After the 'F' character, specify the number of categories of the dependent variables. Then enter the table sample by sample, completing the first sample before beginning the second sample. If you have more than one table (i.e. more than one block), complete one table before beginning another. The following data sequence could be used for the 2×2 frequency table 7.8:

    10000   DATA   2, 1
    10001   DATA   'F'
    10002   DATA   2
    10003   DATA   13, 6, 1, 6

Once the data are in, we need to tell the program what we want it to do. A single character, in quotes, will do this. An 'A' initiates a specific test and should immediately be followed by a list of the coefficients to be used, e.g.

    10010   DATA   'A', 1, 2, 3

The letter 'B' initiates a nonspecific test and 'C' initiates a test of correlation or concordance. 'D' sets the program off in search of a new problem. 'E' will terminate execution. More than one option can be called.

This is a no-frills program which is certainly not user-proof. The most likely problems will occur with data input, and this shows up in the form of messages like

data type mismatch error

caused by trying to read a number into a string variable or reading a letter into a numeric variable. The solution is simply to check the data sequence for errors. The program also prints out the data as they are read in, specifying the sample and block number for each value. These should be checked against the user's intentions and may again give a clue to the problem. The data input conventions are also summarized more formally as follows.

Summary of data input conventions

(1) Data are communicated to the program through numbered DATA statements. To avoid confusion with old data always use consecutive line numbers beginning at 10000.

(2) For raw scores with *equal cell frequencies* use the following format:

```
10000   DATA   (samples), (blocks)
10001   DATA   'R'
10002   DATA   (number of scores in each cell)
10003   DATA   data separated by commas
   ⋮       ⋮           ⋮
```

(3) For raw scores with *unequal cell frequencies*:

```
10000   DATA   (samples), (blocks)
10001   DATA   'R'
10002   DATA   0 (means unequal cell frequencies)
10003   DATA   (number of scores in first cell)
10004   DATA   data in first cell, separated by commas
10005   DATA   (number of scores in second cell)
10006   DATA   data in second cell, separated by commas
   ⋮       ⋮           ⋮
```

(4) For frequency tables use the following format, noting that the number of blocks in this case refers to the number of frequency tables:

```
10000   DATA   (samples), (blocks)
10001   DATA   'F'
10002   DATA   (number of categories)
10003   DATA   (frequencies in sample 1, table 1)
```

```
10004   DATA   (frequencies in sample 2, table 1)
  :        :          :
        DATA   (frequencies in sample 1, table 2)
           :          :
```

(5) The following two rules govern the order of individual scores:

(a) *All* scores from block 1 are input before *any* scores from block 2, and so on.

(b) (Within a given block) *all* scores from sample 1 are input before *any* scores from sample 2, and so on.

(6) Following data input, select the kind of test required using the following options:

'A'   specific test, followed by a list of coefficients
'B'   nonspecific test
'C'   correlation/concordance
'D'   new problem

Examples

Specific test:

```
10020   DATA   'A'
10021   DATA   1, 2, 3, 4 etc.
10022   DATA   'D'
```

Nonspecific test:

```
10020   DATA   'B', 'D'
```

Correlation/concordance:

```
10020   DATA   'C', 'D'
```

Tests can be concatenated thus:

```
10020   DATA   'A'
10021   DATA   1, 2, 3, 4
10022   DATA   'B', 'D'
```

Note that the program cannot handle correlation problems presented in the form of a frequency table as in section 11.3. This is left to the more enterprising reader as an exercise!

Varying conventions for BASIC

Each new computer system seems to have at least one or two important differences in the details of its BASIC syntax. The program is written in the simplest possible subset of BASIC to avoid many of these problems. Even so, some conversion will certainly be required before implementation on your own machine. The version in section 15.2 was written for a PRIME mainframe computer, and some commonly occurring variations are as follows:

PRIME	Other	
**	↑ or Λ	(exponentiation)
'	"	(quotes)
:	;	(PRINT statement separators)

A GLOBAL EDIT is often all that is necessary to put things right. Another potential problem comes with statements of the following type:

IF (condition) THEN (line number)

which may need to be rewritten as

IF (condition) THEN GOTO (line number)

Note that the comment statements at the end of each line (following the semi-colons) will be rejected by most computers and should not be used in the program proper.

The listing in section 15.3 is taken from a BBC micro and illustrates some of these minor differences.

15.2 Detailed description of the program

This section is given for researchers wishing to get to know the program well, perhaps with a view to improving it or adapting it for a particular purpose. It is not necessary to read this section if the researcher merely intends to use the program as it stands. The complete listing in section 15.3 and the introduction in section 15.1 is all he needs. The program is presented in the form of modules rather than line by line:

data input routines
data statements - examples
rank sums from raw scores
rank sums from frequency tables
nonspecific test
specific test
evaluating p from H and Z
main program.

Figure 15.1 gives a program summary.

The program notation is kept as close to the notation in table 6.1 as possible; for example B(I) corresponds to the variable b_i and so on. The major exception

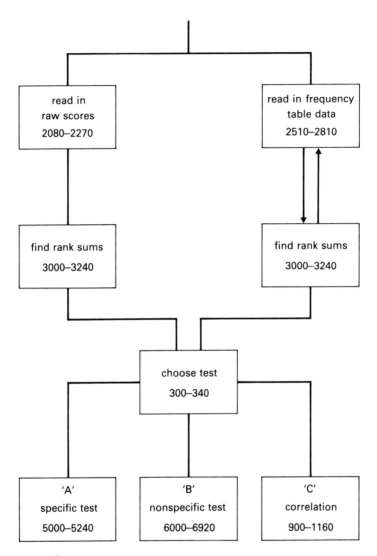

Figure 15.1 Program summary

is the dimension C(I, J), which corresponds to n_{ij}. We can define the main arrays
in the following dimension statements:

```
10   DIM R(k)      ; R_j
20   DIM N(k)      ; n_j
30   DIM L(k)      ; λ_j
40   DIM E(k)      ; E(R_j)
```

50	DIM U(k)	; $[R_j - E(R_j)]/\sigma_j$
60	DIM T(m)	; T_i
70	DIM B(m)	; b_i
80	DIM X(m, a)	; m blocks of a raw scores
90	DIM C(m, k)	; n_{ij}, cell frequencies
100	DIM F(q, k)	; $q \times k$ frequency table
120	DIM S(k, k)	; $\sigma_{jj}^2{}'$, variance–covariance matrix
120	DIM V(k)	; for matrix inversion routine

Set k to the maximum number of samples. Note that k may be very large for correlation questions. On small computers k may need to be reset at the beginning of such computations.

Set m to the maximum number of blocks.
Set q to the maximum number of categories in a frequency table.
Set a to the maximum number of scores in a block.

Data input routines

Machines vary enormously in the available set of input instructions. The most primitive instruction READ will be used here. This instruction requires that all input values be available in DATA statements before the RUN instruction is issued. First of all we specify the number of samples and blocks:

```
210    GOSUB 2010
2010   READ K, M
2020   PRINT 'SAMPLES ': K:' BLOCKS ': M
```

Now we indicate whether the data are in the form of raw scores or frequency tables:

```
2030   READ R$
```

The string 'F' indicates frequency tables and the string 'R' indicates raw scores. Frequency table input is dealt with later.

```
2040   IF R$='F' THEN 2510
```

With raw scores we need to indicate whether cell frequencies are all the same – in which case the cell frequency needs to be read in once only. If C is 0 this is taken to indicate that the cell frequencies will vary and need to be specified for each cell. If C is any other positive value it will be taken as constant at that value:

```
2080   READ C                        ; cell frequency
```

The raw scores are now read in and printed for inspection. If the cell frequencies vary then the scores in each cell are preceded by a number which specifies how many scores there are in that cell. The cell frequencies are stored in C(m, k).

```
2090   FOR I=1 TO M                           ; for each block
2100   PRINT
2110   PRINT TAB(10) : 'BLOCK ' : I
2120   B(I)=0                                  ; bᵢ, block frequencies
2130   A=1                                     ; score counter
2140   FOR J=1 TO K                            ; for each sample
2150   PRINT
2160   PRINT 'SAMPLE ' : J :
2170   C(I, J)=C                               ; standard cell frequency
2180   IF C=0 THEN READ C(I, J)                ; variable cell frequency
2190   B(I)=B(I)+C(I, J)                       ; bᵢ = Σnᵢⱼ
2195   IF C(I, J)=0 THEN 2250                  ; empty cell
2200   FOR S=1 TO C(I, J)                      ; for each score in the cell
2210   READ X(I, A)                            ; ith block, ath score
2220   PRINT TAB(15+ (S−1) * 6) : X(I, A)
2230   A=A+1                                   ; score counter
2240   NEXT S
2250   NEXT J                                  ; next sample
2260   NEXT I                                  ; next block
2265   PRINT
2270   RETURN                                  ; to main program
```

- Line 2120: b_i, block frequencies
- Line 2190: $b_i = \Sigma n_{ij}$

If we are dealing with frequencies, we should note the number of categories in each table and return to the main program:

```
2510   READ Q
2530   RETURN
```

We shall read each table separately later and immediately rank to avoid having to store a three-dimensional array (f_{ijq}). The routine to read the tables will be:

```
2600   REM          READ IN A SINGLE FREQUENCY TABLE
2605   C=0                      ; cell frequency not assumed equal
2610   PRINT

2630   FOR J=1 TO K             ; for each sample
2660   FOR P=1 TO Q             ; for each category

2670   READ F (P, J)            ; observed frequency

2690   NEXT P                   ; next sample

2700   NEXT J                   ; next category
2800   PRINT                    ; prepare for next row
2810   RETURN                   ; to frequency table routine
```

Note that all frequencies from the first sample are input before those for the second sample, and so on.

Data statements - examples

Section 10.2

```
10000   DATA 3, 5                    ; three samples, five blocks
10001   DATA 'R'                     ; raw score input
10002   DATA 1                       ; single score per cell throughout
10003   DATA 1.4, 2.2, 2.3           ; block 1
10004   DATA 1.5, 1.9, 2.2           ; block 2
10005   DATA 1.2, 1.9, 2.5           ; block 3
10006   DATA 1.3, 2.0, 2.6           ; block 4
10007   DATA 2.5, 1.6, 1.7           ; block 5
```

Section 9.2

```
10000   REM          EXAMPLE 9.2
10001   DATA 3, 1                    ; three samples, one block
10002   DATA 'R'                     ; raw score input
10003   DATA 0                       ; variable cell frequency
10004   DATA 5                       ; five scores in sample 1
10005   DATA 96, 128, 83, 61, 101    ; data
10006   DATA 5                       ; five scores in sample 2
10007   DATA 82, 124, 132, 135, 109  ; data
10008   DATA 4                       ; four scores in sample 3
10009   DATA 115, 149, 166, 147      ; data
```

Section 9.7

```
10000   REM          EXAMPLE 9.7
10001   DATA 4, 1                    ; four samples, 1 block
10002   DATA 'F'                     ; frequency table input
10003   DATA 5                       ; five categories
10004   DATA 13, 10, 3, 2, 0         ; sample 1
10005   DATA 15, 13, 8, 5, 0         ; sample 2
10006   DATA 22, 19, 8, 3, 0         ; sample 3
10007   DATA 26, 20, 3, 0, 1         ; sample 4
```

Rank sums from raw scores

Initialize variables which will be later used as accumulators:

```
3000   REM          RANK RAW SCORES
3010   FOR J=1 TO K      ; for each sample
3020   N(J)=0            ; n_j
3030   R(J)=0            ; R_j
```

```
3034   NEXT J

3035   FOR I=1 TO M        ; for each block
3036   T(I)=0              ; Tᵢ
3037   NEXT I

3038   N=0                 ; N
```

We now find the rank for each score and add it into the appropriate sample rank sum. Note that the rank is not stored, having no further value. The rank of a score is found by adding the number of smaller scores (L) to half the number of same scores (T, including self) and adding a further 0.5.

```
3039   PRINT
3040   FOR I=1 TO M                           ; for each block
3041   A=0                                    ; counts the scores in the block
3042   PRINT 'BLOCK':I
3043   FOR J=1 TO K                           ; for each sample
3045   N(J)=N(J)+C(I, J)                      ; find total sample frequency
3050   N=N+C(I, J)                            ; find total frequency
3060   PRINT 'SAMPLE':J
3065   IF C(I, J)=0 THEN 3210                 ; empty cell
3070   FOR P=1 TO C(I, J)                     ; for each score in the cell
3080   A=A+1                                  ; score count
3090   X=X(I, A)                              ; score to be ranked
3100   T=0                                    ; scores tying with X
3110   L=0                                    ; scores less than X

3120   FOR B=1 TO B(I)                        ; compare with every score in
                                                the block
3130   IF X(I, A)>X(I, B) THEN L=L+1          ; smaller score
3140   IF X(I, A)=X(I, B) THEN T=T+1          ; tied score
3150   NEXT B
3160   R=L+T/2+0.5                            ; rank of X(I, A)
3170   PRINT 'SCORE':X(I, A):'RANK':R
3180   R(J)=R(J)+R                            ; rank sum
3190   T(I)=T(I)+T*T−1                        ; for tie correction factor
3200   NEXT P

3210   NEXT J                                 ; next sample

3200   T(I)=1−T(I)/(B(I)*B(I)*B(I)−           ; Tᵢ = 1 − Σ(t³−t)/(b³−b)
       B(I))
3226   PRINT 'T=':T(I)
3227   PRINT
3230   NEXT I                                 ; next block
3240   RETURN
```

Rank sums from frequency tables

Initialize variables which will later be used as accumulators:

```
4000   REM              RANKS FROM FREQUENCIES
4010   FOR J=1 TO K
4020   N(J)=0                    ; n_j
4030   R(J)=0                    ; R_j
4040   FOR I=1 TO M
4050   C(I, J)=0                 ; n_ij
4060   NEXT I

4070   NEXT J
4075   N=0
```

Now read in each frequency table and find the shared rank (R) for each category. Multiply the shared rank by the category frequency and add to the appropriate rank sum:

```
4080   FOR I=1 TO M              ; for each block
4090   T(I)=0                    ; the correction factor T_i
4100   B(I)=0                    ; block frequency b_i

4110   GOSUB 2600               ; read in the ith frequency table
4120   FOR P=1 TO Q             ; pth category in the ith table
4130   G=0                      ; Σt, cumulative category frequency
4135   PRINT 'CATEGORY':P:
4140   FOR J=1 TO K             ; for each sample
4150   F=F(P, J)                ; temporary store to avoid repeated
                                   array use
4155   PRINT TAB(15+(J−1)*5):F:
4160   C(I, J)=C(I, J)+F        ; accumulate cell frequency
4170   N(J)=N(J)+F              ; accumulate sample frequency
4180   B(I)=B(I)+F              ; accumulate block frequency
4190   G=G+F                    ; accumulate category frequency
4195   N=N+F                    ; find total frequency
4200   NEXT J                   ; next sample
4210   R=B(I)−G/2+0.5           ; shared rank for pth category
4220   T(I)=T(I)+G*G*G−G        ; for tie correction
4225   PRINT '------':R
4230   FOR J=1 TO K             ; across all samples
4240   R(J)=R(J)+R*F(P, J)      ; accumulate rank sum
4250   NEXT J                   ; next sample

4260   NEXT P                   ; next category
4270   T(I)=1−T(I)/(B(I)*B(I)*B(I)   ; T_i
       −B(I))
```

```
4275   PRINT 'T=':T(I)
4280   NEXT I                    ; next block
4290   RETURN                    ; to main program
```

Nonspecific test

There are two methods for computing H. The slow method will always work eventually but can be very slow for large numbers of samples. We use the slow method for unequal sample sizes and more than one block:

```
6000   REM                       NONSPECIFIC TEST
6010   IF M=1 THEN 6050          ; single block
6020   IF C=0 THEN 6510          ; unequal cell frequency
```

The quick method computes H thus:

$$H = \frac{12N \sum_{j=1}^{k} \frac{1}{n_j} [R_j - E(R_j)]^2}{\sum_{i=1}^{m} b_i^2 (b_i + 1) \, T_i} \tag{15.1}$$

where

$$E(R_j) = \tfrac{1}{2} \sum_{i=1}^{m} n_{ij}(b_i + 1) \tag{15.2}$$

```
6050   FOR J=1 TO K             ; for each sample
6060   E(J)=0                   ; E(R_j)
6070   FOR I=1 TO M             ; for each block
6080   E(J)=E(J)+0.5*C(I, J)*   ; E(R_j), see (15.2)
       (B(I)+1)
6090   NEXT I                   ; next block
6100   NEXT J                   ; next sample
6110   U=0
6120   FOR J=1 to K             ; for each sample
6130   U=U+((R(J)−E(J))**2)/N(J) ; [R_j − E(R_j)]^2/n_j
6140   NEXT J
6150   V=0                      ; V = Σb_i^2(b_i + 1) T_i
6160   FOR I=1 TO M             ; for each block
6170   V=V+B(I)*B(I)*(B(I)+1)
       *T(I)
6180   NEXT I                   ; next block
```

```
6190   H=12 * N * U/V                    ; see (15.1) above
6200   RETURN
```

The slow method begins by computing the variance-covariance matrix $\sigma_{jj'}^2$, where (for $j \neq j'$)

$$\sigma_{jj'}^2 = -\tfrac{1}{12} \sum_{j=1}^{m} n_{ij} n_{ij'} (b_i + 1) T_i \tag{15.3}$$

and for $j = j'$

$$\sigma_{jj'}^2 = \tfrac{1}{12} \sum_{i=1}^{m} n_{ij} (b_i - n_{ij})(b_i + 1) T_i \tag{15.4}$$

This matrix is then converted to a correlation matrix

$$r_{jj'} = \sigma_{jj'}^2 / \sqrt{(\sigma_{jj}^2 \sigma_{j'j'}^2)} \tag{15.5}$$

to simplify the job of inverting the matrix.

Only the first $k - 1$ rows and columns of the matrix are used.

```
6510   K1=K−1                           ; number of rows and columns used
6520   FOR P=1 TO K1                    ; compute variance-covariance matrix
6530   FOR Q=1 TO K1
6540   S(P, Q)=0
6550   FOR I=1 TO M                     ; compute $\sigma_{jj'}^2$ using (15.3) and
                                          (15.4)
6560   IF P=Q THEN S(P, Q)=S(P, Q)
       +C(I, P) * (B(I)−C(I, P)) * (B(I)
       +1) * T(I)/12
6570   IF P⟨ ⟩Q THEN S(P, Q)=S(P, Q)
       +C(I, P) * C(I, Q) * (B(I)+1) *
       T(I)/12
6572   IF P=Q THEN V(P)=S(P, Q)   ; V(P) = $\sigma_{jj}^2$
6580   NEXT I
6590   NEXT Q
6600   NEXT P
6610   FOR P=1 TO K1                    ; convert S( ) to correlation matrix
6620   FOR Q=1 TO K1
6630   S(P, Q)=S(P, Q)/SQR(V(P) *
       V(P))
6640   NEXT Q
6650   NEXT P
```

Compute

$$u_j = [R_j - E(R_j)]/\sqrt{\sigma_{jj}^2}$$

which is a standardized form of U to match the correlation matrix.

```
6660   FOR J=1 TO K1
6670   E(J)=0
6680   FOR I=1 TO M
6690   E(J)=E(J)+0.5*C(I, J)*        ; E(R_j)
       (B(I)+1)
6700   NEXT I
6710   U(J)=(R(J)−E(J))/SQR(V(J))   ; u_j/σ_jj
6720   NEXT J
```

The next step is to invert the correlation matrix:

```
6730   FOR I=1 TO K1                 ; this is a standard inversion routine
6740   X=S(I, I)
6750   S(I, I)=1
6760   FOR J=1 TO K1
6770   S(I, J)=S(I, J)/X
6780   NEXT J
6790   FOR H=1 TO K1
6800   IF H=I THEN 6850
6810   X=S(H, I)
6815   S(H, I)=0
6820   FOR J=1 TO K1
6830   S(H, J)=S(H, J)−X*S(I, J)
6840   NEXT J
6850   NEXT H
6860   NEXT I
```

We can now compute H:

$$H=UV^{-1}U'$$

where U is the vector of u_j and V is the matrix of the correlations $r_{jj'}$. Note that both U and V are standardized versions of their corresponding entities in chapter 14, but the result is unaffected.

```
6865   H=0
6870   FOR P=1 TO K1
6880   FOR Q=1 TO K1
6890   H=H+U(P)*S(P, Q)*U(Q)        ; H = UV^{-1}U'
6900   NEXT Q
```

```
6910   NEXT P
6920   RETURN                        ; to main program
```

Specific test

The k coefficients are assumed to be next in the DATA lists.

```
5000   L(1)=1                        ; λ₁ = 1 for two-sample test
5005   L(2)=0                        ; λ₂ = 0
5010   IF K=2 THEN 5080              ; if this is a two-sample test
5020   PRINT'COEFFICIENTS FOR SPECIFIC TEST'
5040   FOR J=1 TO K                  ; if not, read coefficients
5050   READ L(J)                     ; λⱼ
5060   PRINT TAB((J−1)*6):L(J):
5070   NEXT J
5075   PRINT
```

The lines above correspond to:

$$L = \sum_{j=1}^{k} \lambda_j R_j$$

We can now compute:

$$L = \sum_{j=1}^{k} \lambda_j R_j$$

$$E(L) = \tfrac{1}{2} \sum_{i=1}^{m} (b_i + 1) \sum_{j=1}^{k} n_{ij} \lambda_j$$

$$\mathrm{var}(L) = \tfrac{1}{12} \sum_{i=1}^{m} (b_i + 1) \left[b_i \sum_{j=1}^{k} n_{ij} \lambda_j^2 - \left(\sum_{j=1}^{k} n_{ij} \lambda_j \right)^2 \right] T_i$$

```
5080   L=0                          ; L
5090   E=0                          ; E(L)
5100   V=0                          ; var(L)
5105   FOR J=1 TO K                 ; for each sample
5107   L=L+L(J)*R(J)                ; L = ΣλⱼRⱼ
5109   NEXT J                       ; next sample
5110   FOR I=1 TO M                 ; for each block
5120   E1=0                         ; Σnᵢⱼλⱼ²
5130   E2=0                         ; Σnᵢⱼλⱼ
5140   FOR J=1 TO K
5160   E1=E1+L(J)*L(J)*C(I, J)
5170   E2=E2+L(J)*C(I, J)
5180   NEXT J                       ; next sample
5190   E=E+(B(I)+1)*E2/2            ; E(L)
5200   V=V+(B(I)+1)*(B(I)*E1−       ; var(L)
       E2*E2)*T(I)/12
5210   NEXT I                       ; next block
```

Z can now be computed:

$$Z = \frac{L - E(L)}{\sqrt{\text{var}(L)}}$$

```
5220   Z=(L−E)/SQR(V)
5225   PRINT 'L=':L
5227   PRINT 'E(L)=':E:'VAR=':V:' ROOT (VAR)=':SQR(V)
5230   PRINT 'Z=':Z
5240   RETURN
```

Evaluating p from H and Z

For this purpose we have a useful formula for converting H to an approximately normally distributed value with a mean of 0 and a variance of 1 (Wilson and Hilferty, 1931):

$$Z = \frac{\left(\dfrac{H}{k-1}\right)^{1/3} - \left[1 - \dfrac{2}{9(k-1)}\right]}{\sqrt{[2/9(k-1)]}}$$

```
7000   REM EVALUATE P FROM H AND Z
7010   Z=((H/(K−1)) * *0.3333−1+2/(9 * (K−1)))/SQR(2/(9 * (K−1)))
```

Z can now be converted to the required probability using another approximation which, it should be noted, is only accurate for $p < 0.1$:

$$p = 0.5/(1 + 0.196\,854Z + 0.115\,194Z^2 + 0.000\,344Z^3 + 0.019\,527Z^4)^4$$

```
7030   P=0.5/(1+0.196854 * Z+0.115194 * Z * *2+0.000344 * Z * *3
       +0.019527 * Z * *4) * *4
7040   RETURN                    ; to main program
```

Main program

The individual subroutines may be strung together using the following main program, if required:

```
210   GOSUB 2000                 ; data input
```

If raw score data input was used, then we need the raw score ranking routine:

```
230   IF R$='R' THEN GOSUB 3000  ; raw score ranking
```

Otherwise use the frequency table routine which both inputs and ranks:

```
250   IF R$='F' THEN GOSUB 4000  ; frequency table ranking
```

We may then print out the sample rank means for inspection:

```
250   PRINT
251   PRINT 'SAMPLE FREQUENCY RANK SUM RANK MEAN'
253   FOR J = 1 TO K
254   PRINT J, N(J), R(J), R(J)/N(J)
255   NEXT J
256   PRINT
257   PRINT
```

At this point we must choose from the following menu:

(a) specific test
(b) nonspecific test
(c) correlation and concordance
(d) stop.

The appropriate character must be the next string in the DATA sequence.

```
260   REM    SELECT TEST
300   READ M$
310   IF M$ = 'A' THEN 500
320   IF M$ = 'B' THEN 700
330   IF M$ = 'C' THEN 900
340   IF M$ = 'D' THEN 210        ; next problem
350   STOP                        ; any other character terminates
```

Note that the selection from the menu must precede the coefficients in the DATA statements. For section 9.2, we might use

```
10010   DATA 'A'                  ; choose a specific test
10011   DATA 1, 2, 3              ; coefficients for specific test
```

Firstly, the specific test:

```
500   PRINT
510   PRINT 'SPECIFIC TEST'
520   GOSUB 5000                  ; specific test
535   IF Z < 1 THEN PRINT 'NOT
      SIGNIFICANT'
540   IF Z < 1 THEN 260           ; choose next option (see above)
```

Convert Z to a probability statement:

```
550   GOSUB 7030                  ; Z → p conversion
570   PRINT 'P =':P:
580   IF P > 0.05 PRINT 'NOT
      SIGNIFICANT'
```

```
590   IF P<=0.05 PRINT
      'SIGNIFICANT'
600   GOTO 260                    ; choose next option
```

Secondly, the nonspecific test:

```
700   PRINT
710   PRINT 'NONSPECIFIC TEST'
720   GOSUB 6000                  ; nonspecific test
730   PRINT 'H=':H:'WITH':K—1:'DEGREES OF FREEDOM':
755   IF H<K—1 THEN PRINT 'NOT SIGNIFICANT'
757   IF H<K—1 THEN 260
760   GOSUB 7000                  ; H → p conversion
780   PRINT 'P=':P
790   IF P>0.05 PRINT 'NOT
      SIGNIFICANT'
800   IF P<=0.05 PRINT 'SIGNIFICANT'
810   GOTO 260                    ; try another option
```

Thirdly, correlation/concordance. We must precede this with a nonspecific test to find H without printing it:

```
900   REM      CORRELATION/CONCORDANCE
920   GOSUB 6000                  ; nonspecific test, find H
```

If there are more than two blocks, then we are dealing with concordance rather than simple correlation:

```
940   IF M>2 THEN 1110            ; concordance
960   IF C<>1 THEN PRINT 'TEST INAPPROPRIATE'
970   IF C<>1 THEN 260
```

We can find ρ from H:

$$\rho = H/(k - 1) - 1$$

```
980   R0=H/(K—1)—1               ; Spearman's rho
990   PRINT 'SPEARMANS RHO=':R0
```

If this is a specific test we can compute

$$Z = \sqrt{[\rho^2 (k - 1)]}$$

```
1000  Z=SQR(R0*R0*(K—1))
1010  GOSUB 7030                  ; convert Z to p
1030  PRINT 'SPECIFIC TEST Z=':Z:' P=':P
```

For a nonspecific test, double P:

```
1035   Z=ABS (Z)                        ; ignore sign of Z
1037   GOSUB 7030
1040   PRINT 'NONSPECIFIC TEST
       H=':Z**2:' P=':2*P
1050   GOTO 260                         ; select new option
```

For concordance compute:

$$W = H/m(k-1)$$

```
1110   W=H/(M*(K-1))
1120   PRINT 'KENDALLS W=':W
1130   GOSUB 7000                       ; convert H → p
1150   PRINT 'H=':H:' P=':P:
       'WITH':K-1:'DF'
1160   GOTO 260                         ; select new option
```

15.3   Complete program listing

This listing is taken from a BBC microcomputer and, when compared with section 15.2, illustrates some of the minor variations which are to be found among BASIC dialects. It also illustrates the use of dynamic dimensioning of variables. The maximum values for K (samples), M (blocks), Q (categories of the dependent variable) and A (number of scores per block) are all set in line 2. The program then sets the array dimensions accordingly. In section 15.2 these were set explicitly. If the variable storage area is relatively small in your computer, you must be prepared to redimension the arrays for different problems.

```
  2   K=22 : M=5 : Q=5 : A=22
 10   DIM R(K)
 20   DIM N(K)
 30   DIM L(K)
 40   DIM E(K)
 50   DIM U(K)
 55   DIM V(K)
 60   DIM T(M)
 70   DIM B(M)
 80   DIM X(M,A)
 90   DIM C(M,K)
100   DIM F(Q,K)
110   DIM S(K,K)
210   GOSUB 2010
230   IF R$="R" THEN GOSUB 3000
250   IF R$="F" THEN GOSUB 4000
251   PRINT
252   PRINT "SAMPLE  FREQ  RANK SUM  RANK MEAN"
253   FOR J=1 TO K
254       PRINT TAB(1);J;TAB(10);N(J);TAB(17);R(J);TAB(27);R(J)/N(J)
255   NEXT J
```

```
256    PRINT
257    PRINT
260    REM                    SELECT TEST
300    READ M$
310    IF M$="A" THEN 500
320    IF M$="B" THEN 700
330    IF M$="C" THEN 900
340    IF M$="D" THEN 210
350    STOP
500    PRINT
510    PRINT "           SPECIFIC TEST"
520    GOSUB 5000
535    IF Z<1 THEN PRINT "NOT SIGNIFICANT"
540    IF Z<1 THEN 260
550    GOSUB 7030
570     PRINT "P= ";P;
580    IF P>0.05 PRINT " NOT SIGNIFICANT"
590    IF P<=0.05 PRINT "    SIGNIFICANT"
600    GOTO 260
700    PRINT
710     PRINT "         NONSPECIFIC TEST"
720    GOSUB 6000
730    PRINT "H= ";H;" WITH ";K-1;" DEGREES OF FREEDOM"
755    IF H<K-1 THEN PRINT " NOT SIGNIFICANT"
757    IF H<K-1 THEN 260
760    GOSUB 7000
780    PRINT "P= ";P;
790    IF P>0.05 PRINT " NOT SIGNIFICANT"
800    IF P<=0.05 PRINT " SIGNIFICANT"
810    GOTO 260
900    REM                CORRELATION / CONCORDANCE
920    GOSUB 6000
940    IF M>2 THEN 1110
960    IF C<>1 THEN PRINT "CORRELATION TEST INAPPROPRIATE"
970    IF C<>1 THEN 260
980    RO=H/(K-1)-1
990    PRINT "SPEARMAN'S RHO= ";RO
1000   Z=SQR(RO*RO*(K-1))
1010   GOSUB 7030
1030   PRINT "SPECIFIC TEST Z= ";Z;" P= ";P
1035   Z=ABS(Z)
1037   GOSUB 7030
1040   PRINT "NONSPECIFIC TEST   H= ";Z^2;" P= ";2*P
1050   GOTO 260
1110   W=H/(M*(K-1))
1120   PRINT "KENDALL'S W= ";W
1130   GOSUB 7000
1150   PRINT "H= ";H;" P= ";P;" WITH ";K-1;" DF"
1160   RETURN
2010   READ K,M
2020   PRINT "SAMPLES ";K;"    BLOCKS ";M
2030   READ R$
2040   IF R$="F" THEN 2510
2080   READ C
2090   FOR I= 1 TO M
2100     PRINT
2110     PRINT TAB(10);"BLOCK   ";I
2120     B(I)=0
2130     A=1
2140     FOR J=1 TO K
2150       PRINT
2160       PRINT "SAMPLE    ";J;
2170       C(I,J)=C
2180       IF C=0 THEN READ C(I,J)
2190       B(I)=B(I)+C(I,J)
2195       IF C(I,J)=0 THEN 2250
2200       FOR S=1 TO C(I,J)
2210         READ X(I,A)
2220         PRINT TAB(15+(S-1)*6);X(I,A);
2230         A=A+1
2240         NEXT S
2250       NEXT J
2260     NEXT I
```

```
2265    PRINT
2270    RETURN
2510    READ Q
2530    RETURN
2600    REM         READ IN SINGLE FREQUENCY TABLE
2605    C=0
2610    PRINT
2630    FOR J=1 TO K
2660      FOR P=1 TO Q
2670        READ F(P,J)
2690        NEXT P
2700      NEXT J
2800    PRINT
2810    RETURN
3000    REM         RANKING RAW SCORES
3010    FOR J=1 TO K
3020      N(J)=0
3030      R(J)=0
3034      NEXT J
3035    FOR I= 1 TO M
3036      T(I)=0
3037      NEXT I
3038    N=0
3039    PRINT
3040    FOR I=1 TO M
3041      A=0
3042      PRINT "BLOCK ";I
3043      FOR J=1 TO K
3045        N(J)=N(J)+C(I,J)
3050        N=N+C(I,J)
3060        PRINT "SAMPLE ";J
3065        IF C(I,J)=0THEN 3210
3070        FOR P=1 TO C(I,J)
3080          A=A+1
3090          X=X(I,A)
3100          T=0
3110          L=0
3120          FOR B=1 TO B(I)
3130            IF X(I,A) > X(I,B) THEN L=L+1
3140            IF X(I,A) = X(I,B) THEN T=T+1
3150            NEXT B
3160          R=L+T/2 +0.5
3170          PRINT "SCORE ";X(I,A);"    RANK ";R
3180          R(J)=R(J)+R
3190          T(I)=T(I)+T*T-1
3200          NEXT P
3210        NEXT J
3220      T(I)=1-T(I)/(B(I)*B(I)*B(I)-B(I))
3226      PRINT "T= ";T(I)
3227      PRINT
3230      NEXT I
3240    RETURN
4000    REM         RANKS FROM FREQUENCIES
4010    FOR J=1 TO K
4020      N(J)=0
4030      R(J)=0
4040      FOR I=1 TO M
4050        C(I,J)=0
4060        NEXT I
4070      NEXT J
4075    N=0
4080    FOR I=1 TO M
4090      T(I)=0
4100      B(I)=0
4110      GOSUB 2600
4120      FOR P=1 TO Q
4130        G=0
4135        PRINT "CATEGORY ";P;
4140        FORJ=1 TO K
4150          F=F(P,J)
4155          PRINT TAB(15+(J-1)*5);F(P,J);
4160          C(I,J)=C(I,J)+F
4170          N(J)=N(J)+F
```

```
4180        B(I)=B(I)+F
4190        G=G+F
4195        N=N+F
4200        NEXT J
4210      R=B(I)-G/2+0.5
4220      T(I)=T(I)+G*G*G-G
4225      PRINT "----- ";R
4230      FOR J=1 TO K
4240        R(J)=R(J)+R*F(P,J)
4250        NEXT J
4260      NEXT P
4270    T(I)=1-T(I)/(B(I)*B(I)*B(I)-B(I))
4275    PRINT "T= ";T(I)
4280    NEXT I
4290  RETURN
5000  REM                SPECIFIC TEST
5002  L(1)=1
5005  L(2)=0
5010  IF K=2 THEN 5080
5020  PRINT "COEFFICIENTS FOR SPECIFIC TEST"
5040  FOR J=1 TO K
5050    READ L(J)
5060    PRINT TAB((J-1)*6);L(J);
5070    NEXT J
5075  PRINT
5080  L=0
5090  E=0
5100  V=0
5105  FOR J=1 TO K
5107    L=L+L(J)*R(J)
5109    NEXT J
5110  FOR I=1 TO M
5120    E1=0
5130    E2=0
5140    FOR J=1 TO K
5160      E1=E1+L(J)*L(J)*C(I,J)
5170      E2=E2+L(J)*C(I,J)
5180      NEXT J
5190    E=E+(B(I)+1)*E2/2
5200    V=V+(B(I)+1)*(B(I)*E1-E2*E2)*T(I)/12
5210    NEXT I
5220  Z=(L-E)/SQR(V)
5225  PRINT "L= ";L
5227  PRINT "E(L)= ";E;"  VAR= ";V;"  ROOT(VAR)= ";SQR(V)
5230  PRINT "Z= ";Z
5240  RETURN
6000  REM               NON SPECIFIC TEST
6010  IF M=1 THEN 6050
6020  IF C=0 THEN 6510
6050  FOR J=1 TO K
6060    E(J)=0
6070    FOR I=1 TO M
6080      E(J)=E(J)+0.5*C(I,J)*(B(I)+1)
6090      NEXT I
6100    NEXT J
6110  U=0
6120  FOR J=1 TO K
6130    U=U+((R(J)-E(J))^2)/N(J)
6140    NEXT J
6150  V=0
6160  FOR I=1 TO M
6170    V=V+B(I)*B(I)*(B(I)+1)*T(I)
6180    NEXT I
6190  H=12*N*U/V
6200  RETURN
6510  K1=K-1
6520  FOR P=1 TO K1
6530    FOR Q=1TO K1
6540      S(P,Q)=0
6550      FOR I=1 TO M
6560        IF P=Q THEN S(P,Q)=S(P,Q)+C(I,P)*(B(I)-C(I,P))*(B(I)+1)*T(I)/12
6570        IF P<>Q THEN S(P,Q)=S(P,Q)-C(I,P)*C(I,Q)*(B(I)+1)*T(I)/12
6572        IF P=Q THEN V(P)=S(P,Q)
```

```
6580          NEXT I
6590         NEXT Q
6600        NEXT P
6610   FOR P= 1 TO K1
6620      FOR Q=1 TO K1
6630        S(P,Q)=S(P,Q)/SQR(V(P)*V(Q))
6640      NEXT Q
6650     NEXT P
6660   FOR J=1 TO K1
6670     E(J)=0
6680      FOR I=1 TO M
6690        E(J)=E(J)+0.5*C(I,J)*(B(I)+1)
6700      NEXT I
6710     U(J)=(R(J)-E(J))/SQR(V(J))
6720   NEXT J
6730   FOR I=1 TO K1
6740     X=S(I,I)
6750     S(I,I)=1
6760      FOR J=1 TO K1
6770        S(I,J)=S(I,J)/X
6780      NEXT J
6790      FOR H=1 TO K1
6800        IF H=I THEN 6850
6810        X=S(H,I)
6815        S(H,I)=0
6820        FOR J=1 TO K1
6830          S(H,J)=S(H,J)-X*S(I,J)
6840        NEXT J
6850      NEXT H
6860    NEXT I
6865   H=0
6870   FOR P=1 TO K1
6880      FOR Q=1 TO K1
6890        H=H+U(P)*S(P,Q)*U(Q)
6900      NEXT Q
6910    NEXT P
6920   RETURN
7000   REM         EVALUATE P FROM H AND Z
7010   Z=((H/(K-1)^3-1+2/(9*(K-1)))/SQR(2/9*(K-1)))
7030   P=.5/(1+.196854*Z+.115194*Z^2+.000344*Z^3+.019527*Z^4)^4
7040   RETURN
9999    REM EXAMPLE 10.2
10000 DATA 3, 5
10001 DATA "R"
10002 DATA 1
10003 DATA 1.4, 2.2, 2.3
10004 DATA 1.5, 1.9, 2.2
10005 DATA 1.2, 1.9, 2.5
10006 DATA 1.3, 2.0, 2.6
10007 DATA 2.5, 1.6, 1.7
10008 DATA "A", 1, 2, 3, "D"
11000 REM EXAMPLE 9.2
11001 DATA 3,1
11002 DATA "R"
11003 DATA 0
11004 DATA 5
11005 DATA 96, 128, 83, 61, 101
11006 DATA 5
11007 DATA 82, 124, 132, 135, 109
11008 DATA 4
11009 DATA 115, 149, 166, 147
11010 DATA "A", 1, 2, 3, "D"
12000 REM EXAMPLE 9.7
12001 DATA 4, 1
12002 DATA "F"
12003 DATA 5
12004 DATA 13, 10, 3, 2, 0
12005 DATA 15, 13, 8, 5, 0
12006 DATA 22, 19, 8, 3, 0
12007 DATA 26, 20, 3, 0, 1
12008 DATA "B", "D"
13000   REM  EXAMPLE 11.2
```

```
13001    DATA 22, 2
13002    DATA "R"
13003    DATA 1
13004    DATA .75, 0, 1, .77, 4.11, .44
13005    DATA 3.75, 3.44, 1.63, 2.33, 3.43
13006    DATA 2, 2.5, 3, 20.2, .87, .55
13007    DATA 1.33, 9.75, 3.33, 5.33, 5
13008    DATA 14, 12, 17, 15, 9, 14, 20
13009    DATA 10, 15, 13, 16, 5, 13, 17
13010    DATA 22, 15, 7, 6, 12, 17, 13, 12
13011    DATA "C"
14001    DATA "E"
>
```

Appendix Critical value tables

Table A Critical values of the normal distribution. See chapter 2

Notes
(1) Estimates of z must be greater than or equal to the critical value to be significant at the appropriate level.
(2) Negative values of z may indicate that a hypothesis is badly expressed in terms of the coefficients used.

5%	2.5%	1.0%	0.1%
$z \geqslant 1.64$	1.96	2.33	3.10

Table B Critical values of the chi-square distribution. See chapter 2

Notes
(1) Estimates of chi-square must be greater than or equal to the critical value to be significant at the appropriate level.
(2) For simple designs, the degrees of freedom are usually one less than the number of samples: $df = k - 1$.
(3) When there are more than 15 degrees of freedom, compute

$$z = \sqrt{(2\chi^2)} - \sqrt{(df - 1)}$$

and evaluate z by consulting table A.

Table B   continued

Degrees of freedom	5%	2.5%	1.0%	0.1%
1	$\chi^2 \geqslant 3.8$	5.0	6.6	10.8
2	6.0	7.4	9.2	13.8
3	7.8	9.3	11.3	16.3
4	9.5	11.1	13.3	18.5
5	11.1	12.8	15.1	20.5
6	12.6	14.5	16.8	22.5
7	14.1	16.0	18.5	24.3
8	15.5	17.5	20.1	26.1
9	16.9	19.0	21.7	27.9
10	18.3	20.5	23.2	29.6
11	19.7	21.9	24.7	31.3
12	21.0	23.3	26.2	32.9
13	22.4	24.7	27.7	34.5
14	23.7	26.1	29.1	36.1
15	25.0	27.5	30.6	37.7

Table C   Critical values of z for problems involving multiple comparisons. See section 12.1

Notes
Q is the number of comparisons being simultaneously tested.

Q	5%	2.5%	1.0%	0.1%
1	$z \geqslant 1.64$	1.96	2.33	3.10
2	1.96	2.24	2.58	3.29
3	2.13	2.39	2.72	3.40
4	2.24	2.50	2.81	3.48
5	2.33	2.58	2.88	3.54
6	2.39	2.64	2.94	3.59
7	2.46	2.69	3.00	3.62
8	2.50	2.74	3.04	3.68
9	2.54	2.77	3.07	3.69
10	2.58	2.81	3.10	3.71

Table D   Values of x^2 and $x^3 - x$

Table D   continued

x	x^2	$(x^3 - x)$	x	x^2	$(x^3 - x)$
1	1	0	41	1 681	68 880
2	4	6	42	1 764	74 046
3	9	24	43	1 849	79 464
4	16	60	44	1 936	85 140
5	25	120	45	2 025	91 080
6	36	210	46	2 116	97 290
7	49	336	47	2 209	103 776
8	64	504	48	2 304	110 544
9	81	720	49	2 401	117 600
10	100	990	50	2 500	124 950
11	121	1 320	51	2 601	132 600
12	144	1 716	52	2 704	140 556
13	169	2 184	53	2 809	148 824
14	196	2 730	54	2 916	157 410
15	225	3 360	55	3 025	166 320
16	256	4 080	56	3 136	175 560
17	289	4 896	57	3 249	185 136
18	324	5 814	58	3 364	195 054
19	361	6 840	59	3 481	205 320
20	400	7 980	60	3 600	215 940
21	441	9 240	61	3 721	226 920
22	484	10 626	62	3 841	238 266
23	529	12 144	63	3 969	249 984
24	576	13 800	64	4 096	262 080
25	625	15 600	65	4 225	274 560
26	676	17 550	66	4 356	287 430
27	729	19 656	67	4 489	300 696
28	784	21 924	68	4 624	314 364
29	841	24 360	69	4 761	328 440
30	900	26 970	70	4 900	342 930
31	961	29 760	71	5 041	357 840
32	1 024	32 736	72	5 184	373 176
33	1 089	35 904	73	5 329	388 944
34	1 156	39 270	74	5 476	405 150
35	1 225	42 840	75	5 625	421 800
36	1 296	46 620	76	5 776	438 900
37	1 369	50 616	77	5 929	456 456
38	1 444	54 834	78	6 084	474 474
39	1 521	59 280	79	6 241	492 960
40	1 600	63 960	80	6 400	511 920

Table D	continued	
x	x^2	(x^3-x)
81	6 561	531 360
82	6 724	551 286
83	6 889	571 704
84	7 056	592 620
85	7 225	614 040
86	7 396	635 970
87	7 569	658 416
88	7 744	681 384
89	7 921	704 880
90	8 100	728 910
91	8 281	753 480
92	8 464	778 596
93	8 649	804 264
94	8 836	830 490
95	9 025	857 280
96	9 216	884 640
97	9 409	912 576
98	9 604	941 094
99	9 801	970 200
100	10 000	999 900
101	10 201	1 030 200
102	10 404	1 061 106
103	10 609	1 092 624
104	10 816	1 124 760
105	11 025	1 157 520
106	11 236	1 190 910
107	11 449	1 224 936
108	11 664	1 259 604
109	11 881	1 294 920
110	12 100	1 330 890
111	12 321	1 367 520
112	12 544	1 404 816
113	12 769	1 442 784
114	12 996	1 481 430
115	13 225	1 520 760
116	13 456	1 560 780
117	13 689	1 601 496
118	13 924	1 642 914
119	14 161	1 685 040
120	14 400	1 727 880

Table D	continued	
x	x^2	(x^3-x)
121	14 641	1 771 440
122	14 884	1 815 726
123	15 129	1 860 744
124	15 376	1 906 500
125	15 625	1 953 000
126	15 876	2 000 250
127	16 129	2 048 256
128	16 384	2 097 024
129	16 641	2 146 560
130	16 900	2 196 870
131	17 161	2 247 960
132	17 424	2 299 836
133	17 689	2 352 504
134	17 956	2 405 970
135	18 225	2 460 240
136	18 496	2 515 320
137	18 769	2 571 216
138	19 044	2 627 934
139	19 321	2 685 480
140	19 600	2 743 860
141	19 881	2 803 080
142	20 164	2 863 146
143	20 449	2 924 064
144	20 736	2 985 840
145	21 025	3 048 480
146	21 316	3 111 990
147	21 609	3 176 376
148	21 904	3 241 644
149	22 201	3 307 800
150	22 500	3 374 850
151	22 801	3 442 800
152	23 104	3 511 656
153	23 409	3 581 424
154	23 716	3 652 110
155	24 025	3 723 720
156	24 336	3 796 260
157	24 649	3 869 736
158	24 964	3 944 154
159	25 281	4 019 520
160	25 600	4 095 840

Table D   continued

x	x^2	(x^3-x)
161	25 921	4 173 120
162	26 244	4 251 366
163	26 569	4 330 584
164	26 896	4 410 780
165	27 225	4 491 960
166	27 556	4 574 130
167	27 889	4 657 296
168	28 224	4 741 464
169	28 561	4 826 640
170	28 900	4 912 830
171	29 241	5 000 040
172	29 584	5 088 276
173	29 929	5 177 544
174	30 276	5 267 850
175	30 625	5 359 200
176	30 976	5 451 600
177	31 329	5 545 056
178	31 684	5 639 574
179	32 041	5 735 160
180	32 400	5 831 820
181	32 761	5 929 560
182	33 124	6 028 386
183	33 489	6 128 304
184	33 856	6 229 320
185	34 225	6 331 440
186	34 596	6 434 670
187	34 969	6 539 016
188	35 344	6 644 484
189	35 721	6 751 080
190	36 100	6 858 810
191	36 481	6 967 680
192	36 864	7 077 696
193	37 249	7 188 864
194	37 636	7 301 190
195	38 025	7 414 680
196	38 416	7 529 340
197	38 809	7 645 176
198	39 204	7 762 194
199	39 601	7 880 400
200	40 000	7 999 800

Table D   continued

x	x^2	(x^3-x)
201	40 401	8 120 400
202	40 804	8 242 206
203	41 209	8 365 224
204	41 616	8 489 460
205	42 025	8 614 920
206	42 436	8 741 610
207	42 849	8 869 536
208	43 264	8 998 704
209	43 681	9 129 120
210	44 100	9 260 790
211	44 521	9 393 720
212	44 944	9 527 916
213	45 369	9 663 384
214	45 796	9 800 130
215	46 225	9 938 160
216	46 656	10 077 480
217	47 089	10 218 096
218	47 524	10 360 014
219	47 961	10 503 240
220	48 400	10 647 780
211	48 841	10 793 640
222	49 284	10 940 826
223	49 729	11 089 344
224	50 176	11 239 200
225	50 625	11 390 400
226	51 076	11 542 950
227	51 529	11 696 856
228	51 984	11 852 124
229	52 441	12 008 760
230	52 900	12 166 770
231	53 361	12 326 160
232	53 824	12 486 936
233	54 289	12 649 104
234	54 756	12 812 670
235	55 225	12 977 640
236	55 696	13 144 020
237	56 169	13 311 816
238	56 644	13 481 034
239	57 121	13 651 680
240	57 600	13 823 760

Table D   continued

Table D   continued

x	x^2	(x^3-x)
241	58 081	13 997 280
242	58 564	14 172 246
243	59 049	14 348 664
244	59 536	14 526 540
245	60 025	14 705 880
246	60 516	14 886 690
247	61 009	15 068 976
248	61 504	15 252 744
249	62 001	15 438 000
250	62 500	15 624 750
251	63 001	15 813 000
252	63 504	16 002 756
253	64 009	16 194 024
254	64 516	16 386 810
255	65 025	16 581 120
256	65 536	16 776 960
257	66 049	16 974 336
258	66 564	17 173 254
259	67 081	17 373 720
260	67 600	17 575 740
261	68 121	17 779 320
262	68 644	17 984 466
263	69 169	18 191 184
264	69 696	18 399 480
265	70 225	18 609 360
266	70 756	18 820 830
267	71 289	19 033 896
268	71 824	19 248 564
269	72 361	19 464 840
270	72 900	19 682 730
271	73 441	19 902 240
272	73 984	20 123 376
273	74 529	20 346 144
274	75 076	20 570 550
275	75 625	20 796 600
276	76 176	21 024 300
277	76 729	21 253 656
278	77 284	21 484 674
279	77 841	21 717 360
280	78 400	21 951 720

x	x^2	(x^3-x)
281	78 961	22 187 760
282	79 524	22 425 486
283	80 089	22 664 904
284	80 656	22 906 020
285	81 225	23 148 840
286	81 796	23 393 370
287	82 369	23 639 616
288	82 944	23 887 584
289	83 521	24 137 280
290	84 100	24 388 710
291	84 681	24 641 880
292	85 264	24 896 796
293	85 849	25 153 464
294	86 436	25 411 890
295	87 025	25 672 080
296	87 616	25 934 040
297	88 209	26 197 776
298	88 804	26 463 294
299	89 401	26 730 600
300	90 000	26 999 700
301	90 601	27 270 600
302	91 204	27 543 306
303	91 809	27 817 824
304	92 416	28 094 160
305	93 025	28 372 320
306	93 636	28 652 310
307	94 249	28 934 136
308	94 864	29 217 804
309	95 481	29 503 320
310	96 100	29 790 690
311	96 721	30 079 920
312	97 344	30 371 016
313	97 969	30 663 984
314	98 596	30 958 830
315	99 225	31 255 560
316	99 856	31 554 180
317	100 489	31 854 696
318	101 124	32 157 114
319	101 761	32 461 440
320	102 400	32 767 680

Table D   continued

x	x^2	(x^3-x)
321	103 041	33 075 840
322	103 684	33 385 926
323	104 329	33 697 944
324	104 976	34 011 900
325	105 625	34 327 800
326	106 276	34 645 650
327	106 929	34 965 456
328	107 584	35 287 224
329	108 241	35 610 960
330	108 900	35 936 670
331	109 561	36 264 360
332	110 224	36 594 036
333	110 889	36 925 704
334	111 556	37 259 370
335	112 225	37 595 040
336	112 896	37 932 720
337	113 569	38 272 416
338	114 244	38 614 134
339	114 921	38 957 880
340	115 600	39 303 660
341	116 281	39 651 480
342	116 964	40 001 346
343	117 649	40 353 264
344	118 336	40 707 240
345	119 025	41 063 280
346	119 716	41 421 390
347	120 409	41 781 576
348	121 104	42 143 844
349	121 801	42 508 200
350	122 500	42 874 650
351	123 201	43 243 200
352	123 904	43 613 856
353	124 609	43 986 624
354	125 316	44 361 510
355	126 025	44 738 520
356	126 736	45 117 660
357	127 449	45 498 936
358	128 164	45 882 354
359	128 881	46 267 920
360	129 600	46 655 640

Table D   continued

x	x^2	(x^3-x)
361	130 321	47 045 520
362	131 044	47 437 566
363	131 769	47 831 784
364	132 496	48 228 180
365	133 225	48 626 760
366	133 956	49 027 530
367	134 689	49 430 496
368	135 424	49 835 664
369	136 161	50 243 040
370	136 900	50 652 630
371	137 641	51 064 440
372	138 384	51 478 476
373	139 129	51 894 744
374	139 876	52 313 250
375	140 625	52 734 000
376	141 376	53 157 000
377	142 129	53 582 256
378	142 884	54 009 774
379	143 641	54 439 560
380	144 400	54 871 620
381	145 161	55 305 960
382	145 924	55 742 586
383	146 689	56 181 504
384	147 456	56 622 720
385	148 225	57 066 240
386	148 996	57 512 070
387	149 769	57 960 216
388	150 544	58 410 684
389	151 321	58 863 480
390	152 100	59 318 610
391	152 881	59 776 080
392	153 664	60 235 896
393	154 449	60 698 064
394	155 236	61 162 590
395	156 025	61 629 480
396	156 816	62 098 740
397	157 609	62 570 376
398	158 404	63 044 394
399	159 201	63 520 800
400	160 000	63 999 600

Table E   Critical values of L for the exact test for two unrelated samples. See section 7.2

Notes
(1) L must be greater than or equal to the critical value to be significant at the appropriate level.
(2) The percentage values quoted are either exact or slight overestimates.

n_1	n_2	5%	2.5%	1.0%	0.1%
20	1	$L \geqslant$ 21	—	—	—
5	2	13	—	—	—
6	2	15	—	—	—
7	2	17	—	—	—
8	2	18	19	—	—
9	2	20	21	—	—
10	2	22	23	—	—
11	2	24	25	—	—
12	2	25	26	—	—
13	2	27	28	29	—
14	2	29	30	31	—
15	2	30	32	33	—
16	2	32	34	35	—
17	2	34	35	37	—
18	2	35	37	39	—
19	2	37	39	40	—
20	2	39	41	42	—
4	3	18	—	—	—
5	3	20	21	—	—
6	3	22	23	—	—
7	3	25	26	27	—
8	3	27	28	30	—
9	3	30	31	32	—
10	3	32	33	35	—
11	3	34	36	38	—
12	3	37	38	40	—
13	3	39	41	43	—
14	3	41	43	46	—
15	3	44	46	48	—
16	3	46	48	51	—
17	3	48	51	53	57
18	3	51	53	56	60
19	3	53	56	59	63
20	3	55	58	61	66

Table E   continued

n_1	n_2	5%	2.5%	1.0%	0.1%
3	4	$L \geqslant$ 22	—	—	—
4	4	25	26	—	—
5	4	28	29	30	—
6	4	31	32	33	—
7	4	34	35	37	—
8	4	37	38	40	—
9	4	40	42	43	—
10	4	43	45	47	50
11	4	46	48	50	54
12	4	49	51	53	58
13	4	52	54	57	61
14	4	55	57	60	65
15	4	58	60	63	69
16	4	60	63	67	72
17	4	63	67	70	76
18	4	66	70	73	79
19	4	69	73	77	83
20	4	72	76	80	87
2	5	25	—	—	—
3	5	29	30	—	—
4	5	33	34	35	—
5	5	36	38	39	—
6	5	40	42	43	—
7	5	44	45	47	—
8	5	47	49	51	55
9	5	51	53	55	59
10	5	54	57	59	64
11	5	58	61	63	68
12	5	62	64	67	73
13	5	65	68	71	77
14	5	69	72	75	82
15	5	72	76	79	86
16	5	76	80	83	90
17	5	80	83	87	95
18	5	83	87	91	99
19	5	87	91	95	103
20	5	90	95	99	108

Table E continued

n_1	n_2	5%	2.5%	1.0%	0.1%
2	6	$L \geqslant$ 33	—	—	—
3	6	37	38	—	—
4	6	42	43	44	—
5	6	46	48	49	—
6	6	50	52	54	—
7	6	55	57	59	63
8	6	59	61	63	68
9	6	63	65	68	73
10	6	67	70	73	78
11	6	71	74	78	83
12	6	76	79	82	89
13	6	80	83	87	94
14	6	84	88	92	99
15	6	88	92	96	104
16	6	92	96	101	109
17	6	97	101	105	114
18	6	101	105	110	119
19	6	105	110	115	124
20	6	109	114	119	129
2	7	42	—	—	—
3	7	47	48	49	—
4	7	52	53	55	—
5	7	57	58	60	—
6	7	62	64	66	70
7	7	66	69	71	76
8	7	71	74	77	82
9	7	76	79	82	88
10	7	81	84	87	93
11	7	86	89	93	99
12	7	91	94	98	105
13	7	95	99	103	111
14	7	100	104	109	117
15	7	105	109	114	123
16	7	110	114	119	129
17	7	114	119	124	134
18	7	119	124	130	140
19	7	124	129	135	146
20	7	129	134	140	152

Table E   continued

n_1	n_2	5%	2.5%	1.0%	0.1%
2	8	$L \geqslant$ 51	52	–	–
3	8	57	58	60	–
4	8	63	64	66	–
5	8	68	70	72	76
6	8	74	76	78	83
7	8	79	82	85	90
8	8	85	87	91	96
9	8	90	93	97	103
10	8	96	99	103	110
11	8	101	105	109	116
12	8	106	110	115	123
13	8	112	116	120	129
14	8	117	122	126	136
15	8	123	127	132	142
16	8	128	133	138	149
17	8	133	138	144	155
18	8	139	144	150	162
19	8	144	150	156	168
20	8	149	155	162	175
2	9	62	63	–	–
3	9	69	70	71	–
4	9	75	77	78	–
5	9	81	83	85	89
6	9	87	89	92	97
7	9	93	96	99	105
8	9	99	102	106	112
9	9	105	109	112	119
10	9	111	115	119	127
11	9	117	121	126	134
12	9	123	127	132	141
13	9	129	134	139	148
14	9	135	140	145	156
15	9	141	146	152	163
16	9	147	152	158	170
17	9	153	159	165	177
18	9	159	165	171	184
19	9	165	171	178	191
20	9	171	177	185	199

Table E continued

n_1	n_2	5%	2.5%	1.0%	0.1%
2	10	$L \geqslant$ 74	75	–	–
3	10	81	82	84	–
4	10	88	90	92	95
5	10	94	97	99	104
6	10	101	104	107	112
7	10	108	111	114	120
8	10	115	118	122	129
9	10	121	125	129	137
10	10	128	132	136	145
11	10	134	139	143	153
12	10	141	146	151	161
13	10	148	152	158	168
14	10	154	159	165	176
15	10	161	166	172	184
16	10	167	173	179	192
17	10	174	180	187	200
18	10	180	187	194	208
19	10	187	193	201	216
20	10	193	200	208	223
2	11	87	88	–	–
3	11	94	96	98	–
4	11	102	104	106	110
5	11	109	112	114	119
6	11	116	119	123	128
7	11	124	127	131	137
8	11	131	135	139	146
9	11	138	142	147	155
10	11	145	150	154	164
11	11	153	157	162	172
12	11	160	165	170	181
13	11	167	172	178	189
14	11	174	180	186	198
15	11	181	187	194	207
16	11	188	195	201	215
17	11	196	202	209	224
18	11	203	209	217	232
19	11	210	217	225	241
20	11	217	224	233	249

P

Table E   continued

n_1	n_2		5%	2.5%	1.0%	0.1%
2	12	$L \geqslant$ 100	101	–	–
3	12	109	110	112	–
4	12	117	119	121	126
5	12	125	127	130	136
6	12	133	136	139	146
7	12	141	144	148	155
8	12	148	152	157	165
9	12	156	160	165	174
10	12	164	169	174	184
11	12	172	177	182	193
12	12	180	185	191	202
13	12	187	193	199	211
14	12	195	201	208	221
15	12	203	209	216	230
16	12	210	217	224	239
17	12	218	225	233	248
18	12	226	233	241	257
19	12	234	241	250	266
20	12	241	249	258	276
2	13	115	116	117	–
3	13	124	126	128	–
4	13	133	135	138	142
5	13	141	144	147	153
6	13	150	153	157	164
7	13	158	162	166	174
8	13	167	171	175	184
9	13	175	180	185	194
10	13	184	188	194	204
11	13	192	197	203	214
12	13	200	206	212	224
13	13	209	215	221	234
14	13	217	223	230	244
15	13	225	232	239	254
16	13	234	240	248	264
17	13	242	249	257	274
18	13	250	258	266	283
19	13	258	266	275	293
20	13	267	275	284	303

Table E   continued

n_1	n_2		5%	2.5%	1.0%	0.1%
2	14	$L \geqslant$ 131		132	133	–
3	14		140	142	145	–
4	14		150	152	155	160
5	14		159	162	165	172
6	14		168	172	176	183
7	14		177	181	186	194
8	14		186	191	195	205
9	14		195	200	205	216
10	14		204	209	215	226
11	14		213	219	225	237
12	14		222	228	235	248
13	14		231	237	244	258
14	14		240	246	254	269
15	14		249	256	264	279
16	14		258	265	273	290
17	14		266	274	283	300
18	14		275	283	292	311
19	14		284	293	302	321
20	14		293	302	312	331
2	15		147	149	150	–
3	15		158	160	162	–
4	15		168	170	173	179
5	15		177	181	184	191
6	15		187	191	195	203
7	15		197	201	206	215
8	15		207	211	216	226
9	15		216	221	227	238
10	15		226	231	237	249
11	15		235	241	248	261
12	15		245	251	258	272
13	15		254	261	268	283
14	15		264	271	279	294
15	15		273	281	289	305
16	15		283	290	299	317
17	15		292	300	309	328
18	15		302	310	320	339
19	15		311	320	330	350
20	15		320	330	340	361

Table E   continued

n_1	n_2		5%	2.5%	1.0%	0.1%
2	16	$L \geqslant$ 165		167	168	–
3	16		176	178	181	–
4	16		186	189	193	198
5	16		197	201	204	211
6	16		207	211	216	224
7	16		218	222	227	237
8	16		228	233	238	249
9	16		238	243	249	261
10	16		248	254	260	273
11	16		258	265	271	285
12	16		268	275	282	297
13	16		279	285	293	309
14	16		289	296	304	321
15	16		299	306	315	333
16	16		309	317	326	344
17	16		319	327	337	356
18	16		329	338	348	368
19	16		339	348	358	380
20	16		349	358	369	391
2	17		184	185	187	–
3	17		195	198	200	204
4	17		206	210	213	219
5	17		218	221	225	233
6	17		229	233	237	246
7	17		239	244	249	259
8	17		250	255	261	272
9	17		261	267	273	285
10	17		272	278	285	298
11	17		283	289	296	311
12	17		293	300	308	323
13	17		304	311	319	336
14	17		314	322	331	348
15	17		325	333	342	361
16	17		336	344	354	373
17	17		346	355	365	385
18	17		357	366	377	398
19	17		367	377	388	410
20	17		378	388	400	423

Table E   continued

n_1	n_2		5%	2.5%	1.0%	0.1%
2	18	$L \geqslant$	203	205	207	–
3	18		216	218	221	225
4	18		227	231	234	240
5	18		239	243	247	255
6	18		251	255	260	269
7	18		262	267	273	283
8	18		274	279	285	297
9	18		285	291	297	310
10	18		296	303	310	324
11	18		308	314	322	337
12	18		319	326	334	350
13	18		330	338	346	363
14	18		341	349	358	377
15	18		353	361	371	390
16	18		364	373	383	403
17	18		375	384	395	416
18	18		386	396	407	429
19	18		397	407	419	442
20	18		408	419	431	455
2	19		224	226	227	–
3	19		237	240	243	247
4	19		249	253	257	263
5	19		262	266	270	278
6	19		274	279	284	293
7	19		286	291	297	308
8	19		298	304	310	322
9	19		310	316	323	336
10	19		322	328	336	351
11	19		334	341	349	365
12	19		346	353	362	378
13	19		357	365	374	392
14	19		369	378	387	406
15	19		381	390	400	420
16	19		393	402	412	434
17	19		404	414	425	447
18	19		416	426	438	461
19	19		428	438	450	474
20	19		440	451	463	488

Table E   continued

n_1	n_2	5%	2.5%	1.0%	0.1%
1	20	$L \geqslant$ 230	—	—	—
2	20	246	248	249	—
3	20	259	262	265	270
4	20	272	276	280	287
5	20	285	290	294	303
6	20	298	303	308	318
7	20	311	316	322	334
8	20	323	329	336	349
9	20	336	342	350	364
10	20	348	355	363	378
11	20	361	368	377	393
12	20	373	381	390	408
13	20	386	394	403	422
14	20	398	407	417	436
15	20	410	420	430	451
16	20	423	432	443	465
17	20	435	445	457	480
18	20	447	458	470	494
19	20	460	471	483	508
20	20	472	483	496	522

Table F   Critical values of K for the exact test (nonspecific) for two unrelated samples. See section 7.6

Notes
(1) K must be greater than or equal to the critical value to be significant at the appropriate level.
(2) The percentage values quoted are either exact or slight overestimates.

n_1	n_2	5%	2.5%	1.0%	0.1%
8	2	$K \geqslant$ 19	—	—	—
9	2	21	—	—	—
10	2	23	—	—	—
11	2	25	—	—	—
12	2	26	27	—	—
13	2	28	29	—	—
14	2	30	31	—	—

Table F   continued

n_1	n_2	5%	2.5%	1.0%	0.1%
15	2	$K \geqslant 32$	33	–	–
16	2	34	35	–	–
17	2	35	36	–	–
18	2	37	38	–	–
19	2	39	40	41	–
20	2	41	42	43	–
5	3	21	–	–	–
6	3	23	24	–	–
7	3	26	27	–	–
8	3	28	29	–	–
9	3	31	32	33	–
10	3	33	35	36	–
11	3	36	37	39	–
12	3	38	40	41	–
13	3	41	43	44	–
14	3	43	45	47	–
15	3	46	48	49	–
16	3	48	50	52	–
17	3	51	53	55	–
18	3	53	55	58	–
19	3	56	58	60	–
20	3	58	61	63	–
4	4	26	–	–	–
5	4	29	30	–	–
6	4	32	33	34	–
7	4	35	36	38	–
8	4	38	40	41	–
9	4	42	43	45	–
10	4	45	46	48	–
11	4	48	50	52	–
12	4	51	53	55	–
13	4	54	56	59	62
14	4	57	59	62	66
15	4	60	63	65	70
16	4	63	66	69	73
17	4	67	69	72	77
18	4	70	73	76	81
19	4	73	76	79	84
20	4	76	79	82	88

Table F   continued

n_1	n_2	5%	2.5%	1.0%	0.1%
3	5	$K \geqslant 30$	–	–	–
4	5	34	35	–	–
5	5	38	39	40	–
6	5	42	43	44	–
7	5	45	47	49	–
8	5	49	51	53	–
9	5	53	55	57	60
10	5	57	59	61	65
11	5	61	63	65	69
12	5	64	67	69	74
13	5	68	71	73	78
14	5	72	75	78	83
15	5	76	79	82	87
16	5	80	82	86	92
17	5	83	86	90	96
18	5	87	90	94	101
19	5	91	94	98	105
20	5	95	98	102	110
3	6	38	39	–	–
4	6	43	44	45	–
5	6	48	49	50	–
6	6	52	54	55	–
7	6	57	58	60	–
8	6	61	63	65	69
9	6	65	68	70	74
10	6	70	72	75	79
11	6	74	77	80	85
12	6	79	81	84	90
13	6	83	86	89	95
14	6	88	91	94	100
15	6	92	95	99	106
16	6	96	100	104	111
17	6	101	104	108	116
18	6	105	109	113	121
19	6	110	114	118	127
20	6	114	118	123	132

Table F continued

n_1	n_2	5%	2.5%	1.0%	0.1%
3	7	$K \geqslant$ 48	49	–	–
4	7	53	54	56	–
5	7	58	60	62	–
6	7	64	65	67	–
7	7	69	71	73	77
8	7	74	76	78	83
9	7	79	81	84	89
10	7	84	86	89	95
11	7	89	92	95	101
12	7	94	97	100	107
13	7	99	102	106	113
14	7	104	107	111	119
15	7	109	113	117	125
16	7	114	118	122	131
17	7	119	123	128	137
18	7	124	128	133	143
19	7	129	134	139	148
20	7	134	139	144	154
2	8	52	–	–	–
3	8	58	59	–	–
4	8	64	66	67	–
5	8	70	72	74	–
6	8	76	78	80	84
7	8	82	84	86	91
8	8	87	90	93	98
9	8	93	96	99	104
10	8	99	102	105	111
11	8	105	108	111	118
12	8	110	114	117	125
13	8	116	119	123	131
14	8	122	125	130	138
15	8	127	131	136	145
16	8	133	137	142	151
17	8	138	143	148	158
18	8	144	149	154	165
19	8	150	155	160	171
20	8	155	160	166	178

Table F   continued

n_1	n_2	5%	2.5%	1.0%	0.1%
2	9	$K \geqslant 63$	–	–	–
3	9	70	71	72	–
4	9	77	78	80	–
5	9	83	85	87	90
6	9	89	92	94	98
7	9	96	98	101	106
8	9	102	105	108	113
9	9	109	111	115	121
10	9	115	118	122	128
11	9	121	125	128	136
12	9	127	131	135	143
13	9	134	138	142	151
14	9	140	144	149	158
15	9	146	151	156	165
16	9	152	157	162	173
17	9	159	163	169	180
18	9	165	170	176	187
19	9	171	176	183	195
20	9	177	183	189	202
2	10	75	–	–	–
3	10	82	84	85	–
4	10	90	91	93	–
5	10	97	99	101	105
6	10	104	106	109	113
7	10	111	113	116	122
8	10	118	121	124	130
9	10	125	128	132	138
10	10	132	135	139	147
11	10	139	142	147	155
12	10	146	149	154	163
13	10	152	157	161	171
14	10	159	164	169	179
15	10	166	171	176	187
16	10	173	178	184	195
17	10	180	185	191	203
18	10	187	192	198	211
19	10	193	199	206	219
20	10	200	206	213	227

Table F continued

n_1	n_2	5%	2.5%	1.0%	0.1%
2	11	$K \geqslant$ 88	–	–	–
3	11	96	97	99	–
4	11	104	106	108	–
5	11	112	114	116	120
6	11	119	122	.125	130
7	11	127	130	133	139
8	11	135	138	141	148
9	11	142	146	149	157
10	11	150	153	158	166
11	11	157	161	166	175
12	11	165	169	174	183
13	11	172	177	182	192
14	11	180	184	190	201
15	11	187	192	198	210
16	11	195	200	206	218
17	11	202	208	214	227
18	11	209	215	222	236
19	11	217	223	230	244
20	11	224	231	238	253
2	12	101	102	–	–
3	12	110	112	113	–
4	12	119	121	123	–
5	12	127	130	132	137
6	12	136	138	141	147
7	12	144	147	150	157
8	12	152	156	159	167
9	12	160	164	168	176
10	12	169	172	177	186
11	12	177	181	186	195
12	12	185	189	195	205
13	12	193	198	203	214
14	12	201	206	212	224
15	12	209	214	221	233
16	12	217	223	229	243
17	12	225	231	238	252
18	12	233	239	247	261
19	12	241	248	255	271
20	12	249	256	264	280

Table F   continued

n_1	n_2	5%	2.5%	1.0%	0.1%
2	13	$K \geqslant 116$	117	–	–
3	13	126	128	129	–
4	13	135	137	140	143
5	13	144	147	149	154
6	13	153	156	159	165
7	13	162	165	169	176
8	13	171	174	178	186
9	13	180	184	188	197
10	13	188	193	197	207
11	13	197	202	207	217
12	13	206	211	216	227
13	13	215	220	226	237
14	13	223	229	235	248
15	13	232	238	244	258
16	13	240	247	254	268
17	13	249	255	263	278
18	13	258	264	272	288
19	13	266	273	281	298
20	13	275	282	291	308
2	14	132	133	–	–
3	14	142	144	146	–
4	14	152	154	157	161
5	14	162	165	168	173
6	14	172	175	178	184
7	14	181	184	188	196
8	14	191	194	199	207
9	14	200	204	209	218
10	14	209	214	219	229
11	14	219	223	229	240
12	14	228	233	239	251
13	14	237	243	249	262
14	14	246	252	259	272
15	14	256	262	269	283
16	14	265	271	279	294
17	14	274	281	289	304
18	14	283	290	299	315
19	14	293	300	308	326
20	14	302	309	318	336

Table F continued

n_1	n_2	5%	2.5%	1.0%	0.1%
2	15	$K \geqslant 149$	150	–	–
3	15	160	162	163	–
4	15	170	173	175	180
5	15	181	184	187	192
6	15	191	194	198	205
7	15	201	205	209	217
8	15	211	215	220	229
9	15	221	226	231	240
10	15	231	236	241	252
11	15	241	246	252	264
12	15	251	256	263	275
13	15	261	267	273	287
14	15	271	277	284	298
15	15	281	287	294	309
16	15	290	297	305	321
17	15	300	307	315	332
18	15	310	317	326	344
19	15	320	328	336	355
20	15	330	338	347	366
2	16	167	168	–	–
3	16	178	180	182	–
4	16	189	192	195	199
5	16	201	203	207	213
6	16	211	215	219	226
7	16	222	226	230	239
8	16	233	237	242	251
9	16	243	248	253	264
10	16	254	259	265	276
11	16	265	270	276	288
12	16	275	281	287	301
13	16	285	292	299	313
14	16	296	302	310	325
15	16	306	313	321	337
16	16	317	324	332	349
17	16	327	335	343	361
18	16	338	345	354	373
19	16	348	356	366	385
20	16	358	367	377	397

Table F   continued

n_1	n_2	5%	2.5%	1.0%	0.1%
2	17	$K \geqslant$ 185	186	–	–
3	17	198	200	202	–
4	17	210	212	215	220
5	17	221	224	228	234
6	17	233	236	240	248
7	17	244	248	253	262
8	17	255	260	265	275
9	17	267	271	277	288
10	17	278	283	289	301
11	17	289	295	301	314
12	17	300	306	313	327
13	17	311	317	325	340
14	17	322	329	337	352
15	17	333	340	348	365
16	17	344	352	360	378
17	17	355	363	372	391
18	17	366	374	384	403
19	17	377	386	395	416
20	17	388	397	407	428
2	18	205	206	–	–
3	18	218	220	223	–
4	18	231	234	237	242
5	18	243	246	250	257
6	18	255	259	263	271
7	18	267	271	276	286
8	18	279	284	289	300
9	18	291	296	302	313
10	18	303	308	314	327
11	18	314	320	327	341
12	18	326	332	340	354
13	18	338	344	352	368
14	18	349	356	365	381
15	18	361	368	377	395
16	18	373	380	389	408
17	18	384	392	402	421
18	18	396	404	414	434
19	18	407	416	426	448
20	18	419	428	439	461

Table F continued

n_1	n_2	5%	2.5%	1.0%	0.1%
2	19	$K \geqslant$ 226	227	228	–
3	19	240	242	244	–
4	19	253	256	259	264
5	19	266	269	273	280
6	19	279	283	287	296
7	19	291	296	301	310
8	19	304	309	314	325
9	19	316	321	328	340
10	19	328	334	341	354
11	19	341	347	354	368
12	19	353	360	367	383
13	19	365	372	380	397
14	19	378	385	393	411
15	19	390	398	406	425
16	19	402	410	420	439
17	19	414	423	432	453
18	19	426	435	445	467
19	19	438	448	458	481
20	19	451	460	471	494
2	20	248	249	250	–
3	20	262	265	267	–
4	20	276	279	282	288
5	20	290	293	297	305
6	20	303	307	312	321
7	20	316	321	326	336
8	20	329	334	340	352
9	20	342	348	354	367
10	20	355	361	368	382
11	20	368	375	382	397
12	20	381	388	396	412
13	20	394	401	410	427
14	20	407	414	423	441
15	20	420	428	437	456
16	20	432	441	451	471
17	20	445	454	464	485
18	20	458	467	478	500
19	20	471	480	491	514
20	20	483	493	505	529

Table G   Critical values of *L* for the exact test (specific) for two related samples (matched pairs). See section 8.2

Notes
(1) *L* must be greater than or equal to the critical value to be significant at the appropriate level.
(2) The percentage values quoted are either exact or overestimates; if two values in the same row are equal, choose the value to the right.

m'	5%	2.5%	1.0%	0.1%
5	$L \geqslant 10$	—	—	—
6	12	12	—	—
7	14	14	14	—
8	15	16	16	—
9	17	17	18	—
10	19	19	20	20
11	20	21	21	22
12	22	22	23	24
13	23	24	25	26
14	25	26	26	27
15	27	27	28	29
16	28	29	30	31
17	30	30	31	33
18	31	32	33	34
19	33	34	34	36
20	35	35	36	38
21	36	37	38	39
22	38	39	39	41
23	39	40	41	43
24	41	42	43	44
25	43	43	44	46
26	44	45	46	48
27	46	47	47	49
28	47	48	49	51
29	49	50	51	53
30	50	51	52	54

Table H   Critical values of *L* for the exact test (nonspecific) for two related samples (matched pairs). See section 8.6

Note
L must be greater than or equal to the critical value to be significant at the appropriate level.

m'	5%	2.5%	1%	0.1%
6	$L \geqslant 12$			
7	14	14		
8	16	16	16	
9	17	18	18	
10	19	19	20	
11	21	21	22	22
12	22	23	23	24
13	24	24	25	26
14	26	26	27	28
15	27	28	28	29
16	29	29	30	31
17	30	31	32	33
18	32	33	33	35
19	34	34	35	36
20	35	36	37	38
21	37	38	38	40
22	39	39	40	41
23	40	41	42	43
24	42	42	43	45
25	43	44	45	46
26	45	46	46	48
27	47	47	48	50
28	48	49	50	51
29	50	50	51	53
30	51	52	53	55

Table I   Critical values of R_- in the Wilcoxon matched pairs (specific) test

Note
R_- must be less than or equal to the critical value to be significant.

m	5%	2.5%	1%	0.1%
5	$R_- \leqslant 0$			
6	2	0		
7	3	2	0	
8	5	3	1	
9	8	5	3	
10	10	8	5	0
11	13	10	7	1
12	17	13	9	2
13	21	17	12	4
14	25	21	15	6
15	30	25	19	8
16	35	29	23	11
17	41	34	27	14
18	47	40	32	18
19	53	46	37	21
20	60	52	43	26
21	67	58	49	30
22	75	65	55	35
23	83	73	62	40
24	91	81	69	45
25	100	89	76	51
26	110	98	84	58
27	119	107	92	64
28	130	116	101	71
30	151	137	120	86
31	163	147	130	94
32	175	159	140	103
33	187	170	151	112

Table J   Critical values of R_- in the Wilcoxon matched pairs (nonspecific) test. See section 8.7

Note
R_- must be less than or equal to the critical value to be significant.

m	5%	2.5%	1%	0.1%
6	$R_- \leqslant 0$			
7	2	0		
8	3	2	0	
9	5	3	1	
10	8	5	3	
11	10	8	5	0
12	13	10	7	1
13	17	13	9	2
14	21	17	12	4
15	25	20	15	6
16	29	25	19	8
17	34	29	23	11
18	40	34	27	14
19	46	39	32	18
20	52	45	37	21
21	58	51	42	25
22	65	57	48	30
23	73	64	54	35
24	81	72	61	40
25	89	79	68	45
26	98	87	75	51
27	107	96	83	57
28	116	105	91	64
29	126	114	100	71
30	137	124	109	78
31	147	134	118	86
32	159	144	128	94
33	170	155	138	102
34	182	167	148	111
35	195	178	159	120

Table K   Critical values of L in the exact test (specific) for trends among three independent samples. See section 9.2

Notes
(1) This table applies only when the coefficients $\lambda_j = 1, 2, 3$ are used.
(2) Results are not given for cases where any one sample has more than five scores.
(3) The obtained value of L must be greater than or equal to the critical values in the table.

N	n_1	n_2	n_3	5%	2.5%	1%	0.1%
5	1	1	3	$L \geqslant 41$			
	1	2	2	38			
	1	3	1	34			
	2	1	2	36			
	2	2	1	32			
	3	1	1	29			
6	1	1	4	59			
	1	2	3	55	56		
	1	3	2	51	52		
	1	4	1	47			
	2	1	3	53			
	2	2	2	49			
	2	3	1	44			
	3	1	2	46	47		
	3	2	1	41	42		
	4	1	1	38			
7	1	1	5	80	80		
	1	2	4	76	77	77	
	1	3	3	71	72	73	
	1	4	2	67	68	68	
	1	5	1	62	62		
	2	1	4	73	75	75	
	2	2	3	69	70	71	
	2	3	2	64	65	66	
	2	4	1	59	60	60	
	3	1	3	66	67	68	
	3	2	2	61	62	63	
	3	3	1	55	56	57	
	4	2	1	52	53	53	
	5	1	1	48	48		

Table K continued

N	n_1	n_2	n_3	5%	2.5%	1%	0.1%
8	1	2	5	$L \geqslant 99$	100	101	
	1	3	4	94	95	97	
	1	4	3	89	90	92	
	1	5	2	84	85	86	
	2	1	5	96	98	99	
	2	2	4	92	93	94	
	2	3	3	86	88	89	
	2	4	2	81	82	83	
	2	5	1	75	76	77	
	3	1	4	88	90	92	
	3	2	3	83	84	86	
	3	3	2	77	79	80	
	3	4	1	71	72	74	
	4	1	3	79	81	83	
	4	2	2	74	75	76	
	4	3	1	67	68	70	
	5	1	2	69	71	72	
	5	2	1	63	64	65	
9	1	3	5	120	122	123	
	1	4	4	115	116	118	
	1	5	3	109	111	112	
	2	2	5	117	119	121	
	2	3	4	112	113	115	117
	2	4	3	106	108	109	111
	2	5	2	100	101	103	
	3	1	5	113	116	117	
	3	2	4	108	110	112	114
	3	3	3	102	104	106	108
	3	4	2	96	98	99	101
	3	5	1	89	91	92	
	4	1	4	104	106	108	
	4	2	3	98	180	102	104
	4	3	2	92	93	95	97
	4	4	1	85	86	88	
	5	1	3	93	96	97	
	5	2	2	87	89	91	
	5	3	1	80	82	83	

Table K   continued

N	n_1	n_2	n_3	5%	2.5%	1%	0.1%
10	1	4	5	$L \geqslant$ 143	145	147	149
	1	5	4	138	139	141	143
	2	3	5	140	142	144	147
	2	4	4	134	136	138	140
	2	5	3	128	129	131	134
	3	2	5	136	138	140	144
	3	3	4	130	132	134	137
	3	4	3	123	125	127	130
	3	5	2	117	118	120	123
	4	1	5	131	134	137	140
	4	2	4	125	128	130	133
	4	3	3	119	121	123	126
	4	4	2	112	114	116	118
	4	5	1	105	106	108	110
	5	1	4	120	123	126	129
	5	2	3	114	116	118	122
	5	3	2	107	109	111	114
	5	4	1	99	101	103	105
11	1	5	5	169	170	172	176
	2	4	4	165	167	169	173
	2	5	4	158	160	162	166
	3	3	5	160	163	165	169
	3	4	4	153	156	158	162
	3	5	3	146	149	151	154
	4	2	5	155	158	161	166
	4	3	4	149	151	154	158
	4	4	3	141	144	146	150
	4	5	2	134	136	138	142
	5	1	5	150	153	157	162
	5	2	4	143	146	149	154
	5	3	3	136	139	141	145
	5	4	2	129	131	133	137
	5	5	1	121	122	124	128
12	2	5	5	191	194	196	201
	3	4	5	187	189	192	197
	3	5	4	179	182	185	189
	4	3	5	181	184	188	193
	4	4	4	174	177	180	185
	4	5	3	166	169	172	176

Table K continued

N	n_1	n_2	n_3	5%	2.5%	1%	0.1%
	5	2	5	$L \geqslant$ 176	179	183	188
	5	3	4	168	171	175	180
	5	4	3	161	163	166	171
	5	5	2	152	155	157	162
13	3	5	5	215	218	221	226
	4	4	5	209	213	216	222
	4	5	4	201	204	208	213
	5	3	5	203	207	211	217
	5	4	4	195	199	202	208
	5	5	3	189	190	193	198
14	4	5	5	239	243	247	253
	5	4	5	233	237	241	248
	5	5	4	224	228	232	238
15	5	5	5	264	268	273	281

Table L Critical values of K for the exact test (nonspecific) for designs with three unrelated samples

Note
K must be greater than or equal to the critical value to be significant.

n_1	n_2	n_3	5%	2.5%	1%	0.1%
3	2	2	$K \geqslant$ 134.00			
3	3	1	136.00			
3	3	2	194.17	195.33		
3	3	3	267.00	269.67	279.00	
4	2	1				
4	2	2	194.00	195.00		
4	3	1	193.25	197.00		
4	3	2	265.83	270.00	273.33	
4	3	3	355.58	358.92	364.33	
4	4	1	262.25	271.25	275.00	
4	4	2	352.50	360.50	367.00	
4	4	3	457.58	466.33	474.58	494.00
4	4	4	581.00	594.00	606.50	627.00

Table L continued

n_1	n_2	n_3	5%	2.5%	1%	0.1%
5	2	1	$K \geqslant$ 192.00			
5	2	2	263.70	270.00	274.00	
5	3	1	262.20	270.33		
5	3	2	350.63	357.53	365.83	
5	3	3	458.13	465.47	473.87	492.00
5	4	1	348.20	356.20	366.25	
5	4	2	454.00	462.75	475.25	490.50
5	4	3	580.53	590.33	603.78	621.33
5	4	4	722.80	738.20	754.70	776.05
5	5	4	885.35	901.60	919.45	952.97
5	5	5	1073.20	1092.00	1119.60	1155.59

Table M Critical values of L for the exact test (specific) for designs with three or four related samples

Notes
(1) Only the following coefficients may be used for this test: $\lambda_1 = 1, \lambda_2 = 2, \lambda_3 = 3$.
(2) L must be greater than or equal to the critical values to be significant.

k	m	5%	2.5%	1%	0.1%
3	2	$L \geqslant$ 28			
	3	41	42	42	
	4	54	55	55	56
	5	67	68	68	70
	6	79	80	81	83
4	2	58	59	60	
	3	84	86	87	89
	4	111	112	114	117
	5	137	138	140	145

Table N   Critical values of K for the exact test (nonspecific) for designs with three related samples

Note
K must be greater than or equal to the critical value to be significant.

m	5%	2.5%	1%	0.1%
3	$K \geqslant 126$			
4	215	216	216	
5	332	338	342	350
6	474	482	486	504
7	638	642	650	674
8	818	830	840	866
9	1 028	1 044	1 058	1 086

References

Amlaner, C. J., Jr. and McFarland, D. J. (1981) Sleep in the Herring gull (*Larus argentatus*). *Animal Behaviour*, **29**, 551–6.

Benard, A. and Van Elteren, P. L. (1953) A generalisation of the method of *m* rankings. *Proc. Kon. Ned. Ak. van Wet. A$_S$*, **56** or *Indag. Math.*, **15**, 358–69.

Bradley, C. and Meddis, R. (1974) Arousal threshold in dreaming sleep. *Physiological Psychology*, **2**, 109–10.

Bradley, J. V. (1968) *Distribution Free Statistical Tests*, Prentice-Hall Inc., Englewood Cliffs, N.J.

Brunden, M. N. and Mohberg, N. R. (1976) The Benard and Van Elteren statistic and non-parametric computation. *Common. Statist-Simula Computa*, B5 (4), 155–62.

Cochran, W. C. (1950) The comparison of percentages in matched samples. *Biometrika*, **37**, 256–66.

Cohen, D. (1979) The University of Texas at Austin, Department of Psychology. Personal Communication.

Dixon, N. F. and Spitz, L. (1980) The detection and auditory visual desynchrony. *Perception*, **9**, 719–21.

Dixon, W. J. and Mood, A. M. (1946) The statistical sign test, *J. Amer. Statist. Assoc.*, **41**, 556–66.

Dunn, O. J. (1964) Multiple comparisons using rank sums. *Technometrics*, **6**, 241–52.

Durbin, J. (1951) Incomplete blocks in ranking experiments. *The British Journal of Psychology, Statistical Section*, **4**, 85–90.

Fleming, M. L., Unwin, A. R. and Meehan, J. P. (1982) Natural history of alcohol dependence and alcohol related disabilities. *Paper presented to European Medical Informatics Conference*.

Friedman, F. (1937) The use of ranks to avoid the assumption of normality implicit in analysis of variance. *J. Amer. Statist. Ass.* **32**, 675–701.

Fulton, E. J. (1983) Motorcycle conspicuity: evaluation of selected lighting options for daytime use. Internal report to the Transport and Road Research Laboratory by the Institute for Consumer Ergonomics. Loughborough University of Technology.

444

Gazely, D. J. and Stone, P. T. (1981) Visual capacity, lighting and task requirements of partially sighted schoolchildren. Light and Low Vision Report No. 3. Department of Human Sciences, Loughborough University.

Geyer, T. A. W., Grant, R. H., Istance, H. O., McGirl, P. J., Stough, P. G. and Feeney, R. J. (1983) Assessment of the adequacy and problems of wearing suitable respiratory protective equipment for use against asbestos. *Report prepared for the (UK) Health and Safety Executive*. Institute of Consumer Ergonomics, Loughborough.

Gledhill, N. (1976) A study of the interactive effect on performance of thermal stress and motivation. Student undergraduate project. Department of Human Sciences, University of Technology, Loughborough.

Green, D. R. (1982) Probability Concepts in School Pupils age 11–16 years. Unpublished Ph.D. thesis, University of Loughborough.

Horne, J. A. and Pettit, A. N. (1983) Effect of attractive monetary reward on vigilance performance during 72 h total sleep deprivation. *Sleep Research*, **12**, 329.

Howard, M., Miller, C. and Calver, M. (1981) Unpublished undergraduate experiment, Department of Human Sciences, University of Loughborough (contact R. Meddis).

Hunt, S. A. (1979) Hypnosis as obedience behaviour. *British Journal of Social and Clinical Psychology*, **18**, 21–7.

Istance, H. and Howarth, P. (1983) Ergonomic problems in the use of visual display units. *Report to European Coal and Steel Community*, under research contract No. 7206–00–802.

Johnstone, E. C., Lawler, P., Stevens, M., Deakin, J. F. W., Frith, C. D., McPherson, K. and Crow, J. J. (1980) The Northwick Park electroconvulsive therapy trial. *The Lancet*, 1317–20.

Joseph, M. H., Durrant, M. L., Wilkins, D. Y. and Garrow, J. S. (1984) Tryptophan reduces food intake in obese subjects. *Appetite*, submitted for publication.

Kendall, M. G. (1945) *The Advanced Theory of Statistics*, Griffin, London.

Kendall, M. G. and Stuart, A. (1967) *The Advanced Theory of Statistics*, Griffin.

Kruskal, W. H. (1952) A non-parametric test for the several sample problem. *Ann. Math. Statist.*, **23**, 525–540.

Langford, G. W., Meddis, R. and Pearson, A. J. D. (1974) Awakening latency from sleep for meaningful and non-meaningful sleep. *Psychophysiology*, **11**, 1–5.

Liddell, D. (1976) Practical tests of 2×2 contingency tables. *The Statistician*, **25**, 295–304.

Mann, H. B. and Whitney, D. R. (1947) On a test of whether one of two random variables is stochastically larger than the other. *Ann. Math. Statist.*, **18**, 50–60.

Marascuilo, L. A. and McSweeney, M. (1967) Non-parametric post hoc comparisons for trend, *Psychol. Bull.*, **67**, 401–12.

Meddis, R. (1975) A simple two-group test for matched scores with unequal cell frequencies. *British Journal of Psychology*, **66**, 225–227.

Meddis, R. (1980) Unified analysis of variance by ranks. *British Journal of Mathematical and Statistical Psychology*, **33**, 84–98.

Meudell, P., Mayes, A. and Neary, D. (1980) Orienting task effects on the recognition of humerous pictures in amnesics and normal subjects. *Journal of Clinical Neuropyschology*, 2, 75-88.

Mood, A. M. (1950) Introduction to the theory of statistics, New York: McGraw-Hill.

Oakley, D. A. (1979) Neocortex and learning. *Topics in Neurosciences.*

Page, E. B. (1963) Ordered hypotheses for multiple treatments: a significance test for linear ranks. *American Statistical Association Journal*, 216-30.

Parkes, K. R. (1982) Occupational stress among student nurses: a natural experiment. *Journal of Applied Psychology*, 67, 784-96.

Parsons, K. C. (1982) The perception of complex building vibration. *Proceedings of the U.K. Informal Group on Human Response to Vibration, Cricklewood, U.K.*

Poulton, E. C. (1973) Unwanted range effects from using within-subject experimental designs. *Psychological Bulletin*, 80, 113-121.

Reason, J. T. (1968) Relationships between motion sickness susceptibility, the spiral after-effect and loudness summation. *British Journal of Psychology*, 59, 285-93.

Scheffé, H. (1953) A method for judging all contrasts in the analysis of variance. *Biometrika*, 40, 87-104.

Siegel, S. (1956) *Nonparametric Statistics*. McGraw Hill, New York.

Southall, D. and Stone, P. T. (1982) The discomfort glare sensitivity of partially sighted schoolchildren. *Light and Low Vision Report No. 4*, Department of Human Sciences, Loughborough University.

Spearman, C. (1906) A footrule for measuring correlation. *British Journal of Psychology*, 2, 89.

Taub, J. M., Globus, G. G., Phoebus, E. and Drury, R. (1971) Extended sleep and performance. *Nature*, 233, 142-3.

Upton, G. J. G. (1982) A comparison of alternative tests for the 2 X 2 comparative trial. *Journal of the Royal Statistical Society A*, 145, 86-105.

Wilcoxon, F. (1945) Individual comparisons by ranking methods. *Biometrics Bulletin*, 6, 80-83.

Wilcoxon, F. (1946) Individual comparisons of grouped data by ranking methods. *J. Economic Entomology*, 39, 269-70.

Wilcoxon, F. (1947) Probability tables for individual comparisons by ranking methods, *Biometrics*, 3, 119-22.

Wilding, J. M. and Meddis, R. (1972) A note on personality and motion sickness. *British Journal of Psychology*, 63, 619-20.

Wilkins, A. J., Nimmo-Smith, I., Tait, A., McManus, I. C., Della Scala, S., Tilley, A., Arnold, K. and Barrie, M. (1984) A neurological basis for visual discomfort. *Brain*, 107.

Wilson, E. B. and Hilferty, M. M. (1931) The distribution of chi-square. *Proceedings of the National Academy of Science*, 17, 684-88.

Winer, B. J. (1971) *Statistical Principles in Experimental Design*, New York: McGraw-Hill (2nd edn.).

Zeichner, A. and Phil, R. O. (1979) Effects of alcohol and behaviour contingencies on human aggression. *Journal of Abnormal Psychology*, 88, 153-60.

Index

A	Number of cases in the upper left hand cell of a 2×2 frequency matrix ($f_{1,2}$). See section 7.5
B	Number of cases in the upper right hand cell of a 2×2 frequency matrix ($f_{1,2}$). See section 7.5
b_i	Total number of cases in the ith block. See section 5.3
C	Number of cases in the lower left hand cell of a 2×2 frequency matrix ($f_{2,1}$). See section 7.5
d_i	Difference between a pair of ranks, used by Spearman in his method for computing ρ. See section 11.2
D	Number of cases in the lower right hand cell of a 2×2 frequency matrix ($f_{2,2}$). See section 7.5
E, E_{ij}	Expected number of cases in the cell in the ith row and the jth column of a frequency table when H_0 true. See section 9.8
$E(L)$	Expected value of L when H_0 true. The average of L over all permutations of the ranks. See section 3.2
f_{ijq}	Number of cases in the qth category of the dependent variable in the jth sample in the ith block. See sections 5.6 and 10.7
H	Statistic measuring the dispersion of the sample rank means; approximately distributed as chi-square. See section 3.3
H'	H when corrected for ties. See section 3.3
H_0	Null hypothesis. See section 2.9
H_1	Specific alternative hypothesis. See section 2.10
i	Index normally reserved for the block number (e.g. b_i)
j	Index normally reserved for the sample number (e.g. R_j)
k	Number of samples. Note that k is lower case
K	Statistic reflecting differences between the rank means. See section 3.3. Note that K is upper case
L	Statistic measuring agreement between data and a specific hypothesis. See section 3.2.
M	Median; a value exceeded by half of the scores in the data set. See section 2.9
m	Number of blocks. See section 5.3
m'	Number of blocks which are not completely tied. See section 8.2
n_{ij}	Number of cases in the cell in the ith block of the jth sample
n_j	Number of cases in the jth sample $\left(\text{i.e. } n_j = \sum_i^m n_{ij} \right)$
n	Number of cases in each cell when cell frequencies are all equal
N	Total number of cases $\left(\text{i.e. } N = \sum_j^k n_j \right)$
O, O_{ij}	Observed (as opposed to expected) number of cases in the cell in the ith row and jth column of a frequency table. See section 9.8
p	Probability; usually the probability of obtaining the results under discussion, or similarly extreme results, when H_0 is true
q	Index normally reserved for the number of a category of the dependent variable. See section 8.9